D1719817

Genetically Engineered Food

Edited by
Knut J. Heller

Genetically Engineered Food

Methods and Detection

Edited by
Knut J. Heller

WILEY-VCH GmbH & Co. KGaA

Editor:

Prof. Dr. Knut J. Heller
Institut für Mikrobiologie
Bundesanstalt für Milchforschung
Hermann-Weigmann-Str. 1
24103 Kiel
Germany

■ This book was carefully produced. Nevertheless, authors, editors and publisher do not warrant the information contained therein to be free of errors. Readers are advised to keep in mind that statements, data, illustrations, procedural details or other items may inadvertently be inaccurate.

**Library of Congress Card No.: applied for
British Library Cataloguing-in-Publication Data.** A catalogue record for this book is available from the British Library.

Bibliographic information published by Die Deutsche Bibliothek
Die Deutsche Bibliothek lists this publication in the Deutsche Nationalbibliografie; detailed bibliographic data is available in the Internet at <http://dnb.ddb.de>.

© WILEY-VCH Verlag GmbH & Co. KGaA, Weinheim, 2003

All rights reserved (including those of translation into other languages). No part of this book may be reproduced in any form – nor transmitted or translated into a machine language without written permission from the publishers. Registered names, trademarks, etc. used in this book, even when not specifically marked as such, are not to be considered unprotected by law.

Printed in the Federal Republic of Germany.
Printed on acid-free paper.

Composition   Hagedorn Kommunikation GmbH, Viernheim
Printing   betz-druck GmbH, Darmstadt
Bookbinding   Litges & Dopf Buchbinderei GmbH, Heppenheim

ISBN   3-527-30309-X

Preface

A new phase in the era of molecular biology was entered 30 years ago with the construction and successful transformation of the first recombinant DNA-molecule by Cohen and co-workers in 1973. This event marked the birth of genetic engineering which on one hand allowed a very thorough analysis of cellular functions and on the other hand provided the tool for targeted manipulation of the genetic material of cells and organisms. Supported by the development of the efficient chain termination method for DNA sequence analysis by Sanger and co-workers in 1977 and the polymerase chain reaction method for targeted amplification of DNA segments of choice by Mullis and co-workers in 1986, genetic engineering of prokaryotic organisms and later of eukaryotes became soon a task easily performed in many laboratories.

Very soon the potential of genetic engineering for food production was recognised and the first genetically engineered food organisms, the now famous "Flavr Savr" tomato with delayed ripening, was created and approved in the United States of America in 1994. Many other plants like rape, maize, and soy beans followed, and for these the introduction of herbicide resistance was the predominant genetic modification. The development of this new breeding technique initiated e.g. in Europe the introduction of new legislation like the "Council Directive (90/220/EEC) on the deliberate release into the environment of genetically modified organisms" and the "Regulation (258/97/EC) concerning novel foods and novel food ingredients". The rationale behind the introduction of this new legislation was to harmonise legislation concerning free trade, to protect public health and consumer rights, and to duly consider environmental aspects.

One consequence of these regulations was the development of detection methods to identify unambiguously foods produced with the aid of genetic engineering. Today, these methods must be capable of determining the amount of genetically engineered ingredients quantitatively at a level of 1% of the entire amount of the ingredient. This is necessary to differentiate between deliberate application and accidental contamination of the genetically engineered ingredient.

This book addresses in three parts the three different aspects of genetic engineering of foods: in part 1 current applications and future potentials of this breeding technique are discussed, in part 2 the legislation in Europe and the frame it sets for the application of this technique are presented, and in part 3 methods developed to detect foods produced with the aid of genetic engineering are presented and the limits of detection are discussed. The book is by no means comprehensive. The focus concerning detection methods is clearly on those methods detecting DNA. Methods for the detection of protein e.g. are not described separately but are dealt with in different chapters of part 3 whereever it is appropriate. The issues food-safety and consumer-acceptance are not delt with deliberately. Food safety is not a specific issue for novel foods but an issue for food in general. Consumer acceptance for genetically modified foods is an issue of very controversial debates and often the arguements in these debates are not at all scientific. It is the feeling of the editor that taking up the consumer acceptance issue would be an obstacle obstructing the view onto the scientific data presented in the book.

All authors are established and active researchers in their fields. It is their expertise which makes me confident that this book will be a valuable work for anyone interested in novel foods and methods of detection. I am grateful to them for having contributed so excellently to this book.

I also wish to acknowledge the excellent cooperation of Karin Dembowsky and of Andrea Pillmann, both from WILEY-VCH, in the initial and in the finishing phase, respectively.

Finally, I like to thank my family Dagmar, Steffen, and Daniel for their patience and constant support especially during the finishing phase of the book.

Knut J. Heller,
Kiel, April 2003

Contents

List of Contributors *XIV*

Part I   Application and Perspectives   *1*

1   **Transgenic Modification of Production Traits in Farm Animals**   *3*
1.1   The Creation of Transgenic Animals   *3*
1.1.1   Pronuclear DNA Microinjection   *4*
1.1.2   Retroviral Vectors   *5*
1.1.3   Pluripotent Stem Cell Technologies   *6*
1.1.4   Nuclear Transfer using Transgenic Cells   *6*
1.1.5   Gene Transfer in Poultry   *7*
1.1.6   Gene Transfer in Fish   *7*
1.2   Transgenes: Gene Constructs   *8*
1.3   Transgenic Animals with Agricultural Traits   *10*
1.3.1   Improved Growth Rate, Carcass Composition, and Feed Efficiency   *11*
1.3.1.1 Transgenes in mammalian farm animals   *11*
1.3.1.2 Transgenes in fish   *12*
1.3.2   Alteration of Milk Composition   *13*
1.3.3   Improved Animal Health   *15*
1.3.3.1 Additive gene transfer of resistance genes   *15*
1.3.3.2 Gene targeting of susceptibility genes   *16*
1.3.4   Improved Biochemical Pathways   *17*
1.3.5   Improved Wool Production   *17*
1.4   Transgenic Farm Animals and Biosafety Issues   *18*
1.5   Conclusions   *19*

2   **Genetically Modified Plants**   *26*
2.1   Methods to Establish Genetically Modified Plants   *26*
2.1.1   Transformation Methods   *26*
2.1.1.1 *Agrobacterium* transformation   *26*
2.1.1.2 Direct gene transfer   *27*
2.1.2   Tissue Requirements   *28*
2.1.3   Molecular Requirements   *28*

2.1.3.1 Promoter 28
2.1.3.2 Codon usage 29
2.1.3.3 Selectable marker and reporter genes 29
2.2 GM Plants Already on the Market (EU, USA, Canada, Japan) 31
2.2.1 Herbicide Resistance in Soybean, Maize, Oilseed rape, Sugar beet, Rice, and Cotton 33
2.2.2 Insect Resistance in Maize, Potatoes, Tomatoes, and Cotton 35
2.2.3 Virus-resistance, male sterility, delayed fruit ripening, and fatty acid contents in GMPs 37
2.3 GM Plants "in the Pipeline" 40
2.3.1 Input Traits 40
2.3.1.1 Insect resistance in rice, soybean, oilseed rape, eggplant, walnut, grape, and peanut 40
2.3.1.2 Disease resistance in maize, potatoes, fruits, and vegetables 41
2.3.1.3 Tolerance against abiotic stresses 44
2.3.1.4 Improved agronomic properties 46
2.3.2 Traits Affecting Food Quality for Human Nutrition 46
2.3.2.1 Increased carotenoid content in rice and tomato 46
2.3.2.2 Elevated iron level in rice and wheat 47
2.3.2.3 Improved amino acid composition in potato plants 47
2.3.2.4 Reduction in the content of antinutritive factors in cassava 48
2.3.2.5 Production of "low-calorie sugar" in sugar beet 48
2.3.2.6 Seedless fruits and vegetables 48
2.3.3 Traits that Affect Processing 49
2.3.3.1 Altered gluten level in wheat to change baking quality 49
2.3.3.2 Altered grain composition in barley to improve malting quality 49
2.3.4 Traits of Pharmaceutical Interest 50
2.3.4.1 Production of vaccines 50
2.3.4.2 Production of pharmaceuticals 51
2.4 Outlook 52

3 Fermented Food Production using Genetically Modified Yeast and Filamentous Fungi 62
3.1 Introduction 62
3.1.1 Why Do We Ferment Foodstuffs? 62
3.1.2 Fermented Foods of Plant Origin 63
3.1.3 Fermented Foods of Animal Origin 63
3.1.4 Conclusion 65
3.2 Application of Recombinant DNA Methods 65
3.2.1 Recombinant DNA Technology in Yeast 65
3.2.1.1 Vectors 67
3.2.2 Recombinant DNA Technology in Filamentous Fungi 69
3.2.2.1 Fungal transformation 69
3.3 Improved Fermentation Efficiency for Industrial Application 71
3.3.1 Industrial *Saccharomyces* Strains 72

3.3.1.1 Beer 72
3.3.1.2 Wine 74
3.3.1.3 Sake 76
3.3.1.4 Bread 76
3.3.2 Other Industrial Yeast Strains 77
3.3.3 Industrial Filamentous Fungi 78
3.4 Commercial use of Genetically Modified Organisms (GMO) 78
3.5 The Future 79

4 Production of Food Additives using Filamentous Fungi 86
4.1 Filamentous Fungi in Food Production 86
4.1.1 Industrial Applications 87
4.2 Additives for the Food Industry 89
4.3 Design of GMM for Production of Food Additives and Processing Aids 90
4.3.1 Gene Disruption 90
4.3.2 Expression Vectors 92
4.4 Industrial Enzyme Production Processes 95

5 Perspectives of Genetic Engineering of Bacteria used in Food Fermentations 100
5.1 Introduction 100
5.2 Lactic Acid Bacteria 101
5.2.1 *Lactococcus lactis* subsp. *lactis* and subsp. *cremoris* 101
5.2.2 *Lactobacillus* spp. 102
5.2.3 *Streptococcus thermophilus* 103
5.2.4 *Leuconostoc* spp. 103
5.2.5 *Pediococcus* spp. 103
5.2.6 *Oenococcus* spp. 104
5.3 Perspective and Aims 104
5.3.1 Bioconservation 104
5.3.2 Bacteriophage Resistance 105
5.3.3 Exopolysaccharides 107
5.3.4 Proteolysis 108
5.4 Metabolic Engineering of Lactic Acid Bacteria 109
5.5 Stress Responses in Lactic Acid Bacteria 110
5.6 Methods 111
5.6.1 Transformation and Vector Systems 111
5.7 Conclusions 112

Part II Legislation in Europe 119

6 The Legal Situation for Genetically Engineered Food in Europe 121
6.1 Introduction 121
6.1.1 The Need for Regulation 121

6.1.2 The History of the Novel Food Regulation *121*
6.2 Status Quo *122*
6.2.1 The Novel Food Regulation *122*
6.2.1.1 Introduction *122*
6.2.1.2 Scope of application *123*
6.2.1.3 Requirements for novel foods and food ingredients *124*
6.2.1.4 Procedures *124*
6.2.1.5 Labeling *126*
6.2.1.6 Other questions *128*
6.2.2 Problems *129*
6.2.2.1 Negative labeling *129*
6.2.2.2 Compliance with World Trade law *130*
6.2.2.3 Competent authorities in the member states *130*
6.2.4 Supplementary and Replacement Regulation *135*
6.2.4.1 History *135*
6.2.4.2 Scope of application of the Replacement Regulation *135*
6.2.4.3 Requirements of labeling *136*
6.2.5 Relation to *Council Directive No 90/220/EEC of 23 April 1990 on the deliberate release into the environment of genetically modified organisms* (OJ (EC) 1990 No L 117/15; from here on: Deliberate Release Directive) *136*
6.2.6 Supplementary National Provisions in German law: The Novel Foods and Food Ingredients Instrument *136*
6.2.6.1 General rules for novel foods *137*
6.2.6.2 Rules on labeling of genetically modified soya beans and genetically modified maize *137*
6.2.6.3 Availability of negative labeling of foodstuffs made without using procedures of genetic engineering *137*
6.2.6.4 Rules on criminal offences and misdemeanors *139*
6.3 Recent Development at the European Level *139*
6.3.1 Introduction *139*
6.3.2 The Commission Proposal for a "Regulation of the European Parliament and of the Council on Genetically Modified Food and Feed" *140*
6.3.2.1 Objective and definitions *141*
6.3.2.2 Requirements of genetically modified food and feed *141*
6.2.2.3 Labeling of genetically modified food and feed *143*
6.3.2.4 General provisions *144*
6.3.3 The Commission Proposal for a Regulation of the European Parliament and of the Council Concerning Traceability and Labeling of Genetically Modified Organisms and Traceability of Food and Feed Products Produced from Genetically Modified Organisms and Amending Directive 2001/18/EC *144*
6.3.4 Stage of Legislative Procedure *145*

Part III Methods of Detection *147*

7   Detection of Genetic Modifications: Some Basic Considerations *149*
7.1   The Conversion of Genetic Information from DNA to Phenotypes *149*
7.2   DNA, Protein and Phenotypes as Targets for Detection Assays *150*
7.3   Food-grade Modifications *153*
7.4   Detection of Unknown Modifications *154*

8   DNA-based Methods for Detection of Genetic Modifications *155*
8.1   Introduction *155*
8.2   Recent DNA Methodology *156*
8.2.1 Sampling Procedure *156*
8.2.2 Extraction and Purification of DNA *157*
8.3   Specific Detection of Genetic Material *158*
8.3.1 DNA Hybridization-based Detection Technique (Southern Blot) *158*
8.4   Nucleic Acid Amplification Methods using PCR *159*
8.4.1 The Common PCR *159*
8.4.2 Real-time PCR *161*
8.4.3 Important Bioinformatic Considerations *163*
8.5   Alternative and Promising DNA Detection Techniques *164*
8.5.1 Thermal Cycling Procedures *164*
8.5.2 Isothermic Amplification *164*
8.5.3 DNA-micro-arrays *165*
8.5.4 Mass Spectrometry (MS) of DNA *166*
8.5.5 Supplementary Photon-driven Monitoring Methodologies *166*
8.5.6 Novel Biological Monitoring Approaches *167*
8.6   Conclusions and Future Prospects of GMO Detection Applying DNA-analysis *168*

9   Genetic Engineering of Fishes and Methods for Detection *174*
9.1   Introduction *174*
9.2   Development and Production of Transgenic Fish *175*
9.2.1 Structure of Gene Cassettes *176*
9.2.2 Methods of Gene Transfer *177*
9.2.3 Evidence for Gene Transfer and Expression *177*
9.3   Examples of Successful Production of Transgenic Fish *180*
9.3.1 Atlantic Salmon *180*
9.3.2 Pacific Salmon *181*
9.3.3 Tilapia (*O. hornorum* hybrid) *182*
9.3.4 Tilapia (*O. niloticum*) *182*
9.3.5 Carp *183*
9.4   Methods to Detect Processed Transgenic Fish *183*
9.5   Food Safety of Transgenic Fish *183*
9.5.1 The Gene Product *184*
9.5.2 Pleiotropic Effects *184*

10	**Detection Methods for Genetically Modified Crops** *188*
10.1	Introduction *188*
10.2	Isolation of Plant DNA *189*
10.2.1	Sampling *189*
10.2.2	Sample Preparation *190*
10.2.3	DNA Extraction and Analysis *191*
10.3	Detection Strategies *192*
10.3.1	Screening *193*
10.3.2	Specific Detection *195*
10.3.2.1	Example for Qualitative Detection *197*
10.3.3	Quantification *198*
10.3.4	Verification *198*
10.3.5	Validation *199*
10.4	Outlook and Conclusions *200*

11	**Methods to Detect the Application of Genetic Engineering in Composed and Processed Foods** *205*
11.1	Introduction *205*
11.2	Challenges Specific to the Detection of GMO in Composed and Processed Foods *206*
11.3	Degradation of Proteins and DNA *207*
11.3.1	Proteins *207*
11.3.2	DNA *208*
11.4	Analytical Approaches *210*
11.4.1	Protein-based Methods *210*
11.4.2	DNA-based Methods *210*
11.4.2.1	Qualitative PCR *211*
11.4.2.2	Quantitative PCR *215*
11.4.2.3	Competitive PCR *216*
11.4.2.4	Real-time PCR *218*
11.5	Conclusions *223*

12	**Mutations in *Lactococcus lactis*, and their Detection** *231*
12.1	Introduction *231*
12.2	Composition of the Genome of *Lactococcus lactis* *232*
12.3	Flexibility in the Genome of *Lactococcus lactis* *232*
12.3.1	Conjugation *233*
12.3.2	Transduction *234*
12.3.3	Transformation *235*
12.3.4	IS Elements and Transposons *235*
12.3.5	Lactococcal Phage as Sources of Genetic Plasticity *236*
12.3.5.1	An example of natural genetic flexibility: the *Lactococcus lactis* NCDO712 family *236*
12.4	Mutations in *Lactococcus lactis* as a Consequence of Environmental Factors and DNA Metabolism *237*

12.5	Methods to Mutate the Genome of *Lactococcus lactis*	238
12.5.1	Genetic Engineering of *Lactococcus lactis*	238
12.6	Strategies to Detect Genetically Modified *Lactococcus lactis*	241
12.6.1	Sample Preparation	242
12.6.2	DNA-based Procedures	242
12.6.2.1	Southern hybridization	243
12.6.2.2	PCR	243
12.6.3	Nucleotide Sequence-based Procedures	244
12.6.3.1	Micro-arrays	244
12.6.4	Protein-based Procedures	245
12.6.4.1	Western hybridization	245
12.6.4.2	ELISA	245
12.6.4.3	SPR	246
12.6.4.4	Two-dimensional gel electrophoresis and mass spectrometry	246
12.7	Conclusions	247

13 Detection Methods for Genetically Modified Microorganisms used in Food Fermentation Processes *251*

13.1	Introduction	251
13.2	Properties of Microorganisms	252
13.3	Current Methods for Detection of GMM	255
13.3.1	DNA Isolation	256
13.3.2	DNA Stability	257
13.3.3	Organism-specific Detection of GMM	258
13.4	Conclusion	260

Index *263*

List of Contributors

Dr. Torsten Bauer
Institute of Food Technology
University of Hohenheim
Garbenstr. 25
D-70599 Stuttgart
Germany

Prof. Dr. Gottfried Brem
Institute of Animal Breeding
and Genetics
Veterinary University of Vienna
Veterinärsplatz 1
A-1210 Vienna
Austria

Dr. Dr. Ralf Einspanier
Institute of Physiology
Technical University of
Munich-Weihenstephan
Weihenstephaner Berg 3
D-85354 Freising
Germany

Prof. Dr. Karl-Heinz Engel
Technische Universität München
Lehrstuhl für Allgemeine
Lebensmitteltechnologie
Am Forum 2
D-85354 Freising-Weihenstephan
Germany

Dr. Arnold Geis
Institut für Mikrobiologie
Bundesanstalt für Milchforschung
Hermann-Weigmann-Str. 1
D-24103 Kiel
Germany

Prof. Dr. Walter Hammes
Institute of Food Technology
University of Hohenheim
Garbenstr. 25
D-70599 Stuttgart
Germany

Prof. Dr. Knut Heller
Institut für Mikrobiologie
Bundesanstalt für Milchforschung
Hermann-Weigmann-Str. 1
D-24103 Kiel
Germany

Dr. Christian Hertel
Institute of Food Technology
University of Hohenheim
Garbenstr. 25
D-70599 Stuttgart
Germany

Dr. Carsten Hjort
Novozymes, Fungal Discovery Unit
Research and Development
Krogshoejvej 36
DK-2880 Bagsvaerd
Denmark

Prof. Dr. Jan Kok
Department of Genetics
Groningen Biomolecular Sciences
and Biotechnology Institute
University of Groningen
Kerklaan 30
NL-9751 NN Haren
The Netherlands

Prof. Dr. Horst Lörz
Institut für Allgemeine Botanik
der Universität Hamburg
Angewandte Molekularbiologie
der Pflanzen II
Ohnhorststr. 18
D-22609 Hamburg
Germany

Rolf Meyer
Nestec Ltd., Nestlé Research Center
Department of Quality and
Safety Assurance
Vers-chez-les-Blanc
CH-1000 Lausanne 26
Switzerland

Dr. Francisco Moreano
Technische Universität München
Lehrstuhl für Allgemeine
Lebensmitteltechnologie
Am Forum 2
D-85354 Freising-Weihenstephan
Germany

Prof. Dr. Mathias Müller
Institute of Molecular Genetics
and Biotechnology in
Veterinary Medicine
Veterinary University of Vienna
Veterinärsplatz 1
A-1210 Vienna
Austria

Dr. Anke Niederhaus
Technische Universität
FG Mikrobiologie und Genetik
Gustav-Meier-Allee 25
D-13355 Berlin
Germany

Prof. Dr. Hartmut Rehbein
Institut für Fischereitechnik und
Fischqualität
Bundesforschungsanstlat für Fischerei
Palmaille 9
D-22767 Hamburg
Germany

Dr. Stephanie Rief
Institute of Physiology
Technical University of
Munich-Weihenstephan,
Weihenstephaner Berg 3
D-85354 Freising
Germany

Prof. Dr. Ulf Stahl
Technische Universität
FG Mikrobiologie und Genetik
Gustav-Meier-Allee 25
D-13355 Berlin
Germany

Dr. Susanne Stirn
Institut für Allgemeine Botanik
der Universität Hamburg
FG Landwirtschaft und
Pflanzenzüchtung
Ohnhorststr. 18
D-22609 Hamburg
Germany

Prof. Dr. Rudolf Streinz
Universität Bayreuth
Lehrstuhl für Öffentliches Recht,
Völker- und Europarecht
Universitätsstr. 30
D-95440 Bayreuth
Germany

Dr. Bertus van den Burg
IMEnz Bioengineering B. V.
L. J. Zielstraweg 1
NL-9713 GX Groningen
The Netherlands

Part I
Application and Perspectives

1
Transgenic Modification of Production Traits in Farm Animals

Mathias Müller and Gottfried Brem

"Genetic engineering" is the umbrella term for procedures that result in a directed alteration in the genotype of an organism. The combined use of molecular genetics, DNA recombination and reproductive biology allows the generation of transgenic animals. For animals, the term "transgenic" originally referred to the stable introduction of new genetic material into the germ-line [1, 2]. This definition of transgenic animals must be extended with respect to two aspects. First, further developments of genetic engineering of animals allow not only additive gene transfer (gain of function) but also deletive gene transfer (knockout, loss of function) and replacement gene transfer (knockin, exchange of function). Second, in addition to germ-line integration of transgenes, somatic gene transfer approaches result in (mostly transient) gene expression, with the longest duration being a life span and no transmission of the transgenes to the progeny. Although somatic gene transfer experiments in farm animals for production purposes have been performed [3, 4], this technology in animal production is more beneficial for the development of DNA-based vaccines [5]. Here, we will mainly concentrate the discussion on germ-line transgenic animals. The production of transgenic farm animals was first reported in the mid-1980s [6, 7], since when the main progress in exploiting this technology has been made in the establishment of animal models for human diseases [8, 9], the production of heterologous proteins in animals (gene farming) [10], and the production of organs for xenotransplantation [11, 12]. In addition to these biomedical approaches, research has focussed on the improvement of the efficiency and quality of animal production by transgenic means (this review and Ref. [13]).

1.1
The Creation of Transgenic Animals

The main routes to transgenesis in mammals include: (i) microinjection of DNA into the pronucleus of a fertilized oocyte (zygote); (ii) integration of a (retro)viral vector into an early embryo; (iii) incorporation of genetically manipulated pluripotent stem cells into an early embryo; and (iv) transfer of genetically altered nuclei into enucleated oocytes. For additional gene transfer methods, especially sperm-mediated gene transfer, we refer to other reviews [14–16].

1.1.1
Pronuclear DNA Microinjection

Microinjection of foreign DNA into the pronuclei of zygotes is the classic method of gene transfer into farm animals, and since its first reports [6, 7], this technique has accounted for production of the large majority of transgenic farm animals [9, 17–19]. DNA microinjection results in random integration of the foreign DNA into the host genome, and is therefore not suitable for targeted modification of genomes. Despite microinjections being performed at the 1-cell-stage, between 20 and 30% of the founder animals are mosaic and therefore may not transmit the integrated gene construct to their progeny [17, 20]. Random integration of the gene constructs may cause alteration of one or more gene loci. An insertional mutagenesis is recessive and mostly characterized by a recombination event in the kilobase range at the transgene integration site [21]. In mice, approximately 5–15% are affected by this recessive mutations [22]. Except for studies in transgenic rabbits [23, 24], few data have been published on the analysis of homozygous transgenic farm animals, this being mainly due to the long generation intervals. However, there is no reason to doubt the mutagenesis frequencies estimated for mice in other transgenic mammals generated by the same technology. In addition, random integration of the gene constructs may result in varying, aberrant or abolished transgene expression due to effects of the adjacent chromatin overcoming the transgene's regulatory sequences. One possibility of avoiding these integration site-dependent effects is the transfer of large DNA constructs, which are able to form functionally independent chromatin domains [25]. The first successful example of this strategy in livestock was the generation of transgenic rabbits harboring yeast artificial chromosomes (YACs) [26]. An alternative approach to protect transgenes from chromosomal position effects is the use of boundary elements (e.g., insulators, locus control regions, matrix attachment regions) in the gene constructs in order to achieve copy number- and promoter-dependent and position-independent expression of transgenes [27]. Although success following this strategy has been reported, the effects of the elements were not in all cases as expected.

The gene transfer efficiency (transgenic newborns/microinjected zygotes) in general is rather low, especially in large animals. One transgenic animal can be expected after microinjection of 40, 100, 90–110 and 1600 zygotes in mice, pigs, small ruminants and cattle, respectively [19]. Differences in efficiency emphasize fundamental differences in the reproductive biology of species. Hence, a high level of technical skill and experience in embryo collection and embryo transfer are critical for efficient transgenic production, though this applies equally to all gene transfer programs.

As mentioned earlier, the protocols for generating large mammals by DNA microinjection have remained basically unchanged for the past two decades, and little improvement in DNA transfer efficiency has been achieved. Following DNA microinjection, embryos are transferred to synchronized foster mothers. Major progress has been made in the field of embyro transfer which, in all farm animal species, has been facilitated by the development of endoscopy-guided minimally invasive techniques, thereby reducing stress to the foster mothers and maximizing embryo survival and pregnancy rates [28–31]. This embyro transfer technique is also advantageous for the gene transfer methods discussed in the following sections.

1.1.2
Retroviral Vectors

The first germ-line transgenic mice were produced by retroviral infections of early embryos [32]. Retroviruses can be considered natural gene delivery vehicles to mammalian cells. Endogenous retroviruses (ERVs) are a subset of retroelements which represent up to 10% of the mammalian genome [33]. The capability of ERVs to reintegrate into the genome through reverse transcription mechanisms results in continuous insertion of new ERVs into the host genome. The retroviral vectors were, until recently, not considered for farm animal transgenesis. This was due to biosafety concerns and the dependence of most retroviruses on dividing cells for integration into the host genome. Retroviral gene transfer therefore often results in genetic mosaics when developing embryos are infected. With the development of replication-defective retroviral vectors mainly for gene therapy purposes, a powerful tool for gene transfer in mammalian cells has been established [34]. To avoid mosaicism, Chan et al. [35] inoculated bovine oocytes in the final stage of maturation with retroviral vectors. These authors obtained a remarkably high transgenesis rate and, as expected, no mosaic transgenic cattle. A similar approach resulted in the generation of transgenic piglets [36]. One major limitation of retroviral vectors is their limited cloning capacity (<10 kb). However, gene constructs grow increasingly larger in order to omit variegated transgene expression (see above). A second problem with many retroviral vector-mediated transgenics relates to transcriptional shutdown of the transgenes [37]. Lentiviral vectors are a new generation of retroviral vectors which, in contrast to the above-mentioned oncoretrovirus-based vectors, do not undergo transcriptional silencing. In addition, lentiviruses are able to infect both dividing and nondividing cells. Recently, germ-line transmission and expression of transgenes delivered by lentiviral vectors to 1-cell embryos has been reported [38]. This technique of transgenesis is more efficient and cost-effective and technically less demanding than pronuclear injection. The obstacle of the limited size of the constructs to be transferred remains, however. As with gene transfer by pronuclear injection, retrovirus-mediated gene transfer can be only used for additive gene transfer, and also carries the danger of insertional mutagenesis.

1.1.3
Pluripotent Stem Cell Technologies

Pluripotent stem cells are capable of developing into many cell types including germ cells upon fusion with proimplantation embyros (morulae, blastococysts). Pluripotent stem cells can be maintained in tissue culture and genetically manipulated and selected *in vitro* prior to reconstitution of the embryo. In mice, the handling of pluripotent cells has become a routine method for targeted modification of the genome by homologous recombination, i. e., deletive or replacement gene transfers [39]. As yet, many efforts to establish pluripotent stem cells in species other than mice have failed, and possible reasons for this failure are discussed elsewhere [40, 41]. As nuclear transfer using transgenic donor cells (see below) has become an attractive alternative tool for targeted gene transfer, efforts to establish germ-line competent stem cells from farm animals have been reduced.

1.1.4
Nuclear Transfer using Transgenic Cells

Nuclear transfer technology – also known as cloning – comprises the transfer of a donor nucleus (karyoplast) into the cytoplasm of an enucleated zygote or oocyte (cytoplast). Initial nuclear transfer experiments in farm animals used early embryonic stages as nuclear donors [42] (see also Ref. [43] for a review). In breakthrough experiments with sheep it was demonstrated that *in vitro*-cultured differentiated fetal cells [44], and even cells derived from adult tissues [45], could serve as nuclear donor for the reconstitution of enucleated oocytes. Cloning by nuclear transfer has subsequently been achieved in cattle [46–48], goat [49], pigs [50–52], and rabbits [53]. For farm animal transgenesis, a novel tool became available in that cultured cells can be genetically modified by conventional transfection methods prior to their use for nuclear transfer. The first reports on this novel gene transfer technique were the generation of transgenic sheep and cattle by nuclear transfer using transfected and selected fetal fibroblasts [54, 55]. Transgenesis by nuclear transfer of genetically modified cells provides a number of advantages over the other additive gene transfer techniques:

- mosaicism is avoided and germ-line transmission is guaranteed, since all cells of the cloned animal contain the transgene;
- insertional mutagenesis and chromosomal positioning effects can be avoided, since integration and eventually transgene expression can be monitored *in vitro*; and
- the use of male or female cell lines predicts the gender of the transgenic animal [56, 57].

Most importantly, gene transfer by nuclear transfer provides the means for gene targeting in farm animal species [40]. Both the targeted disruption of genes by homologous recombination (deletive gene transfer, knockout) in sheep and pigs

[58–60] and the targeted integration of a gene of interest into a given locus (replacement gene transfer, knockin) [61] have been reported.

Despite these impressive reports and the intriguing advantages of the nuclear transfer technique for the generation of transgenic farm animals, the broad use is not an easy task because: (i) the primary fibroblasts presently used for gene transfers have a limited capacity to divide; (ii) homologous recombination is less frequent in somatic cells than in pluripotent stem cells; and (iii) cloning by nuclear transfer has a low yield which is still diminished when nuclear donor cells are previously cultured [62]. In addition, there is an ongoing debate whether it is possible to overcome abnormalities observed in cloned animals [63, 64]. The abnormalities are not restricted to transgenic cloned animals, suggesting that they originate from the nuclear transfer procedure or the *in vitro* culture conditions. Although healthy clones have been reported [65], improvements in the technology and further investigations of the effects of cloning are required [66].

1.1.5
Gene Transfer in Poultry

Depending on the developmental stage, various strategies are used for the generation of transgenic birds, including DNA microinjection of fertilized ova, retroviral infection of blastodermal cells, and genetic manipulation of primordial germ cells (PGCs) or embryonic stem (ES) cells. As in mammals, the first method developed to transfer genes into birds was through microinjection of DNA into the germinal disc of fertilized ova [67]. Although successful germ-line transmission has been reported [68], the method is labor-intensive, ineffective, and frequently results in mosaicism. Retroviral vectors are able to introduce transgenes into the genome at low but acceptable efficiencies. The first transgenic birds were produced using replication-competent vectors, and thus could not be used for a broad application [69]. The development of replication-defective vectors led to a wide use of this technique in the production of transgenic birds [70] and stable transgene expression [71]. As an alternative, chimeras bearing transfected pluripotent cells originating from the blastoderm, from PGCs or from ES cells have been reported, but have not yet yielded a transgenic bird with germ-line transmission [72, 73].

1.1.6
Gene Transfer in Fish

The techniques for gene transfer into fish have focussed on direct transfer of DNA into gametes or fertilized eggs, and include DNA microinjection, electroporation, retroviral vector infection and biolistic methods [74–77]. Stem cell-based technologies are not available in farmed fish. The creation of transgenic fish is distinguished from gene transfer in mammals or birds because: (i) fish generally undergo external fertilization and no culture or transfer of eggs into recipient females is required; (ii) the eggs of many fish have a tough chorion such that special methods are required to deliver the gene constructs; and (iii) DNA delivery (including that by microinjection) is usually made into the cytoplasm. It is most likely due to the cytoplasmic nature of DNA delivery that a high number of founder transgenic fish are mosaic. Germ-line mosaicism seems also to occur because the frequency of transgene transmission to F_1 is clearly less than at Mendelian ratios. Transmission of the transgenes to later progeny occurs at Mendelian frequencies, indicating the stable integration of the transgenes.

1.2
Transgenes: Gene Constructs

The exogenous DNA integrated into the host genome usually is referred to as gene construct or transgene, and encompasses the elements controlling gene expression (5' promoter region, 3' control regions) and the sequences (cDNA, genomic DNA) encoding the transgene product. The various transgenic sequences used for the differing gene delivery methods and gene transfer programs are summarized in Table 1.1.

For additive gene transfer experiments by DNA microinjection or spermatozoa, the prokaryotic cloning vector sequences are removed from the gene construct. Prokaryotic sequences, and especially their CpG dinucleotide base pairs, may undergo methylation or heterochromatin formation in animal cells, which leads to transgene silencing. It is becoming increasingly clear that silenced transgenes have been recognized as foreign elements by host cellular mechanisms, as are retroviral and transposon sequences [78, 79]. As mentioned above, one obvious way to avoid transgene silencing or chromosomal positioning effects would be to use large gene constructs and the abdication of viral vectors. Therefore, an increasing number of transgenic animals carry gene constructs based on phage (PAC), bacterial (BAC) or yeast (YAC) artificial chromosomes [25]. For expression and replication, these large transgenes are dependent on integration into the host genome. In contrast, mammalian artificial chromosomes (MACs) provide both an independent transcription and replication unit. Studies that originated mainly from human gene therapy programs have resulted in the development of human artificial chromosomes based on episomal viral vectors [80] or engineered minimal chromosomal elements [81–83]. In the future, MACs may be also used in farm animal transgenesis.

Alternatively, gene transfer *in vitro* followed by reconstitution of embryos by nuclear transfer or stem cell technologies allows the targeting of transcription units in the host genome or *in vitro* analysis of the chromosomal integration site. These gene transfer techniques however require methods for identification of the genetically modified cells. The identification of transgenic cells is mostly based on (drug) selectable markers, e.g., antibiotic resistance genes, added to the gene constructs. In plants, the presence of marker genes in the genetically modified organism is the main topic of concern regarding biosafety [84]. By analogy, a genetically modified farm animal that is sold commercially should be free of such genes; this can be achieved by using site-specific recombinases to remove undesirable sequences

Table 1.1. Characteristics of transgenes used for different gene delivery methods and gene transfer programs in farm animals.

Functional consequence of gene transfer	Methods of gene transfer	Sequences not related to the transgene per se	Composition and origin of transgene	Example (see below)
Gain of function	Pronuclear DNA micro-injection; sperm-mediated; physical/chemical methods		• species-specific sequences • cross-species sequences • new combination of promoter and coding sequences (species-specific or cross-species)	• additional copies of casein alleles • humanized milk • mammary gland-specific expression of antibodies
	Artificial nonmammalian chromosomes	PAC-, BAC-, YAC-vector elements	See above	See above
	Mammalian artificial chromosomes	Chromosomal elements	See above	See above
	Retroviral vector	Viral sequences	See above	See above
	Cloning by transfer of genetically modified nuclei	Selectable marker genes[1]	See above	See above
Loss of function	Cloning by transfer of genetically modified nuclei	Selectable marker genes[1]	Deleted or nonsense-mutated species-specific sequences	Generation of PrPc gene-deficient ruminants
Exchange of function	Cloning by transfer of genetically modified nuclei	Selectable marker genes[1]	• Introduction of novel allelic variants • exchange of coding sequences in a transcription unit	• targeted alteration of milk protein genes • replacement of genes

[1] Unwanted sequences may be removed in vitro by site-specific recombinases.

after successful identification of the transgenic cells [85, 86]. However, it should be mentioned that these additional genetic engineering steps have not yet been conducted in farm animals.

Retroviral vector-mediated gene transfer methods bear the advantage of the transgene being actively delivered to the cells and integrated into the host genome. The disadvantage of the system is the above-mentioned transcriptional shutdown and heterochromatin formation of the transgenes due to the presence of the viral sequences.

1.3
Transgenic Animals with Agricultural Traits

A key element to the enhanced production of domesticated species is the development of genetically superior breeding stocks that are tailored to their maintenance conditions, and also to the marketplace. Characteristics that are generally desirable in all species include improvements in growth rates, feed conversion efficiencies, disease resistance, and a capacity to utilize low-cost or nonanimal protein diets. The attempts to improve productivity traits in farm animals by transgenesis can be divided into products designed for the consumer's consumption *per se* and for traits not affecting the food chain in the first place. The first area includes stimulation of growth rates, food conversion and alteration of carcass and milk composition. The second field aims at the improvement of fiber products, enhanced disease resistance and the introduction of novel biochemical pathways. Although the transgene product in this field is not meant to be used as food, the meat or milk of genetically modified animals could be considered for consumption.

Initially, it should be noted that progress on the manipulation of agricultural animal traits has occurred far more slowly than was originally envisaged during the early days of transgene technology. The first reason for this is the finding that most economically important traits are controlled by multiple genes, which are still largely unknown and hence not amenable to manipulation. Even in the case where all genes contributing to a complex trait have been identified, the genetic engineering of this trait would require multiple gene transfers. The second reason is that the low efficiency of gene transfer in farm animals (see above) renders research on trangenesis costly. The third reason is that the ability to regulate expression of transgenes is still far from adequate (see above). Finally, public acceptance of genetically modified organisms in the food chain is – at least in Europe – currently not given.

Compared with mammals and fish, gene transfer experiments in chicken are somewhat limited, though the aims of gene transfer into poultry are basically identical to those used in other farm animals (for reviews, see Refs. [87, 88]).

1.3.1
Improved Growth Rate, Carcass Composition, and Feed Efficiency

1.3.1.1 Transgenes in mammalian farm animals

Among the genetically determined factors regulating growth rate and feed conversion, the genes encoding polypeptides of the growth hormone cascade are of particular interest. The positive acting growth hormone-releasing hormone (GHRH, somatoliberin) and its antagonist, somatotropin release-inhibiting factor (SRIF, somatostatin) control the production of growth hormone (GH, somatotropic hormone (STH), somatotropin). The GH action is highly dependent on the metabolic situation of the organism: low blood glucose levels result in catabolic effects (lipolysis), and a positive energy balance causes anabolic effects which is mainly governed by insulin-like growth factor 1 (IGF-1, somatomedin C). Early studies on farm animal transgenesis were influenced by the results of Palmiter et al. [89] which indicated that mice expressing excess GH grew much faster and bigger than nontransgenic control mice. A number of GH transgenic pigs and sheep were produced with human, bovine, ovine, porcine or rat GH under control of several promotors [90, 91]. GH-transgenic pigs expressing the gene constructs at high levels were found to have faster growth rates and an increased feed efficiency. The most dramatic effect of elevated GH levels in pigs was the reduction in carcass fat as transgenic pigs approached market weight [13, 92], though the constitutive and/or high-level expression of GH in pigs was found to cause a variety of pathologic side effects [13, 93, 94]. Transgenic ruminants (cattle, sheep, goat) carrying growth-promoting genes have been also generated, though no positive effects on either growth performance or carcass composition have been reported [90, 95].

It has been recognized that tight regulation of transgene expression would be required to avoid deleterious effects from continuous exposure of mammals to elevated GH, and so far most efforts to use dietary inducible promoters have failed [90]. Two studies have reported the production of growth-promoting transgenic pig lines. A metal ion-inducible promoter linked to the porcine GH gene was introduced into pigs, and a large number of transgenic founder animals were produced. Transgenic pigs were tested for metal-induced transgene expression, and animals showing high basal levels of transgene expression or plasma GH levels outside the range of nontransgenics were excluded from the study. Following this strategy, negative side effects could be avoided [91]. However, due to the random integration of transgenes by DNA microinjection and a lack of shielding sequences, the transgenic lines showed a high degree of variegated gene expression. In a second study, the expression of IGF-1 was directed to muscle by using a skeletal muscle-specific expression cassette. By avoiding the systemic effects of GH, an increase in carcass leanness and no detrimental side effects were observed [96]. Interestingly, in the context of the use of growth hormone cascade transgenes, somatic gene transfer might suffice the demands for improved growth performance. A somatic gene transfer protocol employing a singular intramuscular injection and electroporation of muscle-specific expression vectors encoding a protease-resistant GHRH resulted in elevated GH and IGF-1 levels which in turn augmented long-term growth without pathologic side effects [4].

An alternative approach to alter growth performance involves the differentiation process of muscle cells themselves. For example, the chicken c-ski proto-oncogene was found to induce myogenic differentiation. This muscle differentiation gene was introduced into pigs and cattle [90, 97] but, as was observed with the growth hormone cascade genes, either none or mainly deleterious effects of the transgene expression were reported. Myostatin, also known as growth and differentiation factor 8, negatively regulates skeletal muscle development, and deletion or nonsense mutations in the myostatin gene are responsible for double muscling in cattle [98–100]. Myostatin-deficient mice, generated by gene targeting, produced twice the muscle mass with reduced carcass fatness as wild-type mice, thus mirroring the phenotype of myostatin-mutant cattle [101]. In an additive gene transfer experiment it was shown that expression of a dominant-negative myostatin transgene also led to increased muscle development, growth performance and carcass weight [102]. Therefore, the myostatin gene is an attractive candidate for both gene targeting and gain of function experiments in farm animals.

1.3.1.2 Transgenes in fish

In relation to the farming of mammals or poultry, aquaculture is still in its infancy. Growth rates of many fish species used are naturally slow, but are currently being enhanced by traditional methods of domestication and selection [103]. Growth-promoting gene transfer programs in fish grossly used GH-based gene constructs, though due to a lack of available piscine sequences the initial experiments were conducted with mammalian GH gene constructs. The effects on growth performance were either not detectable or minimal, however. Subsequent gene transfers using fish GH sequences driven by nonpiscine promoters resulted in growth stimulatory effects in transgenic carp, catfish and tilapia of approximately two-fold increases in weight relative to controls. These experiments provided the first consistent data demonstrating that growth acceleration in fish can be achieved by transgenesis [103]. The subsequent use of all-piscine gene constructs produced fish with up to 40-fold elevated circulating GH levels and a five- to 11-fold increased weight after one year of growth [104, 105]. The pleiotropic effects in the GH-transgenic fish included changed body compositions (50% decreased fat levels), unpredictable variations in food consumption and conversion, and some pathologic side effects [103, 106]. Comparative gene transfer programs demonstrated that GH-transgenes dramatically enhanced the growth of wild, but not domesticated, fish [107]. Thus, it can be noted that in domesticated and selected farm animal species the capacity for further growth enhancement by GH may be restricted by limitations in other physiological pathways. In mammals, this is reflected by the dramatic growth stimulation in GH-transgenic mice, but not in domestic livestock that underwent many centuries of genetic selection (see above).

Salmonids are fish of high economic value which are unable to survive in waters characterized by ice and sub-zero temperatures. Antifreeze proteins (AFP) are produced by a number of fish that inhabit waters at extreme cold temperatures [108]. A possible way to solve the problem of overwintering salmon in sea cages in the northern hemisphere is the transfer of antifreeze protein genes. The AFP-transgenic salmons generated so far express the transgene at levels insufficient to confer freeze resistance [109].

1.3.2
Alteration of Milk Composition

Potential changes in milk composition, or in the primary structures of milk proteins and their presumed beneficial effects on the nutritional, physico-chemical and technological properties of milk and products derived thereof, have been reviewed extensively [110–115].

It should be noted that most transgenic strategies are in the stage of being tested in mouse models, and gene transfer technology in the improvement of milk quality is far from commercial application. Attempts to improve the processing properties of milk include changes in the casein contents and the introduction of modified milk proteins [116]. The introduction of human milk protein genes or the replacement of bovine genes by human genes may play an important role in the production of surrogates for human milk. Bovine and human milk differs considerably, and therefore cows' milk is not an ideal source of food for babies. Bovine milk can be humanized by increasing the whey protein content, e.g., by expressing the antimicrobial proteins lactoferrin and lysozyme [117, 118]. Recently, the large-scale production of human lactoferrin in the milk of transgenic cows has been reported [119].

β-Lactoglobulin is the major heat-labile whey protein of ruminant milk, but does not occur in human milk. Although β-lactoglobulin is not the only bovine milk protein with allergenic properties, it is generally assumed that milk depleted of this protein would be a better source of humanized milk. In principle, the technology for a gene knockout in ruminants is now available, and the depletion of particular proteins from milk has been carried out in mice for β-casein [120] and α-lactalbumin [121, 122]. While β-casein was found to be a nonessential component of the milk protein system, the knockout of α-lactalbumin resulted in the disruption of lactation and lactose synthesis. The biological function of β-lactoglobulin and its contribution to bovine milk physiology is not known, and therefore the side effects of a knockout approach cannot be predicted.

Lactose is the major sugar present in milk and is synthesized by the lactose synthase complex composed of a galactosyltransferase and α-lactalbumin. The majority of the adult population suffers from intestinal disorders as a consequence of lactose maldigestion that results from the physiological down-regulation at weaning of the intestinal lactose-hydrolyzing enzyme. Low-lactose milk *in vivo* has been generated by partial inhibition of the α-lactalbumin gene by a RNA-antisense approach [123] and by the mammary gland-specific expression of an intestinal lactase [124].

Milk appears to be an ideal vehicle for the development of nutraceuticals, i. e., dietary supplements, functional and medical foods [125]. For example, pilot studies have demonstrated the expression of protective immunoglobulins against pathogens of the digestive tract; hence milk could serve for passive immunization programs [126, 127]. Other examples for the use of milk as a nutraceutical are listed in Table 1.2.

In an attempt to increase milk production and lactose content for the benefit of the suckling offspring, bovine α-lactalbumin gene transgenic pigs were generated [128]. However, the transgene was expressed at levels where a boost of pre-weaning growth rates in piglets could not be expected.

Table 1.2. Proposed modifications of milk composition.

Target gene	Gene transfer	Transgene effect	Overall effect
α- and β-caseins	Gain of function	Improved cheese-making properties, increased Ca^{2+} content	Improved technical processing
k-casein	Gain of function	Improved heat stability, decreased micelle size, decreased coagulation	Improved technical processing
Novel phosphorylation and proteolytic sites in caseins	Exchange of function	Increased Ca^{2+} content, improved cheese ripening	Improved technical processing
β-lactoglobulin	Loss of function	Increased temperature stability, improved digestibility, decreased allergenicity	Improved technical processing/humanized milk
Human lysozyme	Gain of function	Antimicrobial, increased cheese yield	Improved technical processing/humanized milk
Human lactoferrin	Gain of function	Antimicrobial	Humanized milk
α-lactalbumin	Reduction of function	Decreased lactose	Improved nutritional value
Lactase	Gain of function	Decreased lactose	Improved nutritional value
Acetyl CoA carboxylase	Reduction of function	Decreased fat content	Improved nutritional value
Immunoglobulin genes	Gain of function	Passive oral immunization	Nutraceutical
Antigen genes	Gain of function	Active oral immunization	Nutraceutical
Phe-free α-lactalbumin	Exchange of function	Source of amino acids for patients suffering from phenylketonuria	Nutraceutical

1.3.3
Improved Animal Health

In contrast to the benefits achieved by selection for production performance, attempts to select for improved disease resistance by conventional breeding programs have not been successful. The reduction of disease susceptibility in livestock would be beneficial in terms of animal welfare, and also of economic importance. An improved health status in animal production results in improvements in both production and reproductive performance. Either somatic or germ-line gene transfer can be applied, though in order to introduce new disease resistance traits in farm animals germ-line transmission is required. The strategies of enhancing disease resistance by transgenesis have been reviewed extensively elsewhere [129–131]. Somatic gene transfer mainly focuses on DNA vaccines (genetic immunization, see above). The integration of stable germ-line transmission (and in some cases the expression of gene constructs) designed to confer disease resistance could be demonstrated, though final proof of the successful generation of resistant farm animals – that is, challenge with an infectious pathogen – has not been reported.

1.3.3.1 Additive gene transfer of resistance genes

The term "intracellular immunization" was originally used for the overexpression in the host of an aberrant form (dominant-negative mutant) of a viral protein that is able to interfere strongly with replication of the wild-type virus. This definition was then extended to all approaches based on intracellular expression of transgene products which inhibit the replication of pathogens in host organisms (see Ref. [129] and references therein). Initial studies in farm animals included the "classical" approach of overexpression of a viral protein in transgenic sheep [132], transgenic rabbits expressing antisense constructs complementary to adenovirus RNA [133], and the transfer of the specific disease resistance gene Mx1 of mice into swine [134]. The mouse Mx1 gene is one of the few examples of a single genetic locus which encodes a disease resistance trait. Mice carrying the autosomal dominant Mx1 allele are resistant to influenza viruses, and transfer of the Mx1 gene into susceptible mice that lacked the Mx1 allele was sufficient to restore virus resistance [135]. Swine are susceptible to influenza and provide a substantial reservoir for swine influenza viruses. Different gene constructs containing the mouse Mx1 cDNA controlled by two constitutive promoters and the inducible murine Mx1 promoter were transferred into swine. Constitutive Mx1 expression was found to be detrimental to the organism, and the inducible Mx1 construct responded to stimuli by RNA synthesis, though to levels insufficient to produce detectable amounts of Mx1 protein. This again highlights the importance of tight transgene regulation for the positive outcome of gene transfer experiments [134].

"Congenital immunization" is defined as transgenic expression and germ-line transmission of a gene encoding an immunoglobulin specific for a pathogen, and therefore providing congenital immunity without prior exposure to that pathogen. The approach was tested in farm animals by expressing the gene constructs encoding mouse monoclonal antibodies in transgenic rabbits, pigs and sheep [136, 137]. Both experiments resulted in transgene expression, but also revealed some unexpected findings, e.g., aberrant sizes of the transgenic antibody, or minimal antigen binding capacity. Following this idea, preformed antibodies against a virus causing a neonatal disease were expressed in the mammary gland of mice (see above). When transferred to farm animals, this could improve the protection of suckling offspring through colostrum-delivered antibodies [138]. It remains to be investigated, however, whether the efforts required to optimize the concept of congenital immunization are justified by its benefits in terms of increasing disease resistance in a certain species. Following this route, one has also to bear in mind that a given infectious pathogen will readily be able to escape the transgenic animal's immunity by changing its antigenic determinants. The transfer of antibody encoding genes into farm animals in general is of great importance for the production of therapeutic antibodies for human medicine [139].

"Extracellular immunization" refers to transgene products which exhibit their antipathogenic function extracellularly. This strategy includes the systemic or local expression of immunomodulatory cytokines and pathogen defense molecules of the innate immune system [131]. Peptide-based antimicrobial defense is an evolutionary ancient mechanism of host response that is found in a wide range of animals, from insects to mammals. The small lytic peptides interact with lipid bilayer membranes to cause osmotic disruption and cell death. Bacterial, protozoan, fungal and damaged eukaryotic cells are most susceptible to disruption [140, 141]. Experiments with transgenic mice and fish demonstrated the power of overexpressing antimicrobial peptides in generating enhanced resistance of the animals against invading microbia [142–144]. Lysostaphin is a potent bacterial peptidoglycan hydrolase with a specific bactericidal activity against *Staphylococcus aureus*, the major contagious pathogen in mastitis. Mammary gland-specific expression of lysostaphin could therefore generate mastitis-resistant cows and reduce the major economic burden of the dairy industry. The proof of principle of mammary gland-expressed lysostaphin conferring protection against staphylococcal infection was provided in mice [145].

1.3.3.2 Gene targeting of susceptibility genes

Another potential means to improve disease resistance is the removal of disease susceptibility genes by homologous recombination in order to create null alleles, or to replace a disease allele by a resistance allele. This concept is discussed for the generation of cattle and sheep that are resistant to spongiform encephalopathies caused by infectious prion particles. Mice devoid of the endogenous prion protein PrPc cannot be infected with the infectious particles, and the loss of function mice show now gross abnormalities [146]. Therefore, targeted inactivation of the gene encoding PrPc in cattle and sheep might create bovine spongiform encephalitis (BSE)- or scrapie-resistant animals. Successful targeting of the PrP locus in sheep fetal fibroblasts for the use of nuclear transport was reported, but no living

PrP-deficient animals were obtained [58]. The physiological role of the endogenous prion protein and the genetic components of susceptibility to the disease are still largely unknown, though future studies will indicate whether fundamental differences between mice and ruminants will prevent the generation of such loss of function animals.

1.3.4
Improved Biochemical Pathways

Transgenesis allows the transfer of genetic information across species barriers. Combined with functional promoter elements, nonmammalian genes can be expressed in farm animals in order to modify intermediary metabolism. To address the problem of manure-based environmental pollution in pig production, transgenic pigs were generated which expressed the *Escherichia coli* phytase gene in the salivary gland. As a result, the transgenic pigs were able to digest the phosphorus in phytate, which is the most abundant source of this element in the pig diet. Consequently, fecal excretion of phosphorus by transgenic pigs was drastically reduced [147].

The introduction of new biochemical pathways that increase the availability of specific nutrients that are currently rate-limiting for animal production is a challenging task, and research programs in ruminants to investigate the transfer of prokaryotic genes to introduce cysteine, threonine and lysine biosynthesis or a functional glyoxylate cycle (to convert the major rumen metabolite acetate to glucose) have been unsuccessful [148].

1.3.5
Improved Wool Production

Improved wool production by the means of transgenesis can be achieved by the abundant supply of cysteine which is required for keratin synthesis. Keratins are the major structural proteins of wool fibers, and cysteine is the rate-limiting amino acid for wool production. Dietary addition of cysteine does not increase wool production due to the digestive degradation of cysteine. However, when bacterial cysteine biosynthesis genes were transferred into sheep, there was no improvement in wool growth because the transgene regulation was not adequate to integrate the novel pathway into the existing biochemical homeostasis [149].

A second approach to improve wool quality aimed at modifying the wool fiber's protein composition. The gene encoding the wool intermediate filament keratin was overexpressed in transgenic sheep, and an alteration in fiber ultrastructure was observed, but the changes were not positive for the processing quality of the wool [150]. Additional investigations are in progress using transgenes encoding keratins rich in amino acids positive for wool quality [151].

Another attempt to improve wool production was the targeted overexpression of IGF-1 in wool follicles [152]. Transgenic sheep showed increased wool production in the first generation, but the effect did not persist to the second generation, indicating that the transgene was not shielded from chromosomal position effects [153].

1.4
Transgenic Farm Animals and Biosafety Issues

The biosafety of transgenics is of concern to the consumer, to the environment, and to the transgenic animals *per se*. The food safety evaluation of transgenic animal products (novel foods, nutraceuticals) is governed by national and international authorities (for reviews, see Refs. [125, 154]). In theory, the meat or milk of transgenic farm animals generated for nonfood purposes could also be consumed. For example, in the case of the "phosphorus-friendly" pig, the transgene is exclusively expressed in the salivary gland; hence the meat contains only the transgene DNA and not the transgene product. For the consumer's safety, a risk assessment relating to the toxicity and transfer of ingested transgenes must be carried out. DNA is an essential part of nutrition, and in general in nontoxic after uptake through the digestive tract. Large alimentary DNA fragments appear to survive gastrointestinal passage, enter the bloodstream and are then found in the nuclei of various cell types [155]. It is clear, however, that this does not lead to vertical gene transfer, and there is no reason why a transgene should integrate into the consumer's genome preferentially compared with nonrecombinant DNA.

The biorisks generated by transgenesis in farm animals are based on the animal species, the gene transfer method, the nature of the transgene, and the fate of the transgenics [156, 157]. Genetically modified organisms are not allowed to multiply uncontrolled in the environment, and large farm animals have little chance to escape and cross-breed with wild animals. However, most concern is warranted for the possible escape of transgenic growth-enhanced fish in a cage culture with access to free waters [158], and subsequent spreading of the transgene into the natural population [159]. Therefore, fish in such cages must be rendered sterile or must be kept in artificial containment facilities.

Physico-chemical gene transfer methods and cloning by nuclear transfer do not hold intrinsic biorisks because the DNA is stably integrated into the host genome. Viral vectors are generally liable to the risk of recombination with wild-type viruses that in turn might create the means to spread the transgene.

The major biorisks due to the transgenes are for the animals themselves (see below). It is favorable if all DNA fragments that are transferred are characterized by sequencing, though such a venture may not always be applicable, especially for large gene constructs such as artificial chromosomes. Transgenic animals designed for human consumption will be devoid of marker genes and other sequences not required for transgene function; however, methods to remove such sequences are available (see above).

A transgenic founder animal has, a priori, certain unknown biological properties and must be tested for stable integration and transmission of the transgene. The F_1 generation can be tested for stable and promoter-specific (i.e., nonectopic) expression of the transgene, while homozygous animals may be produced to prove the freedom of insertional mutagenesis. It is clear that, before the use of transgenic animals for production purposes, the desirable and possibly undesirable transgene effects should be extensively checked using both laboratory and veterinarian methods. In this respect, programs for the systematic risk assessment relating to transgenic farm animal welfare and breeding have been developed [160, 161].

1.5
Conclusions

Gene transfer technologies enable the direct introduction of novel traits into farm animals. The biological performance of the transgene in the animal can be measured in a few generations. Breeding success can be achieved in a shorter time compared with classical breeding programs, although gene transfer can be only carried out in a limited number of animals. Once the positive biological outcome of the artificially added or altered allele in the founder animal's offspring is established, the novel trait must be spread through the production population using conventional strategies. To date, no genetically engineered food obtained from farm animals has reached the marketplace, mainly due to the above-mentioned difficulties in generating transgenic farm animals, the frequent failure to transfer the proof of principle success in basic research models to farm animals, and the lack of public acceptance for novel foods. The reasons for this missing acceptance, and the ethics relating to transgenics, are discussed elsewhere [162–164].

Farm animal transgenics has undoubted importance in the field of biomedicine, yet even if the public's perception were to become positive for transgenic food, and optimized gene transfer technologies were to improve the generation and exploitation of transgenics, most livestock-derived products would still be limited to niche markets. These limitations are the clear result of the above-mentioned problems associated with dissemination of the transgene.

Whilst the transgenics and genomics of farm animals have developed in parallel [165], it is clear that these two important technologies will develop further during the next few years. Ultimately, it is likely that both approaches, when operating in tandem, will lead to the generation of genetically modified farm animals that meet the demands of both productivity and biosafety.

The literature regarding the creation of transgenic animals is complemented by a recent report on transgenic cattle which were generated by nuclear transfer (see 1.1.4) and carry human artificial chromosomes (HACs) encoding the human immunoglobulin locus [Y. Kuroiwa, P. Kasinathan, Y. J. Choi, R. Naeem, K. Tomizuka, E.J. Sullivan, J. G. Knott, A. Duteau, R.A. Goldsby, B.A. Osborne, I. Ishida, and J. M. Robl (**2002**) *Nature Biotechnol.* 20, 889–894]. This is the first report on stable 'transchromosomic' farm animals, which show no chromosomal positioning effects of transgenes observed upon integration into the host genome (see 1.1.1). The experimental design aimed at the production of therapeutic antibodies [Y. Echelard, R. Meade (**2002**) *Nature Biotechnol.* 20, 881–882], the approach, however, can also be applied for the genetic alteration of production traits. Cloning by transfer of transgenic nuclei also generated transgenic cattle, which produce higher levels caseins in the milk [B. Brophy, G. Smolensk, T. Wheeler, D. Wells, P. L'Huiller, G. Laible (2003) *Nature Biotechnol.* 21, 157–16]. This is one of the first successful approaches to improve the functional properties of diary milk (see 1.3.2 and [C. N. Karatzas (2003) *Nature Biotechnol.* 21, 138–139]). Finally, 1.1.5 is complemented by a comprehensive review on recent progress in avian transgenesis [R. Ivarie (2003) *Trends Biotechnol.* 21, 14–19].

References

1 J. W. Gordon, G. A. Scangos, D. J. Plotkin, J. A. Barbosa, F. H. Ruddle (1980) *Proc. Natl. Acad. Sci. USA* 77, 7380–7384.

2 J. W. Gordon, F. H. Ruddle (1981) *Science* 214, 1244–1246.

3 M. Düchler, M. Pengg, S. Schüller, F. Pfneisl, C. Bugingo, G. Brem, E. Wagner, K. Schellander, M. Müller (2002) *J. Gene Med.*, 4, 282–291.

4 R. Draghia-Akli, M. L. Fiorotto, L. A. Hill, P. B. Malone, D. R. Deaver, R. J. Schwartz (1999) *Nature Biotechnol.* 17, 1179–1183.

5 C. W. Beard, P. W. Mason (1998) *Nature Biotechnol.* 16, 1325–1328.

6 G. Brem, B. Brenig, H. M. Goodman, R. C. Selden, F. Graf, B. Kruff, K. Springman, J. Hondele, J. Meyer, E.-L. Winnacker, H. Kräußlich (1985) *Reprod. Dom. Anim.* 20, 251–252.

7 R. E. Hammer, V. G. Pursel, C. E. Rexroad Jr., R. J. Wall, R. D. Palmiter, R. L. Brinster (1985) *Nature* 315, 680–683.

8 R. M. Petters, J. R. Sommer (2000) *Transgenic Res.* 9, 347–351.

9 E. Wolf, W. Schernthaner, V. Zakhartchenko, K. Prelle, M. Stojkovic, G. Brem (2000) *Exp. Physiol.* 85, 615–625.

10 L. M. Houdebine (2000) *Transgenic Res.* 9, 305–320.

11 D. K. Cooper, B. Gollackner, D. H. Sachs (2002) *Annu. Rev. Med.* 53, 133–147.

12 M. Cascalho, J. L. Platt (2001) *Nature Rev. Immunol.* 1, 154–160.

13 V. G. Pursel, C. A. Pinkert, K. F. Miller, D. J. Bolt, R. G. Campbell, R. D. Palmiter, R. L. Brinster, R. E. Hammer (1989) *Science* 244, 1281–1288.

14 R. J. Wall (2002) *Theriogenology* 57, 189–201.

15 F. Gandolfi (1999) *Theriogenology* 53, 127–137.

16 J. M. Robl (1999) *Nature Biotechnol.* 17, 636–637.

17 G. Brem, M. Müller (1994) in *Animals with novel genes*, ed. N. Maclean (Cambridge University Press, Cambridge, UK), pp. 179–244.

18 H. Niemann, W. A. Kues (2000) *Anim. Reprod. Sci.* 60-61, 277–293.

19 R. J. Wall (1996) *Theriogenology* 45, 57–68.

20 R. J. Wall, G. E. J. Seidel (1992) *Theriogenology* 38, 337–357.

21 U. Reichart, R. Kappler, H. Schernthan, E. Wolf, M. Müller, G. Brem, B. Aigner (2000) *Biochem. Biophys. Res. Commun.* 269, 496–501.

22 T. Rijkers, A. Peetz, U. Rüther (1994) *Transgenic Res.* 3, 203–215.

23 N. Zinovieva, C. Lassnig, D. Schams, U. Besenfelder, E. Wolf, S. Müller, L. Frenyo, J. Seregi, M. Müller, G. Brem (1998) *Transgenic Res.* 7, 437–447.

24 B. Aigner, U. Besenfelder, J. Seregi, L. V. Frenyo, T. Sahintoth, G. Brem (1996) *Transgenic Res.* 5, 405–411.

25 P. Giraldo, L. Montoliu (2001) *Transgenic Res.* 10, 83–103.

26 G. Brem, U. Besenfelder, B. Aigner, M. Müller, I. Liebl, G. Schütz, L. Montoliu (1996) *Mol. Reprod. Dev.* 44, 56–62.
27 A. G. West, M. Gaszner, G. Felsenfeld (2002) *Genes Dev.* 16, 271–288.
28 U. Besenfelder, N. Zinovieva, E. Dietrich, B. Sohnrey, W. Holtz, G. Brem (1994) *Vet. Rec.* 135.
29 U. Besenfelder, G. Brem (1993) *J. Reprod. Fertil.* 99, 53–56.
30 U. Besenfelder, J. Mödl, M. Müller, G. Brem (1997) *Theriogenology* 47, 1051–1060.
31 U. Besenfelder, G. Brem (1998) *Theriogenology* 50, 739–745.
32 R. Jaenisch (1976) *Proc. Natl. Acad. Sci. USA* 73, 1260–1264.
33 H. M. Temin (1989) *Genome* 31, 17–22.
34 A. Pfeifer, I. M. Verma (2001) *Annu. Rev. Genomics Hum. Genet.* 2, 177–211.
35 A. W. Chan, E. J. Homan, L. U. Ballou, J. C. Burns, R. D. Bremel (1998) *Proc. Natl. Acad. Sci. USA* 95, 14028–14033.
36 R. A. Cabot, B. Kuhholzer, A. W. Chan, L. Lai, K. W. Park, K. Y. Chong, G. Schatten, C. N. Murphy, L. R. Abeydeera, B. N. Day, R. S. Prather (2001) *Anim. Biotechnol.* 12, 205–214.
37 D. Pannell, J. Ellis (2001) *Rev. Med. Virol.* 11, 205–217.
38 C. Lois, E. J. Hong, S. Pease, E. J. Brown, D. Baltimore (2002) *Science* 295, 868–872.
39 M. R. Capecchi (2000) *Nature Genet.* 26, 159–161.
40 J. A. Piedrahita (2000) *Theriogenology* 53, 105–116.
41 K. Prelle, I. M. Vassiliev, S. G. Vassilieva, E. Wolf, A. M. Wobus (1999) *Cells Tissues Organs* 165, 220–236.
42 S. M. Willadsen (1986) *Nature* 320, 63–65.
43 E. Wolf, V. Zakhartchenko, G. Brem (1998) *J. Biotechnol.* 65, 99–110.
44 K. H. S. Campbell, J. McWhir, W. A. Ritchie, I. Wilmut (1996) *Nature* 380, 64–66.
45 I. Wilmut, A. E. Schnieke, J. McWhir, A. J. Kind, K. H. S. Campbell (1997) *Nature* 385, 810–813.
46 D. N. Wells, P. M. Misica, H. R. Tervit (1999) *Biol. Reprod.* 60, 996–1005.
47 V. Zakhartchenko, V. Alberio, M. Stojkovic, K. Prelle, W. Schernthaner, P. Stojkovic, H. Wenigerkind, R. Wanke, M. Düchler, R. Steinborn, M. Müller, G. Brem, E. Wolf (1999) *Mol. Reprod. Dev.* 54, 264–272.
48 Y. Kato, T. Tani, Y. Sotomaru, K. Kurokawa, J. Kato, H. Doguchi, H. Yasue, Y. Tsunoda (1998) *Science* 282, 2095–2098.
49 A. Baguisi, E. Behboodi, D. T. Mclican, J. S. Pollock, M. M. Destrempes, C. Cammuso, J. L. Williams, S. D. Nims, C. A. Porter, P. Midura, M. J. Palacios, S. L. Ayres, R. S. Denniston, M. L. Hayes, C. A. Ziomek, H. M. Meade, R. A. Godke, W. G. Gavin, E. W. Overstrom, Y. Echelard (1999) *Nature Biotechnol.* 17, 456–461.
50 A. Onishi, M. Iwamoto, T. Akita, S. Mikawa, K. Takeda, T. Awata, H. Hanada, A. C. Perry (2000) *Science* 289, 1188–1190.
51 I. A. Polejaeva, S. H. Chen, T. D. Vaught, R. L. Page, J. Mullins, S. Ball, Y. Dai, J. Boone, S. Walker, D. L. Ayares, A. Colman, K. H. Campbell (2000) *Nature* 407, 86–90.
52 J. Betthauser, E. Forsberg, M. Augenstein, L. Childs, K. Eilertsen, J. Enos, T. Forsythe, P. Golueke, G. Jurgella, R. Koppang, T. Lesmeister, K. Mallon, G. Mell, P. Misica, M. Pace, M. Pfister-Genskow, N. Strelchenko, G. Voelker, S. Watt, S. Thompson, M. Bishop (2000) *Nature Biotechnol.* 18, 1055–1059.
53 P. Chesné, P. G. Adenot, C. Viglietta, M. Baratte, L. Boulanger, J.-P. Renard (2002) *Nature Biotechnol.* 20, 366–369.
54 J. B. Cibelli, S. L. Stice, P. J. Golueke, J. J. Kane, J. Jerry, C. Blackwell, F. A. Ponce de Leon, J. M. Robl (1998) *Science* 280, 1256–1258.
55 A. E. Schnieke, A. J. Kind, W. A. Ritchie, K. Mycock, A. R. Scott, M. Ritchie, I. Wilmut, A. Colman, K. H. S. Campbell (1997) *Science* 278, 2130–2133.
56 S. L. Stice, J. M. Robl, F. A. Ponce de Leon, J. Jerry, P. G. Goulke, J. B. Cibelli, J. J. Kane (1998) *Theriogenology* 49, 129–138.
57 K. Wells, K. Moore, R. Wall (1999) *Nature Biotechnol.* 17, 25–26.

58 C. Denning, S. Burl, A. Ainslie, J. Bracken, A. Dinnyes, J. Fletcher, T. King, M. Ritchie, W. A. Ritchie, M. Rollo, P. de Sousa, A. Travers, I. Wilmut, A. J. Clark (2001) *Nature Biotechnol.* 19, 559–562.
59 L. Lai, D. Kolber-Simonds, K. W. Park, H. T. Cheong, J. L. Greenstein, G. S. Im, M. Samuel, A. Bonk, A. Rieke, B. N. Day, C. N. Murphy, D. B. Carter, R. J. Hawley, R. S. Prather (2002) *Science* 295, 1089–1092.
60 Y. Dai, T. D. Vaught, J. Boone, S. H. Chen, C. J. Phelps, S. Ball, J. A. Monahan, P. M. Jobst, K. J. McCreath, A. E. Lamborn, J. L. Cowell-Lucero, K. D. Wells, A. Colman, I. A. Polejaeva, D. L. Ayares (2002) *Nature Biotechnol.* 20, 251–255.
61 K. J. McCreath, J. Howcroft, K. H. S. Campbell, A. Colman, A. E. Schnieke, A. J. Kind (2000) *Nature* 405, 1066–1069.
62 K. Wells (2001) *Nature Biotechnol.* 19, 529–530.
63 J. B. Cibelli, K. H. Campbell, G. E. Seidel, M. D. West, R. P. Lanza (2002) *Nature Biotechnol.* 20, 13–14.
64 J. P. Renard, Q. Zhou, D. LeBourhis, P. Chavatte-Palmer, I. Hue, Y. Heyman, X. Vignon (2002) *Theriogenology* 57, 203–222.
65 R. P. Lanza, J. B. Cibelli, D. Faber, R. W. Sweeney, B. Henderson, W. Nevala, M. D. West, P. J. Wettstein (2001) *Science* 294, 1893–1894.
66 I. Wilmut (2002) *Nature Med.* 8, 215–216.
67 M. M. Perry, H. M. Sang (1993) *Transgenic Res.* 2, 125–133.
68 J. Love, C. Gribbin, C. Mather, H. Sang (1994) *Biotechnology* 12, 60–63.
69 D. W. Salter, E. J. Smith, S. H. Hughes, S. E. Wright, A. M. Fadly, R. L. Witter, L. B. Crittenden (1986) *Poult. Sci.* 65, 1445–1458.
70 H. Iba (2000) *Dev. Growth Differ.* 42, 213–218.
71 A. J. Harvey, G. Speksnijder, L. R. Baugh, J. A. Morris, R. Ivarie (2002) *Nature Biotechnol.* 19, 396–399.
72 N. Allioli, G. Verdier, C. Legras (1997) *Methods Mol. Biol.* 62, 425–432.
73 B. Pain, P. Chenevier, J. Samarut (1999) *Cells Tissues Organs* 165, 212–219.
74 Z. Y. Zhu, Y. H. Sun (2000) *Cell Res.* 10, 17–27.
75 F. Y. Sin, S. P. Walker, J. E. Symonds, U. K. Mukherjee, J. G. Khoo, I. L. Sin (2000) *Mol. Reprod. Dev.* 56, 285–288.
76 N. Maclean (1998) *Mutat. Res.* 399, 255–266.
77 A. Iyengar, F. Müller, N. Maclean (1996) *Transgenic Res.* 5, 147–166.
78 D. Garrick, S. Fiering, D. I. K. Martin, E. Whitelaw (1998) *Nature Genet.* 18, 56–59.
79 D. R. Dorer (1997) *Transgenic Res.* 6, 3–10.
80 J. M. Vos (1999) *Curr. Opin. Mol. Ther.* 1, 204–215.
81 H. F. Willard (2000) *Science* 290, 1308–1309.
82 R. Saffery, K. H. Choo (2002) *J. Gene Med.* 4, 5–13.
83 G. Hadlaczky (2001) *Curr. Opin. Mol. Ther.* 3, 125–132.
84 B. Hohn, A. A. Levy, H. Puchta (2001) *Curr. Opin. Biotechnol.* 12, 139–143.
85 C. Gorman, C. Bullock (2000) *Curr. Opin. Biotechnol.* 11, 455–460.
86 A. Baer, J. Bode (2001) *Curr. Opin. Biotechnol.* 12, 473–480.
87 A. M. Verrinder Gibbins (1998) *Nature Biotechnol.* 16, 1013–1014.
88 A. M. Verrinder Gibbins (1998) *Anim. Biotechnol.* 9, 173–179.
89 R. D. Palmiter, R. L. Brinster, R. E. Hammer, M. E. Trumbauer, M. G. Rosenfeld, N. C. Birnberg, R. M. Evans (1982) *Nature* 360, 611–615.
90 V. G. Pursel (1998) in *Animal Breeding – Technology for the 21st century*, ed. A. J. Clark (Harwood Academic Publishers, Amsterdam NL), pp. 183–200.
91 M. B. Nottle, H. Nagashima, P. J. Verma, Z. T. Du, C. G. Grupen, S. M. McIlfatrick, R. J. Ashman, M. P. Harding, C. Giannakis, P. L. Wigley, I. G. Lyons, D. T. Harrison, B. G. Luxford, R. G. Campbell, R. J. Crawford, A. J. Robins (1999) in *Transgenic animals in agriculture*, eds. J. D. Murray, G. B. Anderson, A. M. Oberbauer, M. M. McGloughlin (CAB International, Wallingford UK), pp. 145–156.

92 M. B. Solomon, V. G. Pursel, E. W. Paroczay, D. J. Bolt (**1994**) *J. Anim. Sci.* 72, 1242–1246.
93 K. M. Ebert, T. E. Smith, F. C. Buonomo, E. W. Overstrom, M. J. Low (**1990**) *Anim. Biotechnol.* 1, 145–159.
94 C. A. Pinkert, E. J. Galbreath, C. W. Yang, L. J. Striker (**1994**) *Transgenic Res.* 3, 401–405.
95 V. G. Pursel, C. E. J. Rexroad (**1993**) *J. Anim. Sci.* 71 Suppl. 3, 10–19.
96 V. G. Pursel, A. D. Mitchell, R. J. Wall, M. B. Solomon, M. E. Coleman, R. J. Schwartz (**2001**) in *Molecular Farming*, eds. J.-P. Toutant, E. Balzs (INRA Editions, Paris F), pp. 77–86.
97 R. A. Bowen, M. L. Reed, A. Schnieke, G. E. Seidel, J. A. Stacey, W. K. Thomas, O. Kajikawa (**1994**) *Biol. Reprod.* 50, 664–668.
98 A. C. McPherron, S. J. Lee (**1997**) *Proc. Natl. Acad. Sci. USA* 94, 12457–12461.
99 R. Kambadur, M. Sharma, T. P. Smith, J. J. Bass (**1997**) *Genome Res.* 7, 910–916.
100 L. Grobet, L. J. Martin, D. Poncelet, D. Pirottin, B. Brouwers, J. Riquet, A. Schoeberlein, S. Dunner, F. Menissier, J. Massabanda, R. Fries, R. Hanset, M. Georges (**1997**) *Nature Genet.* 17, 71–74.
101 A. C. McPherron, A. M. Lawler, S. J. Lee (**1997**) *Nature* 387, 83–90.
102 J. Yang, T. Ratovitski, J. P. Brady, M. B. Solomon, K. D. Wells, R. J. Wall (**2001**) *Mol. Reprod. Dev.* 60, 351–361.
103 R. A. Dunham, R. H. Devlin (**1999**) in *Transgenic animals in agriculture*, eds. J. D. Murray, G. B. Anderson, A. M. Oberbauer, M. M. McGloughlin (CAB International, Wallingford UK), pp. 209–229.
104 R. H. Devlin, T. Y. Yesaki, C. A. Biagi, E. M. Donaldson, P. Swanson, W.-K. Chan (**1994**) *Nature* 371, 209–210.
105 S. J. Du, Z. Gong, G. L. Fletcher, M. A. Shears, M. J. King, D. R. Idler, C. L. Hew (**1992**) *Biotechnology* 10, 176–181.
106 G. L. Fletcher, S. V. Goddard, M. Shears, A. Sutterlin, C. L. Hew (**2001**) in *Molecular Farming*, eds. J.-P. Toutant, E. Balázs (INRA Editions, Paris F), pp. 57–65.
107 R. H. Devlin, C. A. Biagi, T. Y. Yesaki, D. E. Smailus, J. C. Byatt (**2001**) *Nature* 409, 781–782.
108 G. L. Fletcher, C. L. Hew, P. L. Davies (**2001**) *Annu. Rev. Physiol.* 63, 359–390.
109 C. L. Hew, R. Poon, F. Xiong, S. Gauthier, M. A. Sheras, M. J. King, P. L. Davies, G. L. Fletcher (**1999**) *Transgenic Res.* 8, 405–414.
110 J. D. Murray (**1999**) *Theriogenology* 51, 149–159.
111 E. A. Maga, J. D. Murray (**1995**) *Bio/Technology* 13, 1452–1456.
112 B. Pintado, A. Gutierrez-Adan (**1999**) *Reprod. Nutr. Dev.* 39, 535–544.
113 B. Whitelaw (**1999**) *Nature Biotechnol.* 17, 135–136.
114 I. Wilmut, A. L. Archibald, S. Harris, M. McClenaghan, J. P. Simons, C. B. A. Whitelaw, A. J. Clark (**1990**) *J. Reprod. Fertil.* 41, 135–146.
115 A. J. Clark (**1992**) *J. Cell. Biochem.* 49, 121–127.
116 R. Jiminez-Flores, T. Richardson (**1985**) *J. Diary Sci.* 71, 2640–2654.
117 E. A. Maga, G. B. Anderson, M. C. Huang, J. D. Murray (**1994**) *Transgenic Res.* 3, 36–42.
118 P. Krimpenfort, A. Rademakers, W. Eyestone, A. van der Schans, S. van den Broek, P. Kooiman, E. Kootwijk, G. Platenburg, F. Pieper, R. Strijker, H. DeBoer (**1991**) *Bio/Technology* 9, 844–847.
119 P. H. C. van Berkel, M. M. Welling, M. Geerts, H. A. van Veen, B. Ravensbergen, M. Salaheddine, E. K. J. Pauwels, F. Pieper, J. H. Nuijens, P. H. Nibbering (**2002**) *Nature Biotechnol.* 20, 484–487.
120 S. Kumar, A. R. Clarke, M. L. Hooper, D. S. Horne, A. J. R. Law, J. Leaver, A. Springbett, E. Stevenson, J. P. Simons (**1994**) *Proc. Natl. Acad. Sci. USA* 91, 6138–6142.
121 M. G. Stinnakre, J. L. Vilotte, S. Soulier, J. C. Mercier (**1994**) *Proc. Natl. Acad. Sci. USA* 91, 6544–6548.
122 A. Stacey, A. Schnieke, M. Kerr, A. Scott, C. McKee, I. Cottingham, B. Binas, C. Wilde, A. Colman (**1995**) *Proc. Natl. Acad. Sci. USA* 92, 2835–2839.

123 P. J. L'Huillier, S. Soulier, M. G. Stinnakre, L. Lepourry, S. R. Davis, J. C. Mercier, J. L. Vilotte (**1996**) *Proc. Natl. Acad. Sci. USA* 93, 6698–6703.

124 B. Jost, J. L. Vilotte, I. Duluc, J. L. Rodeau, J. N. Freund (**1999**) *Nature Biotechnol.* 17, 160–164.

125 V. Brower (**1998**) *Nature Biotechnol.* 16, 728–731.

126 A. F. Kolb, L. Pewe, J. Webster, S. Perlman, C. B. Whitelaw, S. G. Siddell (**2001**) *J. Virol.* 75, 2803–2809.

127 J. Castilla, B. Pintado, I. Sola, J. M. Sánchez-Morgado, L. Enjuanes (**1998**) *Nature Biotechnol.* 16, 349–354.

128 G. T. Bleck, B. R. White, D. J. Miller, M. B. Wheeler (**1998**) *J. Anim. Sci.* 76, 3072–3078.

129 M. Müller, G. Brem (**1996**) *J. Biotechnol.* 44, 233–242.

130 M. Müller, G. Brem (**1991**) *Experientia* 47, 923–934.

131 M. Müller, G. Brem (**1998**) *Rev. sci. tech. OIE* 17, 365–378.

132 J. E. Clements, L. B. Hu, L. Lindstrom, A. Powell, C. Rexroad, M. C. Zink (**1996**) *AIDS Res. Hum. Retroviruses* 12, 421–423.

133 L. K. Ernst, V. I. Zakcharchenko, N. M. Suraeva, T. I. Ponomareva, O. I. Miroshnichenko, M. I. Prokof'ev, T. I. Tikchonenko (**1990**) *Theriogenology* 35, 1257–1271.

134 M. Müller, B. Brenig, E. L. Winnacker, G. Brem (**1992**) *Gene* 121, 263–270.

135 H. Arnheiter, S. Skuntz, M. Noteborn, S. Chang, E. Meier (**1990**) *Cell* 62, 51–61.

136 D. Lo, V. Pursel, P. J. Linto, E. Sandgren, R. Behringer, C. Rexroad, R. D. Palmiter, R. L. Brinster (**1991**) *Eur. J. Immunol.* 21, 25–30.

137 U. H. Weidle, H. Lenz, G. Brem (**1991**) *Gene* 98, 185–191.

138 L. J. Saif, M. B. Wheeler (**1998**) *Nature Biotechnol.* 16, 334–335.

139 J. S. Logan (**1993**) *Curr. Opin. Biotechnol.* 4, 591–595.

140 R. M. Epand, H. J. Vogel (**1999**) *Biochim. Biophys. Acta* 1426, 11–28.

141 R. E. W. Hancock, R. Lehrer (**1998**) *Trends Biotechnol.* 16, 82–88.

142 X. Jia, A. Patrzykat, R. H. Devlin, P. A. Ackerman, G. K. Iwama, R. E. Hancock (**2000**) *Appl. Environ. Microbiol.* 66, 1928–1932.

143 W. A. Reed, P. H. Elzer, F. M. Enright, J. M. Jaynes, J. D. Morrey, K. L. White (**1997**) *Transgenic Res.* 6, 337–347.

144 S. Yarus, J. M. Rosen, A. M. Cole, G. Diamond (**1996**) *Proc. Natl. Acad. Sci. USA* 93, 14118–14121.

145 D. E. Kerr, K. Plaut, A. J. Bramley, C. M. Williamson, A. J. Lax, K. Moore, K. D. Wells, R. J. Wall (**2001**) *Nature Biotechnol.* 19, 66–70.

146 H. Bueler, A. Aguzzi, A. Sailer, R. A. Greiner, P. Autenried, M. Aguet, C. Weissmann (**1993**) *Cell* 73, 1339–1347.

147 S. P. Golovan, R. G. Meidinger, A. Ajakaiye, M. Cottrill, M. Z. Wiederkehr, D. J. Barney, C. Plante, J. W. Pollard, M. Z. Fan, M. A. Hayes, J. Laursen, J. P. Hjorth, R. R. Hacker, J. P. Phillips, C. W. Forsberg (**2001**) *Nature Biotechnol.* 19, 741–745.

148 K. A. Ward (**2000**) *Trends Biotechnol.* 18, 99–102.

149 C. S. Bawden, A. V. Sivaprasad, P. J. Verma, S. K. Walker, G. E. Rogers (**1995**) *Transgenic Res.* 4, 87–104.

150 C. S. Bawden, B. C. Powell, S. K. Walker, G. E. Rogers (**1998**) *Transgenic Res.* 7, 273–287.

151 C. S. Bawden, B. C. Powell, S. K. Walker, G. E. Rogers (**1999**) *Exp. Dermatol.* 8, 342–343.

152 S. Damak, H. Y. Su, N. P. Jay, D. W. Bullock (**1996**) *Bio/Technology* 14, 185–188.

153 H.-Y. Su, N. P. Jay, T. S. Gourley, G. W. Kay, S. Damak (**1998**) *Anim. Biotechnol.* 9, 135–147.

154 M. A. Miller, J. C. Matheson III (**1997**) in *Transgenic animals – Generation and use*, ed. L. M. Houdebine (Harwood Academic Publishers, Amsterdam), pp. 563–568.

155 R. Schubbert, D. Renz, B. Schnitz, W. Doerfler (**1997**) *Proc. Natl. Acad. Sci. USA* 94, 961–966.

156 L. M. Houdebine (**1997**) in *Transgenic animals – Generation and use*, ed. L. M. Houdebine (Harwood Academic Publishers, Amsterdam), pp. 559–562.

157 O. Mayo (**2001**) in Molecular Farming, eds. J.-P. Toutant, E. Balázs (INRA Editions, Paris F), pp. 307–314.

158 T. Reichhardt (**2000**) *Nature* 406, 10–12.

159 W. M. Muir, R. D. Howard (**1999**) *Proc. Natl. Acad. Sci. USA* 96, 13853–13856.

160 J. P. Gibson (**1998**) in *Animal Breeding – Technology for the 21st century*, ed. A. J. Clark (Harwood Academic Publishers, Amsterdam NL), pp. 201–214.

161 C. G. Van Reenen, T. II. E. Meuwissen, H. Hopster, K. Oldenbroek, T. A. M. Kruip, H. J. Blokhuis (**2001**) *J. Anim. Sci.* 79, 1763–1779.

162 W. S. Burke (**1998**) *Anim. Biotechnol.* 9, 181–184.

163 K. G. Davies (**2001**) *Trends Biotechnol.* 19, 424–427.

164 J. A. Mench (**1999**) in *Transgenic animals in agriculture*, eds. J. D. Murray, G. B. Anderson, A. M. Oberbauer, M. M. McGloughlin (CAB International, Wallingford UK), pp. 251–268.

165 G. Bulfield (**2000**) *Trends Biotechnol.* 18, 10–13.

2
Genetically Modified Plants

Susanne Stirn and Horst Lörz

2.1
Methods to Establish Genetically Modified Plants

As an introduction to genetically modified plants, we would first like to describe the transformation methods which are currently used, as well as the molecular requirements for the stable introduction and expression of the gene(s) of interest.

2.1.1
Transformation Methods

In general, two different approaches to transfer foreign DNA into plant cells can be distinguished: (i) a vector-mediated transformation method (via *Agrobacterium*); and (ii) direct gene transfer methods.

2.1.1.1 *Agrobacterium* transformation

In the first case, the natural ability of the bacterial phytopathogen *Agrobacterium* to transfer DNA into plant cells is exploited. *Agrobacterium tumefaciens* and *Agrobacterium rhizogenes* are soil microorganisms that incite crown gall tumors and hairy root disease, respectively, on a wide range of dicotyledonous as well as on some monocotyledonous plants. For plant transformation, mainly *A. tumefaciens* is used, and this bacterium contains a large tumor-inducing plasmid (Ti). During infection, a specific segment of the plasmid DNA referred to as T-DNA (transferred DNA) is inserted into the plant genome. The T-DNA contains genes for phytohormones which are responsible for cell proliferation and the formation of the crown gall, as well as genes for the formation of special nutrients (opines) in the plant cell [1]. In order to use *Agrobacterium* as a tool in genetic engineering, these genes have been deleted and replaced by the genes of interest. This was possible since only the 25 bp T-DNA border sequences on the right and left border are needed for DNA transfer. A wide variety of such "disarmed" (nontumor-inducing) vectors have been developed. *Agrobacterium*-mediated transformation is the most commonly used method for most dicotyledonous plants [2]. Advantages include the typical insertion of only one or a few copies of the transgene [3], while another advantage is the transfer of relatively large segments of DNA with only minimal rearrangements [4].

2.1.1.2 Direct gene transfer

For many years, *Agrobacterium* could not be used to transform monocotyledonous and other recalcitrant species, so direct gene transfer methods have been developed as an alternative. These include DNA uptake into protoplasts (protoplast transformation) and the shooting of DNA-coated particles into tissues (particle bombardment).

Transformation of protoplasts In contrast to animal cells, plant cells possess a solid cell wall which is the first barrier to overcome when foreign genes are to be transferred into them. One way to circumvent this barrier is to use plant cells in which the cell wall has been digested enzymatically resulting in so called protoplasts. DNA uptake into protoplasts can then be stimulated by the use of either polyethylene glycol (PEG-transformation) or electric pulses (electroporation). Where an appropriate protoplast-to-plant regeneration system is available, large numbers of transformed clones can be regenerated to fertile transgenic plants [5].

In cases where a barrier to gene transfer has not been detected, virtually every protoplast system has proven transformable, though with considerable differences in the efficiency [5]. However, problems can arise from the regeneration of fertile plants from protoplasts as the regeneration is strongly species- and genotype-dependent, and undesired somaclonal variation can occur due to the relatively long tissue culture phase [6]. One advantage of the protoplast transformation method is its great independence from other patented techniques.

Particle bombardment The new method of using high-velocity microprojectiles to deliver DNA into plant tissue was developed by Sanford and colleagues in 1987 [7]. A particle gun is used to accelerate DNA-coated microprojectiles into cells, past the cell wall and the cell membrane. The microprojectile is small enough (0.5–5 μm) to enter the plant cell without causing too much damage, yet be of large enough mass to penetrate the cell wall and carry an appropriate amount of DNA on its surface [8].

The main advantage of particle bombardment is the absence of biological incompatibilities that are found when using biological vectors. Organelles such as chloroplasts have also been transformed using particle bombardment. Unfortunately, particle bombardment as well as other direct DNA uptake methods often result in complex insertion loci which can cause gene silencing [2, 3].

2.1.2
Tissue Requirements

The ability to cultivate plant tissue *in vitro* is a prerequisite in almost all current transformation protocols. Transformation requires competent (i.e., transformable) cultured cells that are embryogenic or organogenic. Plant cells suitable for regeneration are either cocultivated with *Agrobacterium* or bombarded. For *Agrobacterium* transformation, mostly leaf discs or immature embryos are used for particle bombardment in monocotyledonous plant scutellar tissues.

2.1.3
Molecular Requirements

A typical plant gene consists of a promoter, a coding sequence, a transcription terminator, and a polyadenylation signal. The expression level of the gene is mainly determined by these components, but can also be affected by the surrounding sequences.

2.1.3.1 Promoter
The promoter is the main determinant of the expression pattern in the transgenic plant. Constitutive promoters direct expression in all or almost all tissues, independently of developmental or environmental signals. The promoter directing the synthesis of the cauliflower mosaic virus 35S RNA is the most frequently used constitutive promoter; other constitutive promoters are derived from agrobacterial T-DNA, such as the nopaline synthase (nos) promotor or the octopine synthase (ocs) promotor. Both are mainly used in dicotyledonous plants [9].

In monocotyledonous plants, it is mostly promoters from the rice actin 1 gene (act1) [10] or the maize ubiquitin 1 gene (ubi1) [11] that are used. To enhance expression levels, the first intron of the respective genes have been added to the expression cassettes [12]. In dicotyledonous plants, the addition of introns seems to have less pronounced effects on the expression level. Nevertheless, the insertion of an intron is always necessary when the expression of the gene product in bacteria must be completely avoided (e.g., for genes conferring bacterial resistance; see Section 2.3.1.3).

In cases where transgene expression should be directed to certain tissues or developmental stages, regulated promoters are required. Promoter elements responsible for expression in seeds, tubers, vegetative organs and leaves have been isolated.

Environmental influences can induce gene expression after wounding, heat- or cold stress, or anaerobiosis. The use of natural inducible promoters has the disadvantage of causing pleiotropic effects, since endogenous genes will also be turned on [9]. Chemically induced promoters are favored for the production of pharmaceuticals in transgenic plants, as production of the desired compound can be restricted to a certain time point.

2.1.3.2 Codon usage

Expression of the gene of interest can also be influenced at the level of translation: the codon usage differs between plants and bacteria. Therefore, the removal of rare codons, e. g., codons not frequently used in plant cells, from genes of bacterial origin can enhance expression considerably. This was first described for a modified *Bacillus thuringiensis* toxin gene where expression was enhanced up to 100-fold [13].

2.1.3.3 Selectable marker and reporter genes

In plant transformation systems, the efficacy of stable gene transfer is low; hence systems are required which allow the selection of the transformed cells. A selection system consists of a selective agent (which interferes with the plant metabolism) and a selectable marker gene (which codes for a protein, enabling inactivation or evasion of the selective agent) [14].

The most commonly used selectable marker genes in transgenic plants code for antibiotic or herbicide resistance including the following.

Antibiotic resistance genes

- *nptII* gene: the *neo* or *nptII* gene, isolated from transposon Tn5 from *Escherichia coli* K12 codes for the neomycin phosphotransferase (NPTII). This enzyme detoxifies a range of aminoglycoside antibiotics such as neomycin, kanamycin, and geneticin [15]. These antibiotics are added to the culture medium after the transformation procedure. Only transformed cells or tissues will survive and will be regenerated to plants.
- *hpt* gene: the *hpt* gene has been isolated from *E. coli* and codes for hygromycin phosphotransferase, which detoxifies the antibiotic hygromycin B. Most plant tissues show a higher sensitivity to hygromycin B than to kanamycin or geneticin. In particular, cereals which are resistant to kanamycin and geneticin can be selected with hygromycin [14].
- *bla* gene: the bla_{TEM-1} gene codes for the TEM-1 β-lactamase, the most encountered ampicillin resistance marker in molecular biology. TEM-1 β-lactamase attacks narrow-spectrum cephalosporins and all the anti-Gram-negative-bacterium penicillins, except temocillin. Mostly, the ampicillin resistance in transgenic plants is not used as selectable marker for plant transformation but is a reminiscence of the transformation method. With direct gene transfer methods it is used as a selectable marker during the cloning procedure prior to plant transformation. When present on the plasmid used for transformation, the ampicillin resistance gene is transferred and integrated together with the genes of interest, and remains under control of its prokaryotic promotors [16].
- *aadA*: the *aadA* gene confers resistance to streptomycin and spectinomycin. The gene has been found in association with several transposons (Tn7, Tn21,...) and is ubiquitous among Gram-negative bacteria [15].

Herbicide resistance genes
- *bar/pat* gene: the *bar* gene from *Streptomyces hygroscopicus* and the *pat* gene from *S. viridichromogenes* code for phosphinotricin acetyltransferase (PAT) which confers resistance to the herbicide compound phosphinotricin [17]. It is an inhibitor of glutamine synthetase, a plant enzyme involved in ammonia assimilation. Besides phosphinotricin, the commercially available nonselective herbicides Basta and bialaphos can be used as selective agent. Selection can be applied in the culture medium or by spraying of regenerated plantlets [14].

Reporter genes In contrast to selectable marker genes, reporter genes do not confer resistance to selective agents inhibitory to plant development. Reporter genes code for products which can be detected directly or catalyze reactions whose products are detectable [14].
- *gusA* gene: the *gusA* (*uidA*) gene encodes β-glucuronidase (GUS), which hydrolyzes a wide range of β-glucuronides [18]. Substrates for GUS are available for spectrometric, fluorometric, and histochemical assays. The GUS reporter gene system allows easy quantification with high sensitivity [14]. One disadvantage of this system is its nonviability: examined cells and tissues cannot be regenerated into transgenic plants. Therefore, this system is mainly used to optimize a specific transformation protocol.
- *luc* gene: the luciferase reporter gene assay use bioluminescence reactions with substrate/enzyme combinations which lead to detectable light emission. The *luc* gene originates from the firefly *Photinus pyralis*, and codes for a luciferase which decarboxylates beetle luciferin. In contrast to the GUS reporter system, the luciferase reporter gene is nonlethal to plant cells [19, 20].

Alternative marker and excision systems Due to discussions on the possible risk of a gene transfer of antibiotic resistance genes from transgenic plants to clinically important microorganisms, alternative marker systems have been developed. Additionally, through the use of excision systems the presence of antibiotic or herbicide resistance genes in the final product can be avoided [21].

Alternative markers: Several alternative marker systems have recently been developed. For example, Chua and colleagues [22] successfully selected transgenic plants using an inducible *Agrobacterium* gene encoding for isopentenyltransferase (IPT). This enzyme catalyzes the first step in biosynthesis of cytokinins, a class of phytohormones. Only cells containing this gene are able to form shoots and differentiate into mature plants [22].

As another alternative selection system, phosphomannose isomerase (PMI) was successfully used in transgenic sugar beet and maize. PMI catalyzes the interconversion of mannose-6-phoshate and fructose-6-phoshate. Except for leguminous plants, PMI is absent from most plants and therefore only transgenic plants expressing the *E. coli manA* gene can survive on media containing mannose as a carbon source [23].

Excision systems: Another approach is the use of recombination systems which enable the excision of marker genes in the successfully transformed plant. Four different site-specific recombination systems have been shown to function in plant cells. The best-characterized system in plants is the CRE/lox system of bacteriophage P1: the CRE recombinase enzyme recognizes specifically and catalyzes recombination between two lox sequences. Marker genes flanked by these lox sites can be precisely excised in the presence of the CRE recombinase. One disadvantage of the system was that the CRE protein had to be introduced in the transgenic plant via secondary transformation or sexual crossing with a CRE transformant [24]. When inducible developmental stage promoter or tissue-specific promoter are used, both components of the CRE/lox system can be transferred to one transformant. After selection of the transformants the promoter will be intentionally induced, resulting in excision of the marker gene [25, 26]. This system is also applicable to vegetatively propagated plants.

A recent publication describes the efficient excision of selectable marker genes which are framed by *attP*-sites without the induction of a recombinase. The molecular basis of the reaction is still unknown, but presumably it is due to illegitimate recombination. Further experiments are necessary to reveal the underlying mechanism and to verify its general applicability [27, 28].

2.2
GM Plants Already on the Market (EU, USA, Canada, Japan)

In Europe, 10 genetically modified (crop) plants (GMPs) have been approved for placing on the market according to 90/220 EEC. These approvals covered different uses as applied for by the applicant: in two cases the approvals included use as food and feed, namely Roundup Ready soybean (Monsanto) and insect-resistant maize (Novartis, formerly Ciba Geigy). Nevertheless, only few commercial plantings are carried out in Europe, and in Germany GMPs are grown only on an experimental scale.

Since 1997, GMPs intended for food or feed use must be notified or approved according to the "Novel Food Regulation. Since then, refined oil from herbicide-tolerant oil seed rape, refined oil from herbicide-tolerant and insect-resistant cotton and processed products from herbicide-resistant and insect-resistant maize have been notified (Table 2.1).

Worldwide, GMPs have been grown on 52.6 million hectares according to ISAAA, this being an increase of 19% in comparison with 2000. The rise in area is mainly due to increased plantings of herbicide-resistant soybeans (33.3 million hectares, +7.8%) and herbicide- and insect-resistant cotton (6.8 million hectares, +1.5%). In the case of soybeans, 46% of the world's production is now genetically modified, while the percentage of GMPs in cotton amounts to 20%, in oilseed rape to 11%, and in maize to 7%.

Some 77% of the GMPs contain a herbicide resistance gene, followed by 15% carrying insect resistance genes, and 8% both herbicide- and insect-resistance genes (stacked genes).

The main producers of GMPs are the USA (37.5 million hectares), Argentina (11.8 million hectares), Canada (3.2 million hectares), and China (1.5 million hectares). These areas account for 99% of the world's production of GMPs [32, 33].

Table 2.1. Genetically modified crop plants and products approved in the EU.

Plant (line)	Trait (gene)	Applicant	Approved use	Year
Soybean (GTS-40-3-2)	HR (*CP4 EPSPS*)	Monsanto	Import and processing	1996
Maize (Bt 176)	IR (*cryIAb*), HR (*bar*), ABR (*bla*)	Ciba-Geigy	Cultivation; import and processing	1997
Maize (Bt 11)	IR (*cryIAb*), HR (*pat*)	Northrup King	Import and processing; food additives	1998
Maize (MON 810)	IR (*cryIAb*)	Monsanto	Cultivation; flour, semolina, starch and starch products, oil	1998; 1997
Maize (MON 809)	IR (*cryIAb*), HR (*CP4 EPSPS*, *gox*)	Pioneer	Food additives	1998
Maize (T 25)	HR (*pat*), ABR (partial *bla*)	AgrEvo	Cultivation; refined oil, starch and starch products, fermented or heat-treated products from maize flour	1998
Oil seed rape (MS1 × RF1; MS1 × RF2)	MS/RF (*barnase/barstar*), HR (*bar*), ABR (*npt*II)	PGS	Cultivation; refined oil	1997
Oil seed rape (MS8 × RF3)	MS/RF (*barnase/barstar*), HR (*bar*)	PGS	Refined oil	1999
Oil seed rape (Topas 19/2)	HR (*pat*), ABR (*npt*II)	AgrEvo	Import and processing (feed); refined oil	1995; 1997
Oil seed rape (Falcon GS 40/90)	HR (*pat*)	AgrEvo	Refined oil	1999
Oil seed rape (Liberator L62)	HR (*pat*)	AgrEvo	Refined oil	1999
Oil seed rape (GT 73)	HR (*CP4 EPSPS*, *gox*)	Monsanto	Refined oil	1997
Cotton (MON 1445/1698)	HR (*CP4 EPSPS*), ABR (*npt*II, *aad*)	Monsanto	Refined oil	2002
Cotton (MON 531/757/1076)	IR (*cryIAc*), ABR (*npt*II, *aad*)	Monsanto	Refined oil	2002

Sources: Refs. [29–31]

HR = herbicide resistance (glyphosate resistance *CP4 EPSPS* gene, glyphosate resistance oxidoreductase gene (*gox*), glufosinate resistance *pat* or *bar* gene);
ABR = antibiotic resistance (ampicillin resistance (*bla*), kanamycin resistance (*npt*II)), streptomycin resistance gene (*aad*);
IR = insect resistance (*Bacillus thuringiensis cryIA(b)*);
MS = male sterility (*barnase*);
RF = restorer of fertility (*barstar*)

2.2.1
Herbicide Resistance in Soybean, Maize, Oilseed rape, Sugar beet, Rice, and Cotton

As mentioned above, herbicide resistance is the leading trait in commercialized GMP, with 23 lines having been approved for cultivation and/or food and feed use worldwide (Table 2.2).

Table 2.2. Herbicide-resistant crop plants and products worldwide.

Plant (line)	Trait (gene)	Applicant	Country
Soybean (GTS-40-3-2)	HR (CP4 EPSPS)	Monsanto	USA, EU, Canada, Japan, Argentina, Mexico, Australia, Switzerland, Uruguay, Russia, Korea, South Africa
Soybean (A 2704-12, A 2704-21, A 5547-35)	HR (pat)	AgrEvo	USA, Canada, Japan
Soybean (A 5547-127)	HR (pat), ABR (partial bla)	AgrEvo	USA
Soybean (GU 262)	HR (pat), ABR (partial bla)	AgrEvo	USA
Soybean (W 62, W 98)	HR (bar), RP (gus)	AgrEvo	USA
Maize (GA 21)	HR (maize EPSPS)	Monsanto	USA, Canada, Japan, Argentina, Australia
Maize (NK 603)	HR (CP4 EPSPS)	Monsanto	USA, Canada, Japan
Maize (MON 832)	HR (CP4 EPSPS, gox), ABR (nptII)	Monsanto	Canada
Maize (B 16 = DLL 25)	HR (bar), ABR (bla)	DeKalb	USA, Canada, Japan
Maize (T 14, T 25)	HR (pat), ABR (partial bla)	AgrEvo	USA, EU, Canada, Japan, Argentina
Oilseed rape (GT 73)	HR (CP4 EPSPS, gox)	Monsanto	USA, EU, Canada, Japan, Australia
Oilseed rape (GT 200)	HR (CP4 EPSPS, gox)	Monsanto	Canada, USA
Oilseed rape (HCN 92 = Topas 19/2)	HR (pat), ABR (nptII)	AgrEvo	EU, Canada, Japan, USA
Oilseed rape (HCN 10)	HR (pat)	AgrEvo	USA, Canada, Japan
Oilseed rape (HCN 28 = T 45)	HR (pat)	AgrEvo	USA, Canada, Japan

Table 2.2. (continued)

Plant (line)	Trait (gene)	Applicant	Country
Oilseed rape (Oxy 235)	HR (*bxn*)	Rhone Poulenc	Canada, Japan
Brassica rapa (ZSR 500, 501, 502)	HR (*CP4 EPSPS, gox*)	Monsanto	Canada
Sugar beet (T120-7)	HR (*pat*), ABR (*npt*II)	AgrEvo	USA, Canada, Japan
Sugar beet (GTSB 77)	HR (*CP4 EPSPS, gox*)	Novartis	USA
Rice (LLRICE 06, 62)	HR (*bar*)	AgrEvo	USA
Cotton (MON 1445 / 1698)	HR (*CP4 EPSPS*), ABR (*npt*II, *aad*)	Monsanto	USA, Canada, Japan, Argentina, Australia
Cotton (BXN)	HR (*bxn*), ABR (*npt*II)	Calgene	USA, Canada, Japan
Cotton (19-51A)	HR (*als*)	DuPont	USA
Flax (FP967)	HR (*als*), ABR (*npt*II)	University of Saskatoon	Canada, USA

Sources: Refs. [29, 30]

HR = herbicide resistance (glyphosate resistance *CP4/maize EPSPS* gene, glyphosate resistance oxidoreductase gene (*gox*), glufosinate resistance *pat* or *bar* gene, bromoxynil herbicide resistance nitrilase gene (*bxn*), sulfonylurea resistance *als* gene)

ABR = antibiotic resistance (ampicillin resistance gene *bla*, kanamycin resistance gene *npt*II, streptomycin resistance gene (*aad*));

RP = reporter gene (*gus*)

2.2.2
Insect Resistance in Maize, Potatoes, Tomatoes, and Cotton

Insect resistance is the trait found second most often in approved GMPs. Insect-resistant lines, as well as lines which acquire an insect-resistance gene in combination with a herbicide resistance or virus resistance gene, are listed in Table 2.3.

Table 2.3. Insect-resistant crop plants and products worldwide.

Plant (line)	Trait (gene)	Applicant	Country
Maize (MON 810)	IR (*cryIAb*)	Monsanto	USA, EU, Canada, Japan, Argentina, Australia, Switzerland, South Africa
Maize (MON 802)	IR (*cryIAb*), HR (*CP4 EPSPS, gox*), ABR (*npt*II)	Monsanto	USA, Canada, Japan
Maize (MON 80100)	IR (*cryIAb*), HR (*CP4 EPSPS, gox*), ABR (*npt*II)	Monsanto	USA
Maize (MON 863)	IR (*cryIIIBb2*), ABR (*npt*II)	Monsanto	USA, Japan
Maize (MON 809)	IR (*cryIA(b)*), HR (*CP4 EPSPS, gox*)	Pioneer	USA, EU, Canada, Japan
Maize (Bt 176)	IR (*cryIAb*), HR (*bar*), ABR (*bla*)	Ciba-Geigy	USA, EU, Canada, Japan, Argentina, Switzerland, Australia
Maize (Bt 11)	IR (*cryIAb*), HR (*pat*)	Northrup King	USA, EU, Canada, Japan, Argentina, Switzerland, Australia
Maize (DBT 418)	IR (*cryIAc*, protease inhibitor II (*pin*II)); HR (*bar*), ABR (*bla*)	DeKalb	USA, Canada, Japan, Argentina
Maize (TC1507)	IR (*cryIFa2*), HR (*pat*)	Mycogen/Pioneer	USA, Canada, Japan
Potato (Bt 6, Russet Burbank New Leaf)	IR (*cryIIIA*), ABR (*npt*II)	Monsanto	USA, Canada, Japan
Potato (ATBT04-6 etc., 4 lines Atlantic and Superior New Leaf)	IR (*cryIIIA*), ABR (*npt*II)	Monsanto	USA, Canada, Japan

Table 2.3. (continued)

Plant (line)	Trait (gene)	Applicant	Country
Potato (SEMT 15-15, New Leaf Y)	IR (*cryIIIA*), VR (*coat protein PVY*) ABR (*nptII, aad*)	Monsanto	USA, Canada
Potato (RBTM 21-350, Russet Burbank New Leaf Plus)	IR (*cryIIIA*), VR (*replicase PLRV, helicase PLRV*) ABR (*nptII*)	Monsanto	USA, Canada
Cotton (MON 531, 757, 1076, Bollgard)	IR (*cryIAc*), ABR (*nptII, aad*)	Monsanto	USA, Canada, Japan, Argentina, Mexico, Australia, China, South Africa
Cotton (MON 15985, Bollgard II)	IR (*cryIAc, cryIAb*), ABR (*nptII, aad*), RP (*uidA*)	Monsanto	USA, Australia
Cotton (31807/31808)	IR (*cryIAc*), HR (*bxn*) ABR (*nptII*)	Calgene	USA, Japan
Tomato (5345)	IR (*cryIAc*), ABR (*nptII, aad*)	Monsanto	USA, Canada

Sources: Refs. [29, 30]

IR = insect resistance (delta endotoxin genes of *Bacillus thuringiensis* (*cryIAb, cryIAc, cryIIIA, cryIFa2*; protease inhibitor II (*pinII*))
HR = herbicide resistance (glyphosate resistance *CP4 EPSPS* gene, glyphosate resistance oxidoreductase gene (*gox*), glufosinate resistance *pat* or *bar* gene, bromoxynil herbicide resistance nitrilase gene (*bxn*))
VR = virus resistance (potato virus Y (PVY) coat protein gene, replicase and helicase gene of potato leafroll virus (PLRV), respectively)
ABR = antibiotic resistance (ampicillin resistance (*bla*), kanamycin resistance (*nptII*), streptomycin resistance (*aad*));
RP = reporter gene (beta-D-glucuronidase (*gus*))

2.2.3
Virus-resistance, male sterility, delayed fruit ripening, and fatty acid contents in GMPs

GMPs with virus resistance (potato, squash, and papaya; Table 2.4), male sterility (oilseed rape, maize, and chicory; Table 2.5), delayed fruit ripening (tomato; Table 2.6) as well as modified fatty acid content (soybean, oilseed rape; Table 2.7) have also been approved in some countries.

Table 2.4. Virus-resistant crop plants and products worldwide.

Plant (line)	Trait (gene)	Applicant	Country
Potato (SEMT 15-15, New Leaf Y)	VR (*coat protein PVY*), IR (*cryIIIA*), ABR (*nptII, aad*)	Monsanto	USA, Canada
Potato (RBTM 21-350, Russet Burbank New Leaf Plus)	VR (*replicase PLRV, helicase PLRV*), IR (*cryIIIA*), ABR (*nptII*)	Monsanto	USA, Canada
Squash (CZW-3)	VR (*coat protein CMV, coat protein ZYMV, coat protein WMV 2*), ABR (*nptII*)	Asgrow	USA, Canada
Squash (ZW20)	VR (*coat protein ZYMV, coat protein WMV 2*)	Upjohn	USA, Canada
Papaya (55-1/63-1)	VR (*coat protein PRSV*), RP (*gus*), ABR (*nptII*)	Cornell University	USA

Sources: Refs. [29, 30]

VR = virus resistance (coat protein genes of potato virus Y (PVY), cucumber mosaic virus (CMV), zucchini yellows mosaic virus (ZYMV) and watermelon mosaic virus (WMV) 2, respectively; replicase and helicase genes of potato leafroll virus (PLRV))
IR = insect resistance (delta endotoxin gene of *Bacillus thuringiensis* (cryIIIA))
ABR = antibiotic resistance (kanamycin resistance (*nptII*), streptomycin resistance (*aad*));
RP = reporter gene (*gus*)

Table 2.5. Male sterility in crop plants worldwide.

Plant (line)	Trait (gene)	Applicant	Country
Oilseed rape (MS1, RF1 (PGS1))	MS (*barnase*), FR (*barstar*), HR (*bar*), ABR (*nptII*)	PGS	Canada, EU, Japan, USA
Oilseed rape (MS1, RF2 (PGS2))	MS (*barnase*), FR (*barstar*), HR (*bar*), ABR (*nptII*)	PGS	Canada, EU, Japan, USA
Oilseed rape (MS8 × RF3)	MS (*barnase*), FR (*barstar*), HR (*bar*)	PGS	USA, Canada, Japan
Oilseed rape (PHY14, PHY35)	MS (*barnase*), FR (*barstar*), HR (*bar*)	PGS	Japan
Oilseed rape (PHY36)	MS (*barnase*), FR (*barstar*), HR (*bar*)	PGS	Japan
Maize (MS3)	MS (*barnase*), HR (*bar*), ABR (*bla*)	PGS	USA, Canada
Maize (MS6)	MS (*barnase*), HR (*bar*), ABR (*bla*)	PGS	USA
Maize (676, 678, 680)	MS (*dam*), HR (*pat*)	Pioneer	USA
Chicory (RM3-3, RM3-4, RM3-6)	MS (*barnase*), HR (*bar*), ABR (*nptII*)	Bejo Zaden	USA, EU

Sources: Refs. [29, 30]

MS = male sterility (*barnase* from *Bacillus amyloliquefaciens*, *dam* (DNA adenine methylase from *E. coli*)
FR = fertility restoration (*barstar* from *Bacillus amyloliquefaciens*)
HR = herbicide resistance (glufosinate resistance *bar* gene)
ABR = antibiotic resistance (ampicillin resistance (*bla*), kanamycin resistance (*nptII*))

Table 2.6. Delayed fruit ripening in tomatoes and tomato products worldwide.

Plant (line)	Trait (gene)	Applicant	Country
Tomato (FLAVR SAVR)	DR (*PG*, anti-sense) ABR (*npt*II)	Calgene	USA, Canada, Mexico, Japan
Tomato (B, Da, F)	DR (*PG*, sense or antisense) ABR (*npt*II)	Zeneca	USA, Canada
Tomato (8338)	DR (*ACCd*) ABR (*npt*II)	Monsanto	USA
Tomato (35 1 N)	DR (*sam*-K) ABR (*npt*II)	Agritope	USA
Tomato (1345-4)	DR (*ACC*, sense) ABR (*npt*II)	DNA Plant Technology Corporation	USA, Canada

Sources: Refs. [29, 30]

DR = delayed ripening (polygalacturonase gene in sense or anti-sense direction (*PG*); 1-amino-cyclo-propane-1-carboxylic acid deaminase gene (*ACCd*); S-adenosylmethionine hydrolase gene with Kozak consensus sequence (*sam*-K); aminocyclopropane cyclase synthase gene in sense orientation (*ACC*))
ABR = antibiotic resistance (kanamycin resistance (*npt*II))

Table 2.7. Modified fatty acid content in crop plants and products worldwide.

Plant (line)	Trait (gene)	Applicant	Country
Soybean (G94-1, G94-19, G168, high-oleic soybean)	MFA (*GmFad2-1*), ABR (*bla*), RG (*gus*)	DuPont	USA, Canada, Japan, Australia
Oilseed rape (23-18-17, 23-198, high-laurate canola)	MFA (*BayTE*), ABR (*npt*II)	Calgene	USA, Canada

Sources: Refs. [29, 30]

MFA = modified fatty acid content (delta 12 desaturase gene (*GmFad2-1*); thioesterase gene from *Umbellaria californica* (*Bay TE*)
ABR = antibiotic resistance (ampicillin resistance (*bla*), kanamycin resistance (*npt*II))
RP = reporter gene (*gus*)

2.3
GM Plants "in the Pipeline"

2.3.1
Input Traits

In the following chapters, selected examples will be presented of studies being conducted in the genetic engineering of crop plants. First, attempts to reduce the production costs of crop plants are summarized. These studies refer to crop plants with genetically engineered resistance against insects, diseases, abiotic stresses, and improved agronomic properties.

2.3.1.1 Insect resistance in rice, soybean, oilseed rape, eggplant, walnut, grape, and peanut

The best-characterized insecticidal proteins are the delta-endotoxins of *Bacillus thurigiensis* (*B.t.*-toxins), which have been used as biopesticides in agriculture (also organic farming), forestry, and as a mosquito vector control for many years. *B.t.* insecticidal activity is highly specific in that the endotoxins are nontoxic to nontarget insects, birds, and mammals.

Besides the commercially available *B.t.*-maize, -potato, -cotton and -tomato lines (see Table 2.3), insect-resistant rice (*cryIAb* and *cryIAc*) [34, 35], soybeans (*cryIAc*) [36], oilseed rape (*cryIAc*) [37] and eggplants (*cryIIIB*) [38] have been developed. Field trials have also been performed in the USA with insect-resistant sunflower, lettuce, grapefruit, sugarcane, apple, walnut, grape- and peanut [39].

In order to control corn rootworm infections, Mycogen and Pioneer Hi-Bred are developing transgenic maize which contain two novel proteins from *Bacillus thuringiensis* strain PS149B1. These proteins belong to another class of insecticidal proteins with no homology to the delta-endotoxins (Bt PS149B1 toxins) [41].

Another approach to achieve insect-resistant plants is to use plant defense proteins such as proteinase inhibitors [42, 43], lectins [44] or alpha-amylase inhibitor [45]. Lectins are thought to confer resistance towards sap-sucking insects (e.g., the rice brown planthopper) which act as vector for virus transmission [44]. Since the publication of Ewen and Pusztai, the possible toxic effects of the snowdrop lectin on mammals have been broadly discussed [46, 47].

Morton and colleagues transferred the alpha-amylase inhibitor 1 gene from beans to peas and conducted field trials with the insect-resistant pea plants, and reported high insect mortality in transgenic plants [45]. One advantage of using alpha-amylase inhibitor genes is the long history of human consumption of beans and the fact, that bean alpha-amylase has no effect on starch digestion in humans [45].

In an attempt to confer broad-spectrum resistance to storage pests, maize was transformed with the chicken avidin gene [48]. The mode of action of avidin is to cause a deficiency of the vitamin biotin in insects; hence, a thorough safety testing for human consumption must be carried out before the commercial use of biotin maize.

2.3.1.2 Disease resistance in maize, potatoes, fruits, and vegetables

Virus resistance One strategy by which to obtain virus-resistant plants is to transfer genes from the pathogen itself into the plant (pathogen-derived resistance). The most widely used approach is to express the virus coat protein in transgenic plants. In theory, the expression of viral genes disrupts viral infection or symptom development. All but one of the commercially available virus-resistant plants contain viral coat proteins (see Table 2.4), and this technique is extended to other plants such as rice [49], plum tree [50], tomato [51], pea [52], and peanut [53]. Additionally, field trials have been performed in the USA with coat protein-mediated virus-resistant wheat, soybean, sugarcane, cucumber, sweet potato, and grapefruit [39].

Another form of pathogen-derived resistance is the use of viral replicase genes (or RNA-dependent RNA polymerase genes), which presumably act via post-transcriptional gene silencing. This technique has been used to confer resistance to potato leafroll virus in potato (see Table 2.4), to barley yellow dwarf virus in oats, cucumber mosaic virus in tomato, rice tungro spherical virus in rice, and wheat streak mosaic virus in wheat [54–57].

Since varying degrees of virus resistance have been obtained with coat protein-mediated resistance, approaches have been performed to ameliorate resistance against cucumber mosaic virus via satellite RNA, especially in tomato [58, 59]. However, this approach has raised controversy as a single point mutation in the satellite RNA can transform it into a harmful necrogenic form [60].

In order to protect plants against more than one virus, ribosome-inactivating proteins (RIPs) have been expressed in transgenic plants. RIPs are strong inhibitors of protein synthesis and, depending on the plant species from which they originate, they show different levels of toxicity against different hosts. Poke weed antiviral protein (PAP) confers resistance to PVX and PVY in transgenic potatoes, while PAPII confers resistance to TMV, PVX and fungal infections in tobacco, respectively [61, 62].

On a more experimental scale are approaches to achieve virus resistance using antibodies against the virus coat protein. Such antibodies are able to neutralize virus infection, presumably by interacting with newly synthesized coat protein and disrupting viral particle formation [63, 64]. Like RIPs, broad-spectrum antibodies might be used to protect plants against a wider range of viruses, as has been demonstrated in the case of poty viruses [64].

Fungal resistance Fungal resistance can be conferred via the activation of specific self-defense mechanisms in the plant. One of the mechanisms is the so-called hypersensitive response (HR), which enables plants to enclose the pathogen in the infected area by the formation of necrotic lesions. HR induces many defense-related signal molecules such as salicylic acid, ethylene and phytoalexin. HR is also characterized by an accumulation of pathogenesis-related (PR) proteins that includes fungal cell wall-degrading enzymes, antimicrobial peptides, thionins, lipid transfer proteins, and proteinase inhibitors [65].

In rice, the introduction of chitinase and thaumatin-like protein, respectively, led to increased resistance to sheath blight (*Rhizoctonia solani*) [66–68]. Enhanced resistance to the rice blast fungus *Magnaporthe grisea* was observed upon constitutive expression of chitinase and defense-related protein genes in transgenic rice [69, 70].

Pathogenesis-related proteins from plants have been used to confer fungal resistance in alfalfa [71], cucumber [72], oilseed rape [73], tomatoes [74], wheat [75–77], grape vine [78], and oranges [79].

Other antifungal genes of plant origin comprise genes for RIPs [80], genes for phytoalexins [81–84], and anthocyanin genes [85]. An example of an antifungal gene from nonplant sources which has been transferred to plants is the human lysozyme gene [86].

Individual PR-proteins, however, have a narrow spectrum of antifungal activity, and need to function collectively in order to provide a modest but long-term resistance. Therefore, research is currently focusing on genes from mycoparasitic fungi as a source for improving resistance to fungal pathogens. An endochitinase of the mycoparasitic fungus *Trichoderma harzianum* has been transferred to tobacco and potato, and has been reported to confer a high level and a broad spectrum of resistance [87]. When transferred to apple, however, the endochitinase of *Trichoderma harzianum* increased resistance to apple scab but also reduced plant growth [88].

A similar approach has been taken by transferring an antifungal protein from a virus that persistently infects *Ustilago maydis*, to wheat. Transgenic wheat plants exhibited increased resistance against stinking mut (*Tilletia tritici*) [89].

For a comprehensive survey on the different approaches to achieve fungal resistance in transgenic plants, see Tables 1 and 2 in Ref. [90].

Bacterial resistance Resistance to bacterial infections is not yet as developed as virus- and fungal resistance, due partly to the fact that bacterial diseases are a major problem in only some crop plants such as potato, tomato, rice and certain fruit trees.

The most efficient form of protection is genetic resistance, which is based on single dominant or semidominant genes. These R genes usually confer race-specific resistance, and their effectiveness is based on their interaction with complementary pathogen avirulence (AV) genes in the pathogen, the so called gene-for-gene interaction [91]. Resistance to bacterial blight caused by *Xanthomonas oryzae* pv. *oryzae* was achieved by transferring the resistance gene *Xa21* from a wild rice species to the elite indica rice variety 'IR72' [92]. Accordingly, the resistance gene *Bs2* from pepper was transferred to tomato, which then exhibited resistance to bacterial spot disease [93].

The tomato disease resistance gene *Pto* specifies race-specific resistance to *Pseudomonas syringae* pv. *tomato* carrying the *avrPto* gene. By overexpressing *Pto* a race-nonspecific resistance was observed in transgenic tomatoes [94].

The resistance based on single dominant gene expression always bears the danger of early evolution of counter-resistance in the pathogen as a result of the emergence of strains that no longer express the specific avirulence gene product [93]. Therefore, new resistance genes are being investigated for use in pyramiding strategies (combination of resistance genes against the same pathogen, but with different targets). One example is the *AP1* gene from sweet pepper which delays the hypersensitive response when expressed in transgenic rice plants, and which can be used in combination with *Xa21* or other resistance genes [95].

Similar to fungal resistance, the overexpression of PR proteins or transfer of PR protein genes from other sources has led to enhanced resistance against bacterial infections: expression of barley lipid transfer protein LTP2 caused enhanced tolerance to bacterial pathogens in transgenic tobacco plants [96].

In several plant species, bifunctional enzymes with lysozyme activity have been detected which are thought to be involved in defense against bacteria. After transfer of the bacteriophage T4 lysozyme gene, transgenic potatoes showed a decreased susceptibility towards *Erwinia carotovora atroseptica* infections [97]. The transfer of the human lysozyme gene to tobacco led to enhanced resistance against both fungal and bacterial diseases [98].

Plant defense responses also involve the production of active oxygen species such as hydrogen peroxide (H_2O_2). This mechanism was exploited by transferring a fungal gene encoding H_2O_2-generating glucose oxidase to potato plants. The transgenic potato tubers exhibited strong resistance to *Erwinia carotovora* subsp. *carotovora* infections causing bacterial soft rot disease, as well as enhanced resistance to potato late blight caused by *Phytophthora infestans* [99].

Insects produce antimicrobial peptides as major defense response to pathogen attack, and these include sarcotoxins, cecropins, and attacins. The latter has been transferred to apples and pears which had improved resistance to *Erwinia amylovora*, causing fire blight [100, 101]. Sarcotoxins appear to confer a higher anti-bacterial activity and a broader spectrum of resistance, as indicated by experiments with transgenic tobacco [65]. Similarly, the expression of synthetic antimicrobial peptide chimeras in transgenic tobacco led to broad-spectrum resistance against both bacterial and fungal pathogens [102].

Nematode resistance In the USA, field tests are being conducted with nematode-resistant carrots, tomatoes, potatoes, and pineapples [39]. In the case of carrot and tomato, cysteine proteinase inhibitors from cowpea and rice, respectively, have been expressed in the transgenic plants. Proteinase inhibitors are an important element of natural plant defense strategy. The cowpea trypsin inhibitor (CpTI), a serine proteinase inhibitor, and oryzacystatin (Oc-I), an inhibitor of cysteine proteinase, have been shown to be effective against proteinases of a cyst nematode [103]. After site-directed mutagenesis of the latter, the modified cystatin showed enhanced efficacy as a transgene against potato cyst nematode [104], and also limited growth of both a cyst and a root-knot nematode when expressed in *Arabidopsis* [103]. Cysteine proteinase inhibitors are of particular interest because they are the only class of proteinase that is not expressed in the digestive system of mammals [105].

2.3.1.3 Tolerance against abiotic stresses

The effects of weather, erosion, and depleted soils expose plants to a variety of stresses. Genetic engineering has been used to provide plants with additional stress response genes in order to counteract these environmental stresses. These genes can be grouped into two categories: (i) genes that respond directly against a particular stress; and (ii) genes that regulate stress gene expression and signal transduction [106]. Transfer or overexpression of both types of genes have been used in transgenic approaches to enhance tolerance against abiotic stresses. In the following section, salt, drought, and cold stress are considered together as on a cellular basis all act as dehydration stress.

Genes conferring dehydration-stress tolerance in transgenic plants Plants react to these stresses by displaying complex, quantitative traits that involve the function of many genes. Expression of these genes lead to the accumulation of low molecular-weight components such as osmolytes, synthesis of late-embryogenesis-abundant (LEA) proteins, and activation of detoxifying enzymes. In transgenic approaches, genes for the enzymes which are responsible for the production of these compounds have been transferred to nontolerant plants. Most of these studies have been performed in model plants such as *Arabidopsis* or tobacco, but some investigations have been extended to food crops such as rice.

Bajaj and colleagues have summarized the results from these studies in a recent review [107]. After transfer of genes encoding enzymes for osmoprotectants such as glycinebetaine [108], proline [109] or putrescine [110], transgenic rice plants exhibited increased tolerance to salt and drought stress. Constitutive overexpression of the oat arginine decarboxylase gene in rice led to severely affected developmental patterns [110]. When the arginine decarboxylase gene was under control of an ABA-inducible promoter, the transgenic rice plants showed an increase in biomass under salinity-stress conditions [111].

The expression of a LEA protein in the transgenic rice plants led to increased tolerance to water deficit and salt stress in the transgenic plants [112]. As it is believed that abiotic stress primarily affects plants through oxidative damage, genes for detoxifying enzymes have been transferred to sensitive plants. McKersie et al. [113] showed that transgenic alfalfa overexpressing Mn superoxide dismutase to reduce free radicals was more tolerant of water deficit and freezing, and had better winter survival rates. It is envisaged that the use of stress-inducible promoters and the introduction of multiple genes will improve dehydration-stress tolerance [107].

Regulatory genes encoding transcription factors The other promising approach to confer tolerance to dehydration stress is to transfer regulatory genes. As the products of these genes regulate gene expression and signal transduction under stress conditions, their overexpression can activate the expression of many stress-tolerance genes simultaneously. The overexpression of the transcription factor DREB1A in transgenic *Arabidopsis* led to increased tolerance to drought, salt and freezing stresses. However, the constitutive overexpression of DREB1A also resulted in severe growth retardation under normal growth conditions. In contrast, the stress-inducible expression of this gene had minimal effects on plant growth, and provided greater tolerance to stress conditions than genes driven by the 35S promoter [107, 114].

Lee and colleagues [115] cloned the functional homologue of the yeast Dbf2 kinase that enhances salt, drought, cold, and heat tolerance upon overexpression in yeast, as well as transgenic plant cells. However, the utilization of this gene to engineer transgenic crops with enhanced stress tolerance has still to be shown [115].

Additional genes that confer tolerance to salt stress The detrimental effects of salt on plants are a consequence of both a water deficit resulting in osmotic stress and the effects of excess sodium ions on key biochemical processes. To tolerate high levels of salts, plants should be able to use ions for osmotic adjustment and internally to distribute these ions to keep sodium away from the cytosol. The first transgenic approaches introduced genes that modulated cation transport systems. Hence, transgenic tomato plants overexpressing a vacuolar Na^+/H^+ antiport were able to grow, flower, and produce fruits in the presence of 200 mM sodium chloride [116].

Overexpression of the *HAL1* gene from yeast in transgenic tomato plants had a positive effect on salt tolerance by reducing K^+ loss and decreasing intracellular Na^+ from the cells under salt stress [117].

Genes conferring tolerance to low iron In arid and semi-arid regions of the world, the soils are alkaline in nature and therefore crop yields are limited by a lack of available iron. Under iron stress, some plants release specific Fe(III)-binding compounds, known as siderophores, which bind the otherwise insoluble Fe(III) and transport it to the root surface. To increase the quantity of siderophores released under conditions of low iron availability, Takahashi et al. [118] transferred two barley genes coding for the enzyme nicotianamine aminotransferase (*naat-A* and *naat-B*) together with the endogenous promoters into rice plants. Transgenic rice plants excreted about 1.8 times more siderophore under iron stress than wild-type rice plants. This relatively small increase allowed transgenic rice plants to withstand iron deprivation remarkably better, resulting in a four-fold increase in grain yield as compared with wild-type plants [118].

Alternately, increasing the rate-limiting step of Fe(III) chelate reduction which reduces iron to the more soluble Fe(II) form might enhance iron uptake in alkaline soils [119].

2.3.1.4 Improved agronomic properties

Acceleration of sprouting time in potato   The ability to control sprouting time in potato tubers is of considerable economic importance to the potato industry. At present, the potato industry uses a range of chemical treatments in order to obtain the desired control. Using a biotechnological approach, the *pyrophosphatase* gene from *E. coli* under the control of the tuber-specific patatin promoter was transferred into potatoes. The *pyrophosphatase* gene was chosen because of the central role of inorganic phosphate in starch degradation and sucrose biosynthesis. It is believed that starch breakdown and subsequent formation of various metabolites is needed for growth of the sprout.

Transgenic potatoes displayed a significantly accelerated sprouting: the transgenic tubers sprouted on average six to seven weeks faster than did control tubers. In addition, after cold storage the majority of transgenic tubers sprouted within one week, whereas the wild-type tubers needed eight weeks or more [120].

Reduction of generation time in citrus   Citrus trees have a long juvenile phase that delays their reproductive development by between six and twenty years. With the aim of accelerating flower time, juvenile citrus seedlings were transformed with the *Arabidopsis LEAFY* (*LFY*) and *APETALA* (*AP1*) genes, which promote flower initiation in *Arabidopsis*. Both types of transgenic citrus produced fertile flowers as early as the first year. These traits are submitted to the offspring as dominant traits, generating trees with a generation time of one year from seed to seed [121]. Constitutive expression of *LFY* also promoted flower initiation in transgenic rice [122].

2.3.2
Traits Affecting Food Quality for Human Nutrition

The following sections provide examples of how the nutritional value of food crops can be improved through genetic engineering, though none of these genetically engineered food plants has yet reached the marketplace. In addition to the safety of the GMPs for health and environment, the effectiveness of this approach in comparison to alternative methods must be shown for these traits.

2.3.2.1 Increased carotenoid content in rice and tomato
One of the most advanced projects to fortify food crops with vitamins is the so-called "golden rice". To enable provitamin A biosynthesis in the rice endosperm, four additional enzymes are required. Immature rice embryos were co-transformed with two *Agrobacterium* constructs containing the *psy* and the *lcy* gene from daffodil, coding for the phytoene synthase and the lycopene β-cyclase, respectively, as well as the *crtI* gene from *Erwinia uredovora*, coding for a bacterial phytoen desaturase. Ten plants harboring all transferred genes were recovered, and all showed a normal vegetative phenotype and were fertile. In most cases, the transformed

endosperms were yellow, indicating carotenoid formation [123]. After transfer of the new trait into locally best adapted varieties, either by traditional breeding or de-novo transformation, it is hoped that vitamin-A deficiency in the developing countries can be prevented [124].

In contrast to rice, tomato plants already produce carotenoids, and tomato products are viewed as the principal dietary source of lycopene and one of the major sources of β-carotene. Since lycopene and β-carotene are considered beneficial to health (reducing chronic conditions such as coronary heart disease and certain cancers), an attempt was made to enhance the carotenoid content and profile of tomato fruits via genetic engineering. Therefore, the bacterial carotenoid gene (*crtI*), which encodes for the phytoene desaturase, has been introduced into tomato plants. Transgenic plants showed an increased β-carotene content (up to 45% of the total carotenoid content), though the total carotenoid content was not elevated [125].

2.3.2.2 Elevated iron level in rice and wheat

As cereal grains are deficient in certain essential mineral nutrients, including iron, several approaches have been used to increase iron accumulation and alter iron metabolism. Since ferritin is a general iron storage protein in all living organisms, ferritin genes have been introduced into rice and wheat plants. Goto et al. [126] generated transgenic rice plants expressing soybean ferritin under the control of the seed-specific rice Glu-B1 promoter. Transgenic rice seeds accumulated up to three times more iron as wild-type seeds [126]. Likewise, increased iron levels were found in transgenic rice seeds expressing bean ferritin under the control of the related Gt-1 promoter [127].

In order to not only increase iron accumulation but also improve its absorption in the human intestine, two approaches have been adopted. First, the level of the main inhibitor of iron absorption, phytic acid, was decreased by the introduction of a heat-tolerant phytase from *Aspergillus fumigatus*. Transgenic rice plants exhibited at least double phytase activity, whereas in one individual transgenic line the phytase activity of the grains increased about 130-fold. Second, as cysteine peptides are considered a major enhancer of iron absorption, the endogenous cysteine-rich metallothionein-like protein gene (*rgMT*) was overexpressed, and the cysteic acid content increased significantly in transgenic seeds. The authors suggested that high-phytase rice, with an increased iron content and rich in cysteine-peptide, has the potential to greatly improve iron supply in rice-eating populations [127, 128].

2.3.2.3 Improved amino acid composition in potato plants

Potatoes contain limited amounts of the essential amino acids lysine, tryptophan, methionine, and cysteine. In order to improve the nutritional value of potatoes, a nonallergenic seed albumin gene (*AmA1*) from *Amaranthus hypochondriacus* was transferred to potato plants. The seed-specific albumin was under the control of a tuber-specific and a constitutive promoter, respectively. In both transgenic lines, a 35–45 % increase in total protein content was reported in transgenic tubers, which corresponded to an increase in most essential amino acids. Additionally, a two-fold increase in tuber number and a 3.0- to 3.5-fold increase in tuber yield was observed [129].

2.3.2.4 Reduction in the content of antinutritive factors in cassava

Cassava is one of the few plants in nature that contains toxic cyanogenic glycosides in the leaves and roots. Sufficient processing of the harvested roots normally renders the cassava safe, although the processing which renders toxic cassava varieties safe also removes certain nutritional value. Therefore, Moller et al. have isolated the genes responsible for cyanogenic glycoside production and are now transforming cassava with anti-sense constructs of the respective genes *CYP79D1* and *CYP79D2* [130].

Another strategy for reducing the cyanide toxicity of cassava roots is to introduce a gene that codes for the enzyme hydroxynitrile lyase (HNL). This enzyme breaks down the major cyanogen acetone cyanohydrin, and is expressed only in leaves. After transformation of the cDNA encoding HNL, the HNL activities in cassava roots were comparable with those in leaves. Field trials will determine whether the expression of HNL in roots in fact reduces the cyanide toxicity of cassava food products [131].

2.3.2.5 Production of "low-calorie sugar" in sugar beet

Koops and colleagues have developed a new sugar beet that produces fructan, a low-calorie sweetener, by inserting a single gene from Jerusalem artichoke that encodes an enzyme for converting sucrose to fructan (1-sucrose:sucrose fructosyl transferase [1-sst]). Short-chain fructans have the same sweetness as sucrose, but provide no calories as humans lack the fructan-degrading enzymes necessary to digest them. Longer-chain fructans form emulsions having a fat-like texture, while fructans also promote the growth of beneficial bacteria present in the gut. Transgenic sugar beet roots produce the same amount of total sugar, but expression of the *1-sst* gene resulted in conversion of more than 90 % of the stored sucrose into fructans. Under greenhouse conditions, the "fructan beets" developed normally and had almost the same amount of root dry weight as normal sugar beets. The yield of 110 µmol g^{-1} freshweight of fructan makes the extraction of these compounds economically interesting [132].

2.3.2.6 Seedless fruits and vegetables

In plants able to develop fruits without fertilization (parthenocarpic fruits), the seeds are absent – a feature that can increase fruit acceptance by consumers. To achieve parthenocarpic development, it is common practice to treat flower buds with synthetic auxinic hormones. To mimic the hormonal effects by genetic engineering, the expression of a gene able to increase auxin content and activity should be induced in the ovule. Rotino and colleagues [133] used the *iaaM* gene from *Pseudomonas syringae* pv. *savastanoi* under control of an ovule-specific promoter (*DefH9*) from *Antirrhinum majus* to induce parthenocarpic development in transgenic tobaccos and eggplants. In transgenic eggplants, fruit setting took place even under environmental conditions prohibitive for the untransformed line. Fruit size and weight were similar to that obtained by pollination in transgenic and control plants [133]. It was envisaged that this approach might also be valuable in other horticultural species such as pepper and tomatoes, in which fruit quality is susceptible to uneven pollination and seed set during lower temperature fluctuations associated with "winter season" production. A similar approach might also be used to produce seedless watermelons rather than the cumbersome triploid seed production system now employed [134].

2.3.3
Traits that Affect Processing

2.3.3.1 Altered gluten level in wheat to change baking quality

The bread-making quality of wheat flour depends primarily on the presence of high molecular-weight (HMW) glutenins. Wheat is unique among cereals in having this property. The glutenin proteins are encoded by six genes, and the total glutenin content of the grain is proportional to the expression of these genes. It has been shown that the quality of dough can be influenced both by the quantity and quality of the expressed genes. When specific glutenin genes were added back to wheat lines missing some of the glutenin genes, the dough-mixing characteristics were improved significantly [135]. This same technique could be applied to wheat lines that have already been optimized for baking, and could result in flour with glutenin levels higher than the current maximum of 10% of total protein [136]. Concordantly, the quality of durum wheat (*Triticum turgidum* L. var. *durum*) for bread and pasta making has been modified by insertion of HMW glutenin subunit genes [137].

2.3.3.2 Altered grain composition in barley to improve malting quality

β-Glucan is the major constituent of the cell wall of the starchy endosperm of barley. These cell wall molecules are very large, water-soluble, and produce viscous worts if not sufficiently reduced in size by hydrolysis. This causes slow filtration of wort and beer, as well as glucan precipitate in the finished product. β-Glucan hydrolysis is likely to be a function of the level of (1-3,1-4)-β-glucanase produced by the aleurone, and how much of the enzyme survives high-temperature kilning and mashing. Doubling the amount of β-glucanase activity is likely to ensure sufficient β-glucan hydrolysis, and this could be achieved either by increasing the amount of enzyme synthesized or by changing the heat stability of β-glucanase [138]. The latter has been achieved by transferring a fungal thermotolerant endo-1,4-β-glucanase to two barley cultivars. The amount of heterologous enzyme has been shown to be sufficient to reduce wort viscosity by decreasing the soluble β-glucan content [139].

In another approach, biochemically active wheat thioredoxin h has been overexpressed in the endosperm of transgenic barley grain. Such overexpression in germinated grain effected an up to 4-fold increase in the activity of the debranching enzyme pullulanase, a rate-limiting enzyme in the breakdown of starch. The breaking down of starch is a key step in the malting process, and tests with transgenic varieties showed that the time required could be reduced by up to a day [140, 141].

2.3.4
Traits of Pharmaceutical Interest

2.3.4.1 Production of vaccines

Vaccines are designed to elicit an immune response without causing disease. Whilst typical vaccines are composed of killed or attenuated disease-causing organisms, recombinant vaccines are desirable as an alternative as they generally cause fewer side effects than occur when the whole organism is delivered. Many candidate proteins have been identified that may function as effective subunit vaccines. Currently, the most common large-scale production of recombinant proteins are genetically engineered bacteria and yeast, due to the ease of manipulation and their rapid growth. However, recombinant proteins overexpressed in microorganisms must be extensively purified to remove host proteins and compounds. Transgenic plants provide an alternative system, with the great practical advantage of producing directly an edible plant tissue for oral immunization. In addition, edible plant-based recombinant vaccines are safe, provide nutrition, and are easy to administer [142, 143].

At present, a great deal of research is focused on the understanding of transgene expression, stability, and processing in plants. Typical experiments investigating in-planta protein expression employ plant model systems such as potato, tomato, and maize. The ideal plant for human vaccination would be bananas, because they are readily eaten by babies, are consumed uncooked, and are indigenous to many developing countries. Since regeneration and growth to maturity can take up to three years in bananas, it is nevertheless necessary to optimize high-level expression of a vaccine antigen in model plants [143].

For viral infections, virus-like particles (VLPs), which form by self-assembly of viral surface proteins, are effective as vaccines. The hepatitis B surface antigen (HbsAg) has successfully been used as a vaccine against hepatitis B virus. The vaccine was first produced in yeast by Merck and SmithKline Beecham, and was the first recombinant subunit vaccine. Hence, it served as a model for a first attempt to produce a plant-based vaccine. HbsAg was expressed in tobacco and potato plants and shown to assemble into VLPs that are similar to the yeast-derived commercial vaccine [144, 145].

In the United States, preliminary clinical trials have been performed, and two out of three volunteers who ate transgenic lettuce carrying a hepatitis B antigen displayed a good systemic response. Likewise, 19 of 20 people who ate a potato vaccine aimed at the Norwalk virus showed an immune response [146].

Mice fed with potato plants transgenic for a cholera toxin B subunit gene (CTB) developed immunity against the bacterial endotoxin [142].

Currently, clinical trials with a plant-based vaccine against a bacterial pathogen are underway in the United States. Arntzen and colleagues inserted a gene for a part of an *Escherichia coli* enterotoxin (LT-B subunit gene) that caused diarrhea in humans, into potato plants, and all but one of the 11 volunteers who ate the transgenic raw potatoes produced antibodies to the toxin [147].

A common problem of vaccine antigens expressed in plants has been the low level of expression. In LT-B-expressing plants this has been overcome by using a "plant codon usage optimized" synthetic gene encoding LT-B and the use of the tuber-specific patatin promoter (Mason et al., 1998, cited in Ref. [143]).

Future considerations must include containment of the transgenes as well as quality control for antigen content. These considerations are particularly important since inappropriate dosing of vaccines can impair their effectiveness, and constant dosing can lead (potentially) to immunological tolerance [143].

2.3.4.2 Production of pharmaceuticals

Examples of human therapeutic proteins include serum proteins such as hemoglobin as blood substitute, interferons as viral protection agents, lysosomal enzymes lacking in patients with Gaucher or Fabry disease, and other proteins such as hirudin, which is effective as anticoagulant (for a summary, see Ref. [148]).

Some recombinant proteins are already available commercially, including human erythropoietin or glucocerebrosidase. Until now, commercial production has used fermentation in *E. coli* and yeast or mammalian cell systems. However, these expression systems have significant limitations: bacteria cannot perform the complex post-translational modifications required for bioactivity of many human proteins and production of proteins such as growth regulators or cell cycle inhibitors would negatively impact the transgenic animal cell culture [148].

Transgenic plants have some remarkable features for the cost-effective bioproduction of proteins for pharmaceutical uses, including: (i) low production costs; (ii) reduced time to market; (iii) unlimited supply; (iv) eukaryotic protein processing; and (v) safety from blood- or animal tissue-borne human pathogens [149]. The production of a range of therapeutic proteins in transgenic plants has proven their capability for production of bioactive human proteins of pharmaceutical value: the proteins appear fully functional and structurally comparable with the analogous proteins produced in animal cell culture [148].

The majority of these pharmaceuticals are produced in model plants that are easy to transform, such as tobacco and potato, by using the constitutive 35S promoter from the cauliflower mosaic virus. However, as high levels of protein accumulation may negatively impact on yield and/or growth of the transgenic plants, inducible or tissue-specific promoters are preferable [148]. Tobacco is an excellent biomass producer, but does produce toxic compounds. Crop-based production systems (e.g., wheat, rice, corn) lack these toxic substances and provide an existing infrastructure for their cultivation, harvest, distribution, and processing [149].

Cereal crops (rice and wheat) were first used as production and storage system for a single-chain antibody against carcinoembryogenic antigen (CEA), a well-characterized tumor-associated marker antigen. In fact, in dried seeds, the antibodies could be stored for at least five months without significant loss of activity of the antibody [150].

CropTech has developed a post-harvest expression system in tobacco that uses an inducible promoter termed MeGa™ promoter. This has been modified from a defense-related gene and is inactive during normal growth and development, yet is highly expressed after mechanical stress. After plant harvest, recombinant protein production is induced and newly synthesized protein (glucocerebrosidase for the treatment of Gaucher disease) is recovered after 8–24 hours (Cramer and Weissenborn, cited in Ref. [148]).

An alternate technique employs targeting of the recombinant protein to the oil bodies of *Brassica napus* seeds. A synthetic gene coding for a hirudin variant was fused to an *Arabidopsis* oleosin gene and transferred to oilseed rape. The recombinant hirudin was correctly targeted and accumulated on the oilbodies of transgenic seeds which, due to their lower density could easily be separated by flotation centrifugation. The functional biopharmaceutical was then released by protease treatment [151].

A recent publication showed the potential for producing recombinant proteins in the guttation fluid of tobacco. Guttation, which is the loss of water and dissolved materials from uninjured plant organs, leads to the production of a guttation fluid. This can be collected throughout the plant's life, thereby providing a continuous and nondestructive system for recombinant protein production [152].

2.4
Outlook

It is likely that, in the future a rapid development of molecular and technological methods will take place, and this will comprise the initial development of new transgenic approaches in model plants such as *Arabidopsis*, followed by the transfer of ready-established methods to most crop plants. The combination of genetic engineering and molecular marker technology with conventional crossings and subsequent selection will enable a rapid progress in plant breeding.

Despite these possibilities, limitations of widespread application can be foreseen in the patent situation, and this will lead to a partial unpredictability in production costs. A hindrance to the worldwide trade – and therefore also to the marketability of genetically modified foods – is founded in the different legal regulations, e. g., in the USA and Europe. Additionally, the lack of acceptance by the public of genetically modified food, especially in central Europe, might delay the commercial potential of transgenic plants, even when there are clear advantages for the consumer.

References

1 Hansen, G., Chilton, M. D. **1999**. Lessons in Gene Transfer to Plants by a Gifted Microbe. In: Hammond, J., McGarvey, P., Yushibov, V. (eds.), *Plant Biotechnology: New Products and Applications*, Springer-Verlag, Berlin, New York, 21–58.

2 Hammond, J. **1999**. Overview: The Many Uses and Applications of Transgenic Plants. In: Hammond, J., McGarvey, P., Yushibov, V. (eds.), *Plant Biotechnology: New Products and Applications*, Springer-Verlag, Berlin, New York, 1–20.

3 Dai, S., Zheng, P., Marmey, P., Zhang, S., Tian, W., Chen, S., Beachy, R. N., Fauquet, C. **2001**. Comparative analysis of transgenic rice plants obtained by Agrobacterium-mediated transformation and particle bombardment. *Mol. Breeding* **7** (1):25–33.

4 Arencibia, A. D., Carmona, E. R., Tellez, P., Chan, M.-T., Yu, S.-M., Trujillo, L. E., Oramas, P. **1998**. An efficient protocol for sugarcane (*Saccharum* spp. L.) transformation mediated by *Agrobacterium tumefaciens*. *Transgenic Res.* **7**:213–222.

5 Potrykus, I. **1995**. Direct Gene Transfer to Protoplasts. In: Potrykus, I. and Spangenberg, G. (eds.), *Gene Transfer to Plants*. Springer Lab Manual, Berlin, 55–57.

6 Arencibia, A. D., Carmona, E. R., Cornide, M. T., Castiglione, S., O'Reilly, J., Chinea, P., Oramas, P., Sala, F. **1999**. Somaclonal variation in insect-resistant sugarcane (Saccharum hybrid) plants produced by electroporation. *Transgenic Res.* **8**:349–360.

7 Sanford, J. C., Klein, T. M., Wolf, E. D., Allen, N. **1987**. Delivery of substances into cells and tissues using a particle bombardment process. *J. Part. Sci. Tech.* **5**:27–37.

8 Finer, J. J., Finer, K. R., Ponappa, T. **1999**. Particle Bombardment Mediated Transformation. In: Hammond, J., McGarvey, P., Yushibov, V. (eds.), *Plant Biotechnology: New Products and Applications*, Springer-Verlag, Berlin, New York, 59–80.

9 Fütterer, J. **1995**. Genetic Markers and Expression Signals. In: Potrykus, I. and Spangenberg, G. (eds.), *Gene Transfer to Plants*. Springer Lab Manual, Berlin, 311–324.

10 Wang, Y., Zhang, W., Cao, J., McElroy, D., Wu, R. **1992**. Characterization of *cis*-acting elements regulating transcription from the promoter of a constitutively active rice actin gene. *Mol. Cell. Biol.* **12**:3399–3406.

11 Christensen, A. H., Sharrock, R. A., Quail, R. H. **1992**. Maize polyubiquitin genes: structure, thermal perturbation of expression and transcript splicing, and promoter activity following transfer to protoplasts by electroporation. *Plant Mol. Biol.* **18**:675–689.

12 McElroy, D., Brettell, R. I. S. **1994**. Foreign gene expression in transgenic cereals. *Trends Biotechnol.* **12**:62–68.

13 Fujimoto, H., Yamamoto, M., Kyozuka, J, Shimamoto, K. **1993**. Insect-resistant rice generated by introduction of a modified δ-endotoxin gene from *Bacillus thuringiensis*. *Bio/Technology* **11**:1151–1155.

14 Schrott, M. **1995**. Selectable Marker and Reporter Genes. In: Potrykus, I. and Spangenberg, G. (eds.), *Gene Transfer to Plants*. Springer Lab Manual, Berlin, 325–336.

15 Collard, J.-M. **1999**. Aminoglycoside resistance. http://www.antibioresistance.be/aminoglycosides.html (accessible February 2003).

16 Moens, W., Collard, J.-M. **2002**. GM Plants containing antibiotic resistance genes. http://www.biosafety.be/ARGMO/GMO_Plants.html (accessible February 2003).

17 De Block, M., Botterman, J., Vandewiele, M., Dockx, J., Thoen, C., Gossele, V., Movva, N. R., Thompson, C., Van Montague, M., Leemans, J. **1987**. Engineering herbicide resistance in plants by expression of a detoxifying enzyme. *EMBO J.* **6**:2513–2518.

18 Jefferson, R. A., Kavanagh, T. A., Bevan, M. W. **1987**. GUS fusions: β-glucuronidase as a sensitive and versatile gene fusion marker in higher plants. *EMBO J.* **6**:39001–3908.

19 Ow, D. W., Wood, K. V., DeLuca, M., de Wet, J. R., Helinski, D. R., Howell, S. H. **1986**. Transient and stable expression of the firefly luciferase gene in plant cells and transgenic plants. *Science* **234**:856–859.

20 Millar, A. J., Short, S. R., Hiratsuka, K., Chua, N. H., Kay, S. A. **1992**. Firefly luciferase as a reporter of regulated gene expression in higher plants. *Plant Mol. Biol. Rep.* **10**:324–337.

21 Stirn, S. **2000**. Antibiotic resistance and horizontal gene transfer. In: *GTZ (Gesellschaft für Technische Zusammenarbeit)* (ed.), Development of Frame Conditions for the Utilisation of Biotechnology and Genetic Engineering. http://www.gtz.de/biotech/dokumente/biotech5.pdf (accessible February 2003).

22 Kunkel, T., Niu, Q.-W., Chan, Y.-S., Chua, N.-H. **1999**. Inducible isopentenyl transferase as a high-efficiency marker for plant transformation. *Nature Biotechnol.* **17**: 916–919.

23 Joersbo, M., Donaldson, I., Petersen, S. G., Brunstedt, J., Okkels, F. T. **1998**. Analysis of mannose selection used for transformation of sugar beet. *Mol. Breeding* **4**:111–117.

24 Odell, J. T., Russell, S. H. **1994**. Use of Site-Specific Recombination Systems in Plants. In: Paszkowski J. (ed.), *Homologous Recombination and Gene Silencing in Plants*. Kluwer Academic Publ., Netherlands, 219–270.

25 Schiemann, J., Weber, A., Hassa, A. **1998**. Überprüfung von Konzepten zur nachträglichen Fremdgen-Eliminierung aus transgenen Pflanzen. In: *Jahresbericht der Biologischen Bundesanstalt für Land- und Forstwirtschaft 1998*, Berlin und Braunschweig, Abstract Nr. 189.

26 Zuo, J., Niu, Q.-W., Møller, S. G., Chua, N.-H. **2001**. Chemical-regulated, site-specific DNA excision in transgenic plants. *Nature Biotechnol.* **19**:157–161.

27 Zubko, E., Scutt, C., Meyer, P. **2000**. Intrachromosomal recombination between attP regions as a tool to remove selectable marker genes from tobacco transgenes. *Nature Biotechnol.* **18**:442–445.

28 Puchta, H. **2000**. Removing selectable marker genes: taking the short cut. *Trends Plant Sci.* **5** (7):273–274.

29 Transgen (Transparenz für Gentechnik bei Lebensmitteln) **2002**. Gentechnisch veränderte Pflanzen: Zulassungen weltweit. http://www.transgen.de (data from December 2002).

30 AGBIOS (Agriculture & Biotechnology Strategies (Canada) Inc.) **2002**. GMO Database. http://www.agbios.com/default.asp (data from December 2002).

31 RKI (Robert-Koch-Institut, Berlin) **2002**. Zentrum Gentechnologie. http://www.rki.de/GENTEC/GENTEC.HTM (data from November 2002).

32 ISAAA (International Service for the Acquisition of Agri-biotech Applications) **2002**. Global Review of Commercialized Transgenic Crops: 2001. http://www.isaaa.org/publications/briefs/Brief_24.htm (January 2002)

33 Transgen **2002**. Anbauflächen gentechnisch veränderter Pflanzen 2001. http://www.transgen.de/Aktuell/ISAAA_2001_preview.html (data from January 2002).

34 Cheng, X., Sardana, R., Kaplan, H., Altosaar, I. **1998**. *Agrobacterium*-transformed rice plants expressing synthetic *cryIA(b)* and *cryIA(c)* genes are highly toxic to striped stem borer and yellow stem borer. *Proc. Natl. Acad. Sci. USA* **95**:2767–2772.

35 Tu, J., Zhang, G., Datta, K., Xu, C., He, Y., Zhang, Q., Khush, G. S., Datta, S. K. **2000**. Field performance of transgenic elite commercial hybrid rice expressing *Bacillus thurigiensis* δ-endotoxin. *Nature Biotechnol.* **18**:1101–1104.

36 Stewart, C. N., Adang, M. J., All, J. N., Boerma, H. R., Cardineau, G., Tucker, D., Parrott, W. A. **1996**. Genetic transformation, recovery, and characterization of fertile soybean transgenic for a synthetic *Bacillus thuringiensis* cryIAc gene. *Plant Physiol.* **112** (1):121–129.

37 Stewart, C. N., Adang, M. J., All, J. N., Raymer, P. L., Ramachandran, S., Parrott, W. A. **1996b**. Insect control and dosage effects in transgenic canola containing a synthetic *Bacillus thuringiensis cryIAc* gene. *Plant Physiol.* **112** (1):115–120.

38 Arpaia, S., Mennella, G., Onofaro, V., Perri, E., Sunseri, F., Rotino, G. L. **1997**. Production of transgenic eggplant (*Solanum melongena* L.) resistant to Colorado Potato Beetle (*Leptinotarsa decemlineata* Say). *Theoret. Appl. Genet.* **95**:329–334.

39 USDA (United States Department of Agriculture) **2002**. Field tests release database for the U. S.. http://www.isb.vt.edu/cfdocs/fieldtests1.cfm (data from February 2002).

40 EPA (Environmental Protection Agency, USA) **2001**. Biopesticides Federal Register Notices – By Date. http://www.epa.gov/pesticides/biopesticides/biop_fr_date.htm (accessible February 2002).

41 Moellenbeck, D. J., Peters, M. L., Bing, J. W. et al. **2001**. Insecticidal proteins from *Bacillus thuringiensis* protect corn from corn root worms. *Nature Biotechnol.* **19**:668–672.

42 Gatehouse, A. M. R., Hilder, V. A., Powell, K. S., Wang, M., Davison, G. M., Gatehouse, L. N., Down, R. E., Edmonds, H. S., Boulter, D., Newell, C. A., Merryweather, A., Hamilton, W. D. O., Gatehouse, J. A. **1994**. Insect-resistant transgenic plants: choosing the gene to do the 'job'. *Biochem. Soc. Trans.* **22**:944–949.

43 Marchetti, S., Delledonne, M., Fogher, C., Chiabà, C., Chiesa, F., Savazzini, F., Giordano, A. **2000**. Soybean Kunitz, C-II and PI-IV inhibitor genes confer different levels of insect resistance to tobacco and potato transgenic plants. *Theoret. Appl. Genet.* **101**: 519–526.

44 Rao, K. V., Rathore, K. S., Hodges, T. K., Fu, X., Stoger, E., Sudhakar, D., Williams, S., Christou, P., Bharathi, M., Bown, D. P., Powell K. S., Spence, J., Gatehouse, A. M. R., Gatehouse, J. A. **1998**. Expression of snowdrop lectin (GNA) in transgenic rice plants confers resistance to rice brown planthopper. *Plant J.* **15**(4): 469–477.

45 Morton, R. L., Schroeder, H. E., Bateman, K. S., Chrispeels, M. J., Armstrong, E,. Higgins, T. J. **2000**. Bean alpha-amylase inhibitor 1 in transgenic peas (*Pisum sativum*) provides complete protection from pea weevil (*Bruchus pisorum*) under field conditions. *Proc. Natl. Acad. Sci. USA* **97**(8):3820–3835.

46 Ewen, S. W. B., Pusztai, A. **1999**. Effects of diets containing genetically modified potatoes expressing *Galanthus nivalis* lectin on rat small intestine. *Lancet* **354**:1353–1354.

47 Kuiper, H. A., Noteborn, H. P. J. N., Peijenburg, A. A. C. M. **1999**. Adequacy of methods for testing the safety of genetically modified food. *Lancet* **354**:1315–1316.

48 Kramer, K. J., Morgan, T. D., Throne, J. E., Dowell, F. E., Bailey, M., Howard, J. A. **2000**. Transgenic avidin maize is resistant to storage insect pests. *Nature Biotechnol.* **18**: 670–674.

49 Hayakawa, T., Zhu, Y., Itoh, K., Izawa, T., Shimamaoto, K., Toriyama, S. **1992**. Genetically engineered rice resistant to rice stripe virus, an insect-transmitted virus. *Proc. Natl. Acad. Sci. USA* **89** (20):9865–9869.

50 Ravelonandro, M., Scorza, R., Bachelier, J. C., Labonne, G., Levy, L., Dam-

steegt, V., Callahan, A. M., Dunez, J. **1997**. Resistance of transgenic *Prunus domestica* to plum pox virus infection. *Plant Dis.* **81**: 1231–1235.

51 Kaniewski, W., Ilardi, V., Tomassoli, L., Mitsky, T., Layton, J., Barba, M. **1999**. Extreme resistance to cucumber mosaic virus (CMV) in transgenic tomato expressing one or two viral coat proteins. *Mol. Breeding* **5**:111–119.

52 Chowrira, G. M., Cavileer, T. D., Gupta, S. K., Lurquin, P. F., Berger, P. H. **1998**. Coat-protein-mediated resistance to pea enation mosaic virus in transgenic *Pisum sativum* L. *Transgenic Res.* **7**:265–271.

53 Magbanua, Z. V., Wilde, H. D., Roberts, J. K., Chowdhury, K., Abad, J., Moyer, J. W., Wetzstein, H. Y., Parrott, W. A. **2000**. Field resistance to tomato spotted wilt virus in transgenic peanut (*Arachis hypogaea* L.) expressing an antisense nucleocapsid gene sequence. *Mol. Breeding* **6**:227–236.

54 Koev, G., Mohan, B. R., Dinesh-Kumar, S. P., Torbert, K. A., Somers, D. A., Miller, W. A. **1998**. Extreme reduction of disease in oats transformed with the 5′ half of the barley yellow dwarf virus-PAV genome. *Phytopathology* **88**:1013–1019.

55 Gal-On, A., Wang, Y., Faure, J. E., Pilowsky, M., Zelcer, A. **1998**. Transgenic resistance to cucumber mosaic virus in tomato: blocking of long-distance movement of the virus in lines harbouring a defective viral replicase. *Phytopathology* **88** (10):1101–1107.

56 Huet, H., Mahendra, S., Wang, J., Sivamani, E., Ong, C. A., Chen, L., de Kochko, A., Beachy, R. N., Fauquet, C. **1999**. Near immunity to rice tungro spherical virus achieved in rice by replicase-mediated resistance strategy. *Phytopathology* **89**: 1022–1027.

57 Sivamani, E., Brey, C. W., Dyer, W. E., Talbert, L. E., Qu, R. **2000**. Resistance to wheat streak mosaic virus in transgenic wheat expressing the viral replicase (Nib) gene. *Mol. Breeding* **6**:469–477.

58 Valanzuolo, S., Catello, S., Colombo, M., Dani, M., Monti, M. M., Uncini, L., Petrone, P., Spingo, P. **1994**. Cucumber Mosaic Virus resistance in transgenic San Marzano tomatoes. *Acta Horticulturae* **376**:377–386.

59 Stommel, J. R., Tousignant, M. E., Wai, T., Pasini, R., Kaper, J. M. **1998**. Viral satellite RNA expression in transgenic tomato confers field tolerance to cucumber mosaic virus. *Plant Dis.* **82** (4):391–396.

60 Tepfer, M. **1993**. Viral genes and transgenic plants. *Bio/Technology* **11**:1125–1132.

61 Lodge, J. K., Kaniewski, W. K., Tumer, N. E. **1993**. Broad-spectrum resistance in transgenic plants expressing pokeweed antiviral protein. *Proc. Natl. Acad. Sci. USA* **90**:7089–7093.

62 Wang, P., Zoubenko, O., Tumer, N. E. **1998**. Reduced toxicity and broad spectrum resistance to viral and fungal infection in transgenic plants expressing pokeweed antiviral protein II. *Plant Mol. Biol.* **38**:957–964.

63 Tavladoraki, P., Benvenuto, E., Trinca, S., De Martinis, D., Cattaneo, A., Galeffi, P. **1993**. Transgenic plants expressing a functional single-chain Fv antibody are specifically protected from virus attack. *Nature* **366**:469–472.

64 Xiao, X. W., Chu, P. W. G., Frenkel, M. J., Tabe, L. M., Shukla, D. D., Hanna, P. J., Higgins, T. J. V., Müller, W. J., Ward, C. W. **2000**. Antibody-mediated improved resistance to ClYVV and PVY infections in transgenic tobacco plants expressing a single-chain variable region antibody. *Mol. Breeding* **6**: 421–431.

65 Mitsuhara, I., Matsufuru, H., Ohshima, M., Kaku, H., Nakajima, Y., Murai, N., Natori, S., Ohashi, Y. **2000**. Induced expression of sarcotoxin IA enhanced host resistance against both bacterial and fungal pathogens in transgenic tobacco. *Mol. Plant-Microbe Interact.* **13** (8):860–868.

66 Lin, W., Anuratha, C. S., Datta, K., Potrykus, K., Muthukrishnan, S., Datta, S. K. **1995**. Genetic engineering of rice for resistance to sheath blight. *Bio/Technology* **13**:686–691.

67 Datta, K., Velazhahan, R., Oliva, N., Ona, I., Mew, T., Khusgh, G. S.,

Muthukrishnan, S., Datta, S. K. **1999**. Over-expression of the cloned rice thaumatin-like protein (PR-5) gene in transgenic rice plants enhances environmental friendly resistance to *Rhizoctonia solani* causing sheath blight. *Theoret. Appl. Genet.* **98**: 1138–1145.

68 Datta, K., Tu, J. M., Oliva, N., Ona, I., Velazhahan, R., Mew, T. W., Muthukrishnan, S., Datta, S. K. **2001**. Enhanced resistance to sheath blight by constitutive expression of infection-related rice chitinase in transgenic elite indica rice cultivars. *Plant Sci.* **160** (3):405–414.

69 Nishizawa, Y., Nishio, Z., Nakazono, K., Soma, M., Nakajima, E., Ugaki, M., Hibi, T. **1999**. Enhanced resistance to blast (*Magnaporthe grisea*) in transgenic Japonica rice by constitutive expression of rice chitinase. *Theoret. Appl. Genet.* **99** (3/4):383–390.

70 Schaffrath, U., Mauch, F., Freydell, E., Schweizer, P., Dudler, R. **2000**. Constitutive expression of the defense-related *Rir1b* gene in transgenic rice plants confers enhanced resistance to the rice blast fungus *Magnaporthe grisea*. *Plant Mol. Biol.* **43**:59–66.

71 Dixon, R. A., Lamb, C. J., Masoud, S., Sewalt, V. J. H., Paiva, N. L. **1996**. Metabolic engineering: prospects for crop improvement through the genetic manipulation of the phenylpropanoid biosynthesis and defense responses – a review. *Gene* **179** (1):61–71.

72 Tabei, Y., Kitade, S., Nishizawa, Y., Kikuchi, N., Kayano, T., Hibi, T., Akutsu, K. **1998**. Transgenic cucumber plants harbouring a rice chitinase gene exhibit enhanced resistance to gray mold (*Botrytis cinerea*). *Plant Cell Rep.* **17** (3):159–164.

73 Wang Y., Nowak, G., Culley, D., Hadwiger, L. A., Fristensky, B. **1999**. Constitutive expression of pea defense gene DRR206 confers resistance to blackleg (*Leptoshaeria maculans*) disease in transgenic canola (*Brassica napus*). *Mol. Plant-Microbe Interact.* **12** (5):410–418.

74 Tabaeizadeh, Z., Agharbaoui, Z., Harrak, H., Poysa, V. **1999**. Transgenic tomato plants expressing a *Lycopersicon chilense* chitinase gene demonstrate improved resistance to *Verticillium dahliae* race 2. *Plant Cell Rep.* **19** (2): 197–202.

75 Bliffeld, M., Mundy, J., Potrykus, I., Fütterer, J. **1999**. Genetic engineering of wheat for increased resistance to powdery mildew disease. *Theoret. Appl. Genet.* **98** (6/7): 1079–1086.

76 Chen, W. P., Chen, P. D., Liu, D. J., Kynast, R., Friebe, B., Velazhahan, R., Muthukrishnan, S., Gill, B. S. **1999**. Development of wheat scab symptoms is delayed in transgenic wheat plants that constitutively express a rice thaumatin-like protein gene. *Theoret. Appl. Genet.* **99** (5):755–760.

77 Oldach, K., Becker, D., Lörz, H. **2001**. Heterologous expression of genes mediating enhanced fungal resistance in transgenic wheat. *Mol. Plant-Microbe Interact.* **14** (7):832–838.

78 Yamamoto, T., Iketani, H., Ieki, H., Nishizawa, Y., Notsuka, K., Hibi, T., Hayashi, T., Matsuta, N. **2000**. Transgenic grapevine plants expressing a rice chitinase with enhanced resistance to fungal pathogens. *Plant Cell Rep.* **19** (7):639–646.

79 Fagoaga, C., Rodrigo, I., Conejero, V., Hinarejos, C., Tuset, J. J., Arnau, J., Pina, J. A., Navarro, L., Pena, L. **2001**. Increased tolerance to *Phytophthora citrophthora* in transgenic orange plants constitutively expressing a tomato pathogenesis related protein PR-5. *Mol. Breeding* **7** (2):175–185.

80 Maddaloni, M., Forlani, F., Balmas, V., Donini, G., Stasse, L., Corazza, L., Motto, M. **1997**. Tolerance to the fungal pathogen *Rhizoctonia solani* AG4 of transgenic tobacco expressing the maize ribosome inactivating protein b-32. *Transgenic Res.* **6** (6):393–402.

81 Hain, R., Reif, H. J., Krause, E., Langbartels, R., Kindl, H., Vornam, B., Wiese, W., Schmelzer, E., Schreier, P. H., Stocker, R. H., Stenzel, K. **1993**. Disease resistance results from foreign phytoalexin expression in a novel plant. *Nature* **361**:153–156.

82 Stark-Lorenzen, P., Nelke, B., Hanssler, G., Mühlbach, H.-P., Thomzik, J. E.

1997. Transfer of a grapevine stilbene synthase gene to rice (*Oryza sativa* L.). *Plant Cell Rep.* **16** (10):668–673.

83 Leckband, G., Lörz, H. **1998**. Transformation and expression of a stilbene synthase gene of *Vitis vinifera* L. in barley and wheat for increased fungal resistance. *Theoret. Appl. Genet.* **96** (8):1004–1012.

84 He, X. Z., Dixon, R. A. **2000**. Genetic manipulation of isoflavone 7-O-methyltransferase enhances biosynthesis of 4′-O-methylated isoflavonoid phytoalexins and disease resistance in alfalfa. *Plant Cell* **12** (9):1689–1702.

85 Gandikota, M., de Kocho, A., Chen, L., Ithal, N., Fauquet, C., Reddy, A. R. **2001**. Development of transgenic rice plants expressing maize anthocyanin genes and increased blast resistance. *Mol. Breeding* **7** (1):73–83.

86 Takaichi, M., Oeda, K. **2000**. Transgenic carrots with enhanced resistance against two major pathogens, *Erysiphe heraclei* and *Alternaria dauci*. *Plant Sci.* **153**:135–144.

87 Lorito, M., Woo, S. L., Fernandez, I. G., Colucci, G., Harman, G. E., Pintor-Toro, J. A., Filippone, E., Muccifora, S., Lawrence, C. B., Zoina, A., Tuzun, S., Scala, F. **1998**. Genes from mycoparasitic fungi as source for improving plant resistance to fungal pathogens. *Proc. Natl. Acad. Sci. USA* **95**:7860–7865.

88 Bolar, J. P., Norelli, J. L., Wong, K. W., Hayes, C. K., Harman, G. E., Aldwinckle, H. S. **2000**. Expression of endochitinase from *Trichoderma harzianum* in transgenic apple increases resistance to apple scab and reduces vigor. *Phytopathology* **90**:72–77.

89 Clausen, M., Kräuter, R., Schachermayr, G., Potrykus, I., Sautter, C. **2000**. Antifungal activity of a virally encoded gene in transgenic wheat. *Nature Biotechnol.* **18**:446–449.

90 Husemann, M., Oldendorf, S., Schütte, G. **2001**. Krankheitsresistenz gegen Pilze und Bakterien. In: Schütte, G., Stirn, S., Beusmann, V. (eds.). *Transgene Nutzpflanzen: Sicherheitsforschung, Risikoabschätzung und Nachgenehmigungs-Monitoring*, Birkhäuser Verlag, Basel, 154–174.

91 Prakash, C. S. **1998**. Regulatory Gene Confers Resistance to Multiple Diseases. ISB News Report, June 1998. http://www.isb.vt.edu/news/1998/news98.jun.html#jun9803 (accessible February 2003).

92 Tu, J., Ona, I., Zhang, Q., Mew, T. W., Khush, G. S., Datta, S. K. **1998**. Transgenic rice variety 'IR72' with Xa21 is resistant to bacterial blight. *Theoret. Appl. Genet.* **97** (1/2): 31–36.

93 Tai, T. H., Dahlbeck, D., Clark, E. T., Gajiwala, P., Pasion, R., Whalen, M. C., Stall, R. E., Staskawicz, B. J. **1999**. Expression of the *Bs2* pepper gene confers resistance to bacterial spot disease in tomato. *Proc. Natl. Acad. Sci. USA* **96** (24):14153–14158.

94 Tang, X., Xie, M., Kim, Y. J., Zhou, J., Klessig, D. F., Martin, G. B. **1999**. Overexpression of *Pto* activates responses and confers broad resistance. *Plant Cell* **11** (1):15–29.

95 Tang, K., Sun, X., Hu, Q., Wu, A., Lin, C.-H., Lin, H.-J., Twyman, R. M., Christou, P., Feng, T. **2001**. Transgenic rice plants expressing the ferredoxin-like protein (AP1) from sweet pepper show enhanced resistance to *Xanthomonas oryzae* pv. *oryzae*. *Plant Sci.* **160**:1035–1042.

96 Molina, A., Garcia-Olmedo, F. **1997**. Enhanced tolerance to bacterial pathogens caused by the transgenic expression of barley lipid transfer protein LPT2. *Plant J.* **12** (3):669–675.

97 Düring, K., Porsch, P., Fladung, M., Lörz, H. **1993**. Transgenic potato plants resistant to the phytopathogenic bacterium *Erwinia carotovora*. *Plant J.* **3**(4):587–598.

98 Nakajima, H., Muranaka, T., Ishige, F., Akutsu, K., Oeda, K. **1997**. Fungal and bacterial disease resistance in transgenic plants expressing human lysozyme. *Plant Cell Rep.* **16**:674–679.

99 Wu, G., Shortt, B. J., Lawrence, E. B., Fitzsimmons, K. C., Shah, D. M. **1995**. Disease resistance conferred by expression of a gene encoding H_2O_2-generating glucose oxidase in trans-

genic potato plants. *Plant Cell* **7**:1357–1368.

100 Norelli, J. L., Aldwinckle, H. S., Deséfano-Beltrán, A., Jaynes, J. M. **1994**. Transgenic 'Malling 26' apple expressing the *attacin E* gene has increased resistance to *Erwinia amylovora*. *Euphytica* **77**:123–128.

101 Reynoird, J. P., Mourges, F., Norelli, J., Aldwinckle, H. S., Brisset, M. N., Chevreau, E. **1999**. First evidence for improved resistance to fire blight in transgenic pear expressing the *attacin E* gene from *Hyalophora crecropia*. *Plant Sci.* **149**:23–31.

102 Osusky, M., Zhou, G., Osuska, L., Hancock, R. E., Kay, W. W., Misra, S. **2000**. Transgenic plants expressing cationic peptide chimeras exhibit broad-spectrum resistance to phytopathogens. *Nature Biotechnol.* **18**:1162–1166.

103 Urwin, P. E., Lilley, C. J., McPherson, M. J., Atkinson, H. J. **1997**. Resistance to both cyst and root-knot nematodes conferred by transgenic *Arabidopsis* expressing a modified plant cystatin. *Plant J.* **12** (2):455–461.

104 Urwin, P. E., Atkinson, H. J., Waller, D. A., McPherson, M. J., **1995**. Engineered oryzacystatin-I expressed in transgenic hairy roots confers resistance to *Globodera pallida*. *Plant J.* **8**:121–131.

105 Atkinson, H. J., Urwin, P. E., Hansen, E., McPherson, M. J. **1995**. Designs for engineered resistance to root-parasitic nematodes. *Trends Biotechnol.* **13**:369–374.

106 Lohr, J. T. **1999**. Another step towards improved stress response. ISB News Report, April 1999. http://www.isb.vt.edu/articles/apr9908.htm (accessible February 2003).

107 Bajaj, S., Targolli, J., Liu, L.-F., Ho, T.-H., Wu, R. **1999**. Transgenic approaches to increase dehydration-stress tolerance in plants. *Mol Breeding* **5**:493–503.

108 Sakamoto, A., Alia, Murata, N. **1998**. Metabolic engineering of rice leading to biosynthesis of glycinebetaine and tolerance to salt and cold. *Plant Mol. Biol.* **38**:1011–1019.

109 Zhu, B., Su, J., Chang, M. C., Verma, D. P. S., Fan, Y. L., Wu, R. **1998**. Overexpression of a Δ^1-pyrroline-5-carboxylate synthetase gene and analysis of tolerance to water- and salt-stress in transgenic rice. *Plant Sci.* **139**:41–48.

110 Capell, T., Escobar, C., Liu, H., Burtin, D., Lepri, O., Christou, P. **1998**. Overexpression of the oat arginine decarboxylase cDNA in transgenic rice (*Oryza sativa*) affects normal development patterns *in vitro* and results in putrescine accumulation in transgenic plants. *Theoret. Appl. Genet.* **97**:246–254.

111 Roy, M., Wu, R. **2001**. Arginine decarboxylase transgene expression and analysis of environmental stress tolerance in transgenic rice. *Plant Sci.* **160**:869–875.

112 Xu, D., Duan, X., Wang, B., Hong, B., Ho, T. H. D., Wu, R. **1996**. Expression of a late embryogenesis abundant protein gene, *HVA1*, from barley confers tolerance to water deficit and salt stress in transgenic rice. *Plant Physiol.* **110**:249–257.

113 McKersie, B. D., Bowley, S. R., Jones, K. S. **1999**. Winter survival of transgenic alfalfa overexpressing superoxide dismutase. *Plant Physiol.* **119**:839–847.

114 Kasuga, M., Liu, Q., Yamaguchi-Shinozaki, K., Shinozaki, K. **1999**. Improving plant drought, salt, and freezing tolerance by gene transfer of a single stress-inducible transcription factor. *Nature Biotechnol.* **17**:287–291.

115 Lee, J. H., van Montague, M., Verbruggen, N. **1999**. A highly conserved kinase is an essential component for stress tolerance in yeast and plant cells. *Proc. Natl. Acad. Sci. USA* **96**:5873–5877.

116 Zhang, H.-X., Blumwald, E. **2001**. Transgenic salt-tolerant tomato plants accumulate salt in the foliage but not in fruit. *Nature Biotechnol.* **19**:765–768.

117 Gisbert, C., Rus, A. M., Bolarin, C., Lopez-Coronado, J. M., Arrilaga, I., Montesinos, C., Caro, M., Serrano, R., Moreno, V. **2000**. The yeast *HAL1* gene improves salt tolerance of transgenic tomato. *Plant Physiol.* **123**:393–402.

118 Takahashi, M., Nakanishi, H., Kawasaki, S., Nishizawa, N. K., Mori, S. 2001. Enhanced tolerance of rice to low iron availability in alkaline soils using barley nicotianamine aminotransferase genes. *Nature Biotechnol.* **19**:466–469.

119 Guerinot, M. L. 2001. Improving rice yields – ironing out the details. *Nature Biotechnol.* **19**:417–418.

120 Farré, E. M., Bachmann, A., Willmitzer, L., Trethewey, R. N. 2001. Acceleration of potato tuber sprouting by the expression of a bacterial pyrophosphatase. *Nature Biotechnol.* **19**:268–272.

121 Pena, L., Martin-Trillo, M., Juarez, J., Pina, J. A., Navarro, L., Martinez-Zapater, J. M. 2001. Constitutive expression of *Arabidopsis LEAFY* or *APETALA1* genes in citrus reduces their generation time. *Nature Biotechnol.* **19**:263–267.

122 He, Z., Zhu, Q., Dabi, T., Li, D., Weigel, D., Lamb, C. 2000. Transformation of rice with the *Arabidopsis* floral regulator *LEAFY* causes early heading. *Transgenic Res.* **9**:223–227.

123 Ye, X., Al-Babili, S., Klöti, A., Zhang, J., Lucca, P., Beyer, P., Potrykus, I. 2000. Engineering the provitamin A (β-carotene) biosynthetic pathway into (carotenoid-free) rice endosperm. *Science* **287**:303–305.

124 Potrykus, I. 2001. Golden Rice and Beyond. *Plant Physiol.* **125**:1157–1161.

125 Römer, S., Fraser, P. D., Kiano, J. W., Shipton, C. A., Misawa, N., Schuch, W., Bramley, P. M. 2000. Elevation of the provitamin A content of transgenic tomato plants. *Nature Biotechnol.* **18**:666–669.

126 Goto, F., Yoshihara, T., Shigemoto, N., Toki, S., Takaiwa, F. 1999. Iron fortification of rice seeds by the soybean ferritin gene. *Nature Biotechnol.* **17**:282–286.

127 Lucca, P., Wünn, J., Hurrell, R. F., Potrykus, I. 2000. Development of iron-rich rice and improvement of its absorption in humans by genetic engineering. *J. Plant Nutr.* **23** (11&12):1983–1988.

128 Lucca, P., Hurrell, R., Potrykus, I. 2001. Genetic engineering approaches to improve the bioavailability and the level of iron in rice grains. *Theoret. Appl. Genet.* **102**:392–397.

129 Chakraborty, S., Chakraborty, N., Datta, A. 2000. Increased nutritive value of transgenic potato by expressing a nonallergenic seed albumin gene from *Amaranthus hypochondriacus*. *Proc. Natl. Acad. Sci. USA* **97** (7):3724–3729.

130 Pratt, T. 1999. The quest for nontoxic cassava. ISB News Report, January 1999. http://www.isb.vt.edu/articles/jan9902.htm (accessible February 2003).

131 Sayre, R. T. 2000. Cyanogen reduction in transgenic cassava. Generation of safer food product for subsistence farmers. ISB News Report, August 2000. http//www.isb.vt.edu/articles/aug0003.htm (accessible February 2003).

132 Sevenier, R., Hall, R. D., van der Meer, I., Hakkert, H. J. C., van Tunen, A. J., Koops, A. J. 1998. High level fructan accumulation in a transgenic sugar beet. *Nature Biotechnol.* **16**:843–846.

133 Rotino, G. L., Perri, E., Zottini, M., Sommer, H., Spena, A. 1997. Genetic engineering of parthenocarpic plants. *Nature Biotechnol.* **15**:1398–1401.

134 Tomes, D. T. 1997. Seedless hopes bode well for winter vegetables. *Nature Biotechnol.* **15**:1344–1345.

135 Barro, F., Rooke, L., Békés, F., Gras, P., Tatham, A. S., Fido, R., Lazzeri, P., Shewry, P. R., Barcelo, P. 1997. Transformation of wheat with high molecular weight subunit genes results in improved functional properties. *Nature Biotechnol.* **15**:1295–1299.

136 Westwood, J. 1997. Transgenic bread is on the rise. ISB News Report, December 1997. http:www.isb.vt.edu/articles/dec9704.htm (accessible February 2003).

137 He, G. Y., Rooke, L., Steele, S., Békés, F., Gras, P., Tatham, A. S., Fido, R., Barcelo, P., Shewry, P. R., Lazzeri, P. A. 1999. Transformation of pasta wheat (*Triticum turgidum* L. var. *durum*) with high molecular weight glutenin subunit genes and modification of dough functionality. *Mol. Breeding* **5**(4):377–386.

138 McElroy, D., Jacobsen, J. **1995**. What's brewing in barley biotechnology? *Bio/Technology* **13**:245–249.

139 Nuutila, A. M., Ritala, A., Skadsen, R. W., Mannonen, L., Kaupinnen, V. **1999**. Expression of fungal thermotolerant endo-1,4-beta-glucanase in transgenic barley seeds during germination. *Plant Mol. Biol.* **41** (6):777–783.

140 Cho, M.-J., Wong, J. H., Marx, C., Jiang, W., Lemaux, P. G., Buchanan, B. B. **1999**. Overexpression of thioredoxin *h* leads to enhanced activity of starch debranching enzyme (pullulanase) in barley grain. *Proc. Natl. Acad. Sci. USA* **96** (25):14641–14646.

141 Pew Initiative on Food and Biotechnology **2001**. Harvest on the horizon: future uses of agricultural biotechnology. http://pewagbiotech.org/research/harvest/harvest.pdf. (accessible February 2003).

142 Arakawa, T., Chong, D. K. X., Langridge, W. H. R. **1998**. Efficacy of a food plant-based oral cholera toxin B subunit vaccine. *Nature Biotechnol.* **16**:292–297.

143 Richter, L., Kipp, P. B. **2000**. Transgenic plants as edible vaccines. In: Hammond, J., McGarvey, P., Yusibov, V. (eds.), *Plant Biotechnology: New Products and Applications*, Springer-Verlag, Berlin, New York, 159–176.

144 Mason, H. S., Lam, D. M. K., Arntzen, C. J. **1992**. Expression of hepatitis B surface antigen in transgenic plants. *Proc. Natl. Acad. Sci. USA* **89**:11745–11749.

145 Thanavala, Y., Yang, Y.-V., Lyons, P., Mason, H. S., Arntzen, C. J. **1995**. Immunogenicity of transgenic plant-derived hepatitis B surface antigen. *Proc. Natl. Acad. Sci. USA* **92**:3358–3361.

146 Langridge, W. H. R. **2000**. Edible vaccines. *Sci. Am.* **283** (3):66–71.

147 Tacket, C. O., Mason, H. S., Losonsk, G., Clements, J. D., Levine, M. M., Arntzen, C. J. **1998**. Immunogenicity in humans of a recombinant bacterial antigen delivered in a transgenic potato. *Nature Med.* **4**:607–609.

148 Cramer, C. L., Boothe, J. G., Oishi, K. K. **2000**. Transgenic plants for therapeutic proteins: Linking upstream and downstream strategies. In: Hammond, J., McGarvey, P., Yusibov, V. (eds.), *Plant Biotechnology: New Products and Applications*, Springer-Verlag, Berlin, New York, 95–118.

149 Fischer, R., Drossard, J., Commandeur, U., Schillberg, S., Emans, N. **1999**. Towards molecular farming in the future: moving from diagnostic protein and antibody production in microbes to plants. *Biotechnol. Appl. Biochem.* **30**:101–108.

150 Stöger, E., Vaquero, C., Torres, E., Sack, M., Nicholson, L., Drossard, J., Williams, S., Keen, D., Perrin, Y., Christou, P., Fischer, R. **2000**. Cereal crops as viable production and storage systems for pharmaceutical scFv antibodies. *Plant Mol. Biol.* **42**:583–590.

151 Parmenter, D. L., Boothe, J. G., Van Rooijen, G. J. H., Yeung, E. C., Moloney, M. M. **1995**. Production of biological active hirudin in plant seeds using oleosin partitioning. *Plant Mol. Biol.* **29**:1167–1180.

152 Komarnytsky, S., Borisjuk, N. V., Borisjuk, L. G., Alam, M. Z., Raskin, I. **2000**. Production of recombinant proteins in tobacco guttation fluid. *Plant Physiol.* **124**:927–933.

3
Fermented Food Production using Genetically Modified Yeast and Filamentous Fungi

Anke Niederhaus and Ulf Stahl

3.1
Introduction

3.1.1
Why Do We Ferment Foodstuffs?

Mankind has been fermenting foodstuffs of plant or animal origin for many thousands of years. In fact, the brewing of beer and making of bread has evolved from a common process which dates back to ancient Babylon some 8,000 years ago [1]. Although information from such a long time ago is not easily accessible, other more recent civilizations have also used beer and bread-making processes, as well as other biotechnological activities during the past millennia [2].

There are several reasons why people have chosen the fermentation process to develop novel foodstuffs. There has always been a need to create something more palatable out of raw foodstuff materials, thereby making them more easy to digest, to improve the taste and, of course, to increase storage and keeping characteristics (referred to nowadays as the 'shelf life') [3]. Thus, the fermentation process not only improves the taste of the raw product but also makes it more nutritious.

Our demands on food have not changed very much over time. The fermentation process can first and foremost be seen as a tool for conserving food and protecting it against spoilage. The use of ethanol or acids gained through the fermentation process prevents the development of unwanted microorganisms and results in a product which is more-or-less preserved, such as beer, wine, or vinegar. The ability of microorganisms to break down indigestible polymers by the use of amylolytic or proteolytic enzymes also contributes to the palatability of the product. Other metabolic activities can reduce toxins in raw foodstuffs, or be responsible for producing vitamins and essential amino and lipid-acids, thereby increasing the food's nutritional value. Moreover, the microorganisms also contribute to the flavor and texture of the food as they secrete metabolites such as aldehydes, acids, esters, ketones and sulfur compounds.

Examples of microhabitat products which have evolved world-wide include soy sauce, miso and tempe from the South-East Asian regions, mold-ripened cheeses and sausages from Europe, and beverages of sour gruel or sugar cane juice originating from South America.

The next two sections will provide a short overview on fermented food, categorized according to the type of raw material involved; that is, whether of plant or animal origin. This chapter only deals with those fermented foods which are produced using yeast or filamentous fungi. Given the enormous amount of information to deal with, our listings will of course be far from complete.

3.1.2
Fermented Foods of Plant Origin

As so many different types of foodstuff of plant origin exist (Table 3.1), the present text will focus only on those which are well known, while the microhabitats examined will be restricted to those of yeast and filamentous fungi.

3.1.3
Fermented Foods of Animal Origin

In general, microorganisms used in producing foodstuffs such as starter cultures must fulfill certain criteria: (i) an absence of toxinogenic and pathogenic potential; (ii) competitiveness; (iii) mycelia of white, yellowish, and ivory color; and (iv) proteolytic and lipolytic activity for aroma formation [3, 16].

Raw foodstuffs of animal origin result in products such as cheeses and meats. Mold-ripened cheeses include blue-veined cheeses such as Roquefort (France), Gorgonzola (Italy), and Stilton (United Kingdom). Roquefort is made only from sheep's milk, whereas Gorgonzola and Stilton are produced from cows' milk. All blue cheeses are made with *Penicillium roqueforti*, which grows inside the product. Examples of mold-ripened soft cheeses having *Penicillium camemberti* growing on the surface are Camembert and Brie [3]. Other cheeses such as Limburger, Tilsiter, Port Salut, Trappist, Brick and the Danish Danbo are also surface-ripened cheeses. The use of yeast is predominant during the initial period of ripening where it is used to raise the pH on the cheese surface, thereby allowing the growth of *Brevibacterium linens* and other bacteria present in the smear. Strains of *Debaryomyces hansenii* (among *Kloeckera* and *Trichosporon* species) in particular are thought to be responsible for the increase in pH by degrading lactic acid [17].

Examples of fermented meat products are Italian or Hungarian salami, where a combination of different microorganisms is involved. Preservation within the product is achieved by decreasing the pH via production of lactic acid by various *Lactobacilli*. Nontoxic *Penicillium*-species such as *Penicillium nalgiovense* are applied to the surface to prevent the growth of mycotoxin-exhibiting molds such as *Aspergillus*.

3.1 Introduction

Table 3.1. Fermented food of plant origin.

Raw material	Product	Predominant MO	References
Barley	Beer	*S. cerevisiae*	4
Sorghum	Pito	*S. cerevisiae*	5
Cereals	Bread	*S. cerevisiae*	6
Rice	Sake	*S. cerevisiae* *Aspergillus oryzae*	7, 8
Maize	Sour maize gruel	*S. cerevisiae* *Lactobacillus confusus*	9
Sugar cane	Cachaca (Aguardente)	*S. cerevisiae*	10, 11
Grape	Wine	*S. cerevisiae*	12
Grape	Champagne	*S. bayanus*	13
Millet	Pombe (beer)	*Schizos. pombe*	13
Palm juice	Ogogoro (gin)	*Schizos. pombe*	13
Soy bean, wheat	Shoyu (Soy sauce)	*Aspergillus* spp. (Koji) *Zygosaccharomyces rouxii* (Moromi)	3, 8
Soy bean, rice	Miso	*Aspergillus oryzae*, Yeast, *Lactobacillus* spp.	3, 8
Soy bean, peanuts, coconut	Tempe	*Rhizopus oligosporus* Other *Rhizopus* spp.	8
Soy bean (milk)	Sufu (Tofu)	*Rhizopus* spp. *Mucor* spp. *Aspergillus* spp.	3
Peanut	Oncom	*Rhizopus oligosporus* *Neurospora sitophila*	3
Cocoa	Chocolate	Yeast, *Acetobacter*	8
Tea extract	Kombucha	*Gluconacetobacter xylinus* *Saccharomyces* spp.	14, 15

Other such meat products are mold-ripened hams, such as 'Bündnerfleisch' from Switzerland and 'Südtiroler Bauernspeck' from Tirol, as well as Italian 'coppa'. It is possible to see once again that local preferences have resulted in different products [3].

3.1.4
Conclusion

The combination of raw foodstuffs with the microflora population can result either in an attractive product ready for commercial exploitation, or in food spoilage. In addition, unexplored habitats contain thousands of unknown species which might be of biotechnological value for food processing. The screening of different sources for microorganisms for potential applications in agriculture and biotechnology can also occur in sophisticated habitats such as the sweet potato flower in Korea [18], extreme environments such as the Brazilian rain forest [19], or ectosymbiontic insects collected in the countryside surrounding Perugia, Italy [20].

These techniques (which of course extend our knowledge of biodiversity) are not the only ones which can be used to improve the fermentation of raw foodstuffs of already known species. Until the 1970s, strain development resulted mainly from selection and mutation, rather than classical breeding. However, during the past 20 years recombinant DNA-technologies have been developed which have allowed the specific improvement and selection of microorganisms within a relatively short time span.

3.2
Application of Recombinant DNA Methods

This chapter focuses on the uses of molecular biology in yeast and filamentous fungi, and in particular on the exploitation of the formation of biomass, the expression of (heterologous) genes of interest, and the enhancement of metabolism with respect to fermentation behavior. All genes referred to in the text are summarized in Table 3.2.

3.2.1
Recombinant DNA Technology in Yeast

When discussing 'yeast', the term relates primarily to the yeast strain *Saccharomyces cerevisiae*, which is very well known in genetic terms. This was the first eukaryotic organism for which the genome has been fully sequenced [21], and these data are readily accessible on the Internet [22]. As *S. cerevisiae* is the yeast most involved in the fermentation of foodstuffs (see Table 3.1), the initial focus of this section will be on its recombinant techniques (for reviews, see Refs. [13, 23–27]), while at a later stage attention will be focused on non-*Saccharomyces* yeasts.

The first protocol on molecular transformation in *S. cerevisiae* was described in 1978 [28], and involves the production of protoplasts with enzymes such as glucuronidase or zymolyase in an osmoticum such as sorbitol. Resuspension takes place in $CaCl_2$, transforming DNA and polyethylene glycol (PEG) is added, and cells are plated out on selective media. One of the main disadvantages of this "spheroplast method" (due to an incomplete removal of the cell wall [29]) is

Table 3.2. Genes of yeast and filamentous fungi and their function.

Gene	Organism	Function
ALD6	S. cerevisiae	Aldehyde dehydrogenase
ALD7	S. cerevisiae	Aldehyde dehydrogenase
AmdS	Aspergillus nidulans	Acetamidase
AOX	Pichia pastoris	Alcohol oxidase
ARO4	S. cerevisiae	Aldolase
ATF1	S. cerevisiae	Alcohol acetyltransferase
CAR1	S. cerevisiae	Arginase
FLO1	S. cerevisiae	Cell-surface protein
GPD1	S. cerevisiae	Glycerin-3-phosphate dehydrogenase
HAP4	S. cerevisiae	Transcription factor
HIS3	S. cerevisiae	Dehydratase in histidine biosynthesis
HSP30	S. cerevisiae	Heat-shock-protein 30
HXT	S. cerevisiae	Permease gene family
ILV5	S. cerevisiae	Acetolactate reductoisomerase
LEU2	S. cerevisiae	Dehydrogenase in leucine biosynthesis
MAL	S. cerevisiae	Maltose permease gene family
MEL1	S. cerevisiae	Melibiase
MET2	S. cerevisiae	Homoserine acetyltransferase
MET3	S. cerevisiae	Sulfate adenyltransferase
MET10	S. cerevisiae	Sulfite-reductase
MET14	S. cerevisiae	Adenyl-sulfatekinase
MET25	S. cerevisiae	Acetyl homoserine-(thiol)-lyase
MIG1	S. cerevisiae	Transcription co-repressor
MIG2	S. cerevisiae	Transcription co-repressor
niaD	A. nidulans	Nitrate reductase
PGK1	S. cerevisiae	Phosphoglyceratekinase
PGU1	S. cerevisiae	Endopolygalacturonase
STA2	S. diastaticus	Glucoamylase
TRP1	S. cerevisiae	Phosphoribosylanthranilate isomerase
URA3	S. cerevisiae	Dexarboxylase

that PEG stimulates cell fusion, which may result in diploids and polyploids [30]. Cell fusion is avoided by the transformation of whole cells treated with lithium salts followed by PEG [31]. The successful uptake of DNA with PEG alone was demonstrated in the same year [32], but transformation was only one-third as efficient compared to treatment with additional lithium acetate [33]. These authors also showed that efficiency of transformation varied for different strains and methods [24], thus emphasizing that strain and method of transformation must be carefully selected. Electroporation was applied to both yeast spheroplasts [34] and intact yeast cells, including PEG in the electroporation buffer [35]. Other protocols involved sorbitol [36] or HEPES buffer [37]. A hybrid technique using elements of both the LiAc/PEG and electroporation protocols has been reported [38]. Other methods for the transformation of yeast cells involve agitation with glass beads [39] or the biolistic approach, which was first developed for plant cells [40] and then successfully transferred to yeast cells [41].

In order to identify positive clones, i.e., successfully transformed cells, the incorporated DNA transfers a marker for selection to the cells [13, 23]. The most commonly used auxotrophic markers are *LEU2*, *HIS3*, *TRP1*, and *URA3*, which are used in corresponding mutant strains auxotrophic for leucine, histidine, tryptophan, and uracil, respectively. Selection takes place on minimal medium without respective nutrient, and is thus complemented by the corresponding marker gene. For applied purposes, the use of an auxotrophic marker is less suitable as industrial strains are often diploid or even polyploid. In this case, dominant markers are available which transfer resistance to antibiotics such as hygromycin B, geneticin/kanamycin, and chloramphenicol to reasonably sensitive strains. Here, wild-type strains can be used and subsequently selected on rich media. Only the combination of the gene for resistance together with a yeast promoter leads to sufficient levels of drug resistance, even in single-copy transformants. Methods for removing marker sequences after integration into the host's genome are already well-known [42], thus avoiding the unwanted transfer of resistance to microorganisms, e.g., to the gut or the environment.

3.2.1.1 Vectors

Vectors used for transformation fall into two classes: (i) those which are maintained as an episomal plasmid which replicates autonomously; and (ii) those which are integrated into the host's genome while being passively inherited. The former tend to be unstable to varying degrees, whereas the latter are inherited with very high stability [25]. Taking the copy number into account, episomal plasmids are more advantageous than integrative vectors.

Extrachromosomal vectors These are shuttle vectors which enable propagation in both *E. coli* and *S. cerevisiae*. Propagation of the recombinant DNA is firstly carried out in *E. coli* and completed in *S. cerevisiae*. These vectors are either based on the *S. cerevisiae* nuclear 2µm plasmid, *ARS*-based plasmids (Autonomously Regulated Sequences) called Yeast Replicating plasmids (YRp), or centromere-containing vectors termed Yeast Centromeric plasmids (YCp). Most used vectors are the 2µm-based, collectively known as Yeast Episomal plasmids (YEp), being multicopy, highly mitotically stable, and freely replicating. The first plasmids constructed [43] contained the whole 2µm plasmid combined with pBR322 for replication in *E. coli*, plus the *LEU2* gene for selection. Nowadays, smaller vectors have been designed containing, for example, only the ORI-STB-region as the minimal portion [25], thereby circumventing recombination with the endogenous 2µm plasmid of yeast observed earlier.

Integrative vectors These contain sequences which show significant homology to the host's genome. The gene of interest (GOI) or the complete expression cassette plus selectable marker is integrated via homologous integration at the target locus. This leads to an increase of the copy-number of a chromosomal gene or incorporation of a heterologous gene into the genome. Vectors contain sequences of sufficient homology to the host's genome so that literally every gene of already completely sequenced strains could be targeted. For these purposes, vectors can either be: (i) used in circular form to accomplish a single cross-over event between the incorporating gene and the target sequence; or (ii) linearized in order to perform single or more cross-over integrations, where duplication of the chromosomal target sequence takes place, so that the vector minus the GOI can subsequently pop out by excisional recombination. This can be successfully achieved when homologous sequences are spread within the vector's sequence, i. e., flanking of unwanted genes such as resistance markers of bacterial origin leads more easily to the excisional recombination. Useful targets are either the chromosomal mutant allele of the selection marker used, or ribosomal and mitochondrial genes appearing in several copies throughout the genome. In the latter case, integration of more than one copy is obtained. Integrative vectors can also be used to disrupt or replace a desired gene. In *S. cerevisiae*, the use of "replacement cassette" vectors permits the deletion of every single gene, e. g., the kanamycin (G418) "replacement cassette" [44]. Homologous recombination between both ends of the cassette and the target gene results in a double crossing-over, leading to the deletion of the gene and subsequent G418-resistance. For commercial reasons, strains should be free of nonessential foreign DNA (as mentioned above). Wach et al. [45] designed a protocol for the integration of linear DNA fragments including the kan^R gene and its subsequent excision by homologous recombination. Puig *et al.* [46] applied this protocol to the development of stable, food-safe recombinant wine yeast strains, maintaining all of its useful wine-making abilities.

Examples of non-*Saccharomyces* yeast of biotechnological importance are *Arxula adeninivorans*, *Candida* spp., *Hansenula polymorpha*, *Kluyveromyces* spp., *Pichia* spp., *Schizosaccharomyces pombe*, *Yarrowia lipolytica*, and *Zygosaccharomyces rouxii* [13]. Compared to *S. cerevisiae*, these so-called *nonconventional* yeast exhibit particular advantages in cloning technology, especially for the production of heterologous proteins, with respect to:

- growth on unusual carbon sources, e. g., *Candida maltosa* assimilates *n*-alkanes and fatty acids [47], and *Yarrowia lipolytica* grows on *n*-paraffins as sole carbon source [48];
- stringently promoted gene expression, e. g., in both the methylotrophic yeast [49] *Hansenula polymorpha* and *Pichia pastoris*, methanol-inducible promoters (AOX: alcohol oxidase) for the production of proteins, which might be toxic, are available;
- lack of hyperglycosylation in *Pichia pastoris* [50] and *Schwanniomyces occidentalis* [51];
- absence of Crabtree-effect in *Pichia pastoris* and *Hansenula polymorpha*, and therefore:
- the ability to grow up to high cell densities, >100 g dry cell wt L^{-1} on simple growth media of *Pichia pastoris*, *Kluyveromyces lactis* and *K. marxianus*.

Transformation systems for nonconventional yeast depend heavily on those developed for *S. cerevisiae*. The most commonly used markers are derived from the biosynthetic pathway of the homologous system, but those from *S. cerevisiae* are also found to be effective. Autonomous replication is usually based on *ARS*-like sequences (e. g., the *PARS* sequences for *Pichia pastoris* or *HARS* sequences for *Hansenula polymorpha*), isolated and selected for the same yeast in which they will be used. In many yeasts, targeted integration or gene disruption can be performed as described earlier [49, 52].

3.2.2
Recombinant DNA Technology in Filamentous Fungi

Several filamentous fungi have been investigated on the genetic level, and all basic tools for incorporating DNA are now available. In particular, the Euascomycetes *Aspergillus nidulans* and *Neurospora crassa* have been studied in great detail as regards fundamental and applied aspects, leading to the development of fungal vectors and genetic manipulation techniques. Subsequently, other Ascomycetes and Deuteromycetes of biotechnological interest have been transformed successfully. However, these protocols cannot be easily transferred to other fungal classes, such as the Basidiomycetes (for reviews, see Refs. [53–56]).

3.2.2.1 **Fungal transformation**
More than twenty years after the first fungal transformation of *N. crassa* [57], methods for transformation still rely in general on those obtained for *S. cerevisiae*, with only minor differences. The uptake of DNA is traditionally achieved via protoplasts exposed to lithium salt [58], with release of protoplasts from mycelium through the use of lytic enzymes such as Novozym 234. Electroporation does not seem to be the method of choice without prior removal of the cell wall, and this somewhat limits the approach [59]. Whole mycelium can either be treated with lithium acetate [60], or through the use of a particle gun (shotgun or biolistic approach), where DNA-coated tungsten balls are shot into the mycelium [61]. Successful transformation of conidia and hyphal tissue has been shown to be mediated by the T-DNA of *Agrobacterium tumefaciens* [62, 63]. Filamentous fungi tested include *A. niger*, *Trichoderma reesei*, *N. crassa*, and *Fusarium venenatum*.

Selection markers A wide range of selection markers are available for hyphal fungi, and, in contrast to yeast, dominant selection systems are preferred. One of the most commonly used systems is characterized by the incorporation of the *amdS* gene of *A. nidulans*, coding for acetamidase which hydrolyzes acetamide to ammonium and acetate, thus providing both a nitrogen and a carbon source. Since the first experiments were conducted in mutant strains of *A. nidulans* [64], other fungi have been investigated, but their growth on acetamide was poor. When wild-type strains instead of mutants were transformed, growth on acetamide increased [65], for example of *Penicillium chrysogenum* and *Trichoderma reesei*. Other selection systems include drug resistance (e. g., to oligomycin or benomyl), coded either by fungal genes or by bacterial genes (cloned together with a fungal promoter), such as phleomycin or hygromycin B. In addition, several selection systems are based on the complementation of defined mutants by the corresponding wild-type gene. Nitrate-reductase mutants (*niaD*) are screened according to their chlorate resistance, and pyrimidine auxotrophs (e.g., orotidine-monophosphate-carboxylase mutants) are selectable with respect to their resistance against the toxic analogue 5-fluoro-orotic acid (5-FOA) [53].

Ectopic integration In contrast to yeast, the standard transformation modus in filamentous fungi is ectopic integration, i.e., nonhomologous integration, which does not require much homology between vector and target DNA. In this case, multiple integration of several vector molecules either at different sites or tandemly arranged in one site is common. This leads to transformants with high copy numbers and considerable mitotic stability [56]. A strict correlation between copy number and gene dose is often not observed in such strains due to the effects of position and titration. The position of integration seems to play a major role in level of expression achieved, whereas limitation of transcription factors available could somewhat downsize the approach [54].

Autonomously replicating vectors The use of autonomously replicating vectors is restricted due to an almost complete lack of natural plasmids in hyphal fungi which could act as a basis for such vectors. The *AMA* (Autonomous Maintenance in Aspergillus) sequence of *A. nidulans* [66] confers autonomous replication of vectors on *Aspergillus* and other fungi, similar to the *ARS* sequence of *S. cerevisiae*. Telomere-sequences, e.g., the short repeats of TTAGGG, have been shown to be effective for direct cloning in *Podospora anserina* when fused to both ends of a linearized vector [67]. For dominant selection, one vector was constructed suitable for transformation of fungi such as *T. reesei, F. oxysporum, Botrytis cinera,* and *Colletotrichum lindemuthanium*, respectively. Nevertheless, plasmids are easily lost when transformants are grown under nonselective conditions.

Homologous integration This is used for targeted gene inactivation (or modification) via either single site integration (where in a single recombination event the whole vector integrates into the genome, thus inactivating the target gene) or via gene replacement (where two crossing over events are required). Nevertheless, this approach is strictly strain-dependent, and the best method of integration must be tested individually. The transformation of *A. awamori* by *Agrobacterium tumefaciens*-mediated homologous recombination was recently reported [68], and this could be a promising method to generate recombinant fungal strains with multiple copies of a gene integrated at a predetermined site in the genome.

3.3
Improved Fermentation Efficiency for Industrial Application

Strain improvement during the past few years has changed radically from previous classical attempts via mutagenesis and screening to a more rational approach such as genetic engineering. This involves the introduction of directed genetic changes through recombinant DNA technology into the cell. In 1991, Bailey defined this approach as "metabolic engineering", that is simply the "technological manifestation of applied molecular biology". Since the first definition of metabolic engineering, this term has broadened its meaning and now deals with various aspects of molecular biology. Three definitions which have been outlined during the past ten years are provided in Table 3.3.

In the following section we will describe different approaches to improving industrial strains by metabolic engineering. Extensive research has, and is being, conducted on laboratory strains with regard to improving features that are of interest to industry, the main aim being to extend these findings to industrial strains [72–75].

Table 3.3. Definitions of "metabolic engineering".

Reference		
69	70	71
Recruiting heterologous activities	Improved production of chemicals already produced by the host organism	Heterologous protein production
AND	AND	AND
Redirecting metabolite flow	Extended substrate range for growth and product formation	Extension of substrate range
	AND	AND
	Addition of new catabolic activities for degradation of toxic materials	Pathways leading to new products
	AND	AND
	Production of chemicals, new to the host genome	Pathways for degradation of xenobiotics
	AND	AND
	Modification of cell properties	Engineering of cellular physiology for process improvement
		AND
		Elimination or reduction of by-product formation
		AND
		Improvement of yield or productivity

Industrial *Saccharomyces* strains allow us to more closely examine the attempts which have been made at genetic modification in order to produce beer, bread, and wine (as examples) more efficiently. The exploitation of other non-*Saccharomyces* yeast and filamentous fungi for the production of food will be investigated later in the chapter.

3.3.1
Industrial *Saccharomyces* Strains

Species of the genus *Saccharomyces* are used for many biologically processed food products, such as bread, spirits, wine, and beer. In order to modify biochemical pathways in industrial yeast, it is essential to acquire extensive knowledge of their genetic structure. Those strains of industrial *Saccharomyces* are genetically closely related (for a review, see Ref. [76]). The most detailed studies have been made on a lager brewer's yeast called *Saccharomyces carlsbergensis* in 1908 by Hansen. These "bottom-fermenting" yeasts are alloploid, i.e., they are hybrids of at least two divergent genomes, namely *S. cerevisiae* and *S. monacensis*.

By contrast, "top-fermenting" ale yeast is a diverse group of polyploid yeast, but is closely related to standard laboratory strains of *S. cerevisiae*, as is the case for "baker's yeast" and "distiller's yeast".

"Wine yeast" are usually seen as isolates of *S. cerevisiae* or *S. bayanus*. The type strain of *S. bayanus* seems to contain a genome originated from both *S. cerevisiae* and *S. monacensis*. Another industrial wine strain was classified as being part of the *S. pastorianus* species, due to the occurrence of two different *MET3* genes. Furthermore, the strain of *S. pastorianus* was found to be identical, or closely related to, *S. carlsbergensis* [77].

When referring to brewer's, wine or baker's yeast in the following text, we are referring to species of *S. cerevisiae* and their closely related subspecies.

3.3.1.1 **Beer**
Beer is produced from barley, hops and water, according to the German "Reinheitsgebot" of 1487 (regulations for beer brewing). It was not however until 1680 that the microscope was developed and that another important ingredient was identified, namely brewer's yeast. Apart from improving the brewing process by technological means, the genetic modification of brewer's yeast has been subject to extensive research during the past fifteen years. Strains have been produced with the ability to ferment a wider range of carbohydrates, with altered flocculation properties, and which produce beers with modified flavors (for reviews, see Refs. [77–80].

Carbohydrate utilization Only mono, di, and trihexoses are utilized by brewer's yeast. In order to metabolize other oligomers, e.g., dextrins or β-glucans, several approaches have been made with regard to the expression of glucoamylase and β-glucanase-genes of various origins. The breaking down of the former unfermen-

Table 3.4. Extension of substrate range in industrial yeast strains (modified after Ref. [80]).

Substrate extension	Achievements	Reference
Starch and dextrins	Expression of α-amylase (*AMY1*) and glucoamylase (*GAM1*) genes of *Schwanniomyces occidentalis*[1,3]	81
β-Glucans	Integration of a β-glucanase gene of barley[1]	82
Malate	Expression of malate permease (*MAE1*) of *Schizosaccharomyces pombe* and malic enzyme of *S. pombe* (*MAE2*) or *Lactococcus lactis* (*mleS*)[2]	83
Melibiose	Introduction of melibiase gene (*MEL1*)[3]	84

[1]Brewer's yeast; [2]wine yeast; [3]baker's yeast.

table wort compounds such as maltotetrose and dextrins leads to a low-carbohydrate beer due to a higher conversion into ethanol. The elimination of β-glucans results in improved filterability of the beer, thus providing an example of not only extension of the substrate range but also an improvement in the process of beer production. Examples and references relating to extension of the substrate range in industrial yeast strains are provided in Table 3.4.

Flocculation A suitable brewer's yeast strain should be able to flocculate, as this is the most cost-effective way of clearing at the end of fermentation. Flocculation in brewer's yeast is provided by a NewFlo phenotype, whereas flocculating laboratory strains exhibit the Flo1 phenotype whose genetic make-up is well known. Flo1 is expressed constitutively, thus in contrast to NewFlo, revealed at the end of exponential growth when starvation is present. Unfortunately, the genetics behind NewFlo still need to be elucidated. The flocculation ability of industrial strains has been improved by inserting the laboratory strain's *FLO1*-gene which encodes a cell-surface protein [85]. Flocculation throughout the fermentation process has been observed, thus resulting in lower cell counts and increased fermentation time. Regulated expression of the flocculation gene *FLO1* has been demonstrated with the help of phase-specific promoter *HSP30*, inducing high gene expression during late stationary phase [86].

Flavor One use for genetic engineering in beer production relates to the development of flavor, mainly to the elimination of unwanted by-products such as diacetyl and hydrogen sulfide, or to the stabilization of flavor compounds such as sulfur dioxide.

Diacetyl, which causes a major off-flavor in finished beer, is a product of the valine biosynthetic pathway. It is formed during fermentation by spontaneous oxidation of α-acetolactate, which diffuses from yeast cells into the fermenting wort. The expensive and time-consuming maturation required by freshly produced "green" beer is necessary as diacetyl is converted only very slowly to the metabolites acetoin and 2,3-butanediol, both of which have a much higher taste threshold than diacetyl. There are many methods to avoid this extensive maturation time by reducing diacetyl formation (see Ref. [78], and references therein). These utilize the introduction of the heterologous gene for α-acetolactate decarboxylase (ALDC), thereby enabling the cell to produce acetoin directly from α-acetolactate and reducing the time required for maturation ("lagering") from weeks to hours. Other attempts to minimize diacetyl formation involve the overexpression of specific genes encoding enzymes of the valine biosynthetic pathway, e.g., the *ILV5* gene for acetolactate reductoisomerase [87].

The construction of a recombinant brewer's yeast strain with a reduced capacity to produce hydrogen sulfide has also been successful, and results in a shorter maturation time and decreased costs for lagering. One promising approach is to place the *MET25* gene of the sulfate metabolic pathway under the control of a constitutive glycolytic promotor and integrate it into the rDNA array of a brewer's yeast strain [88]. Using this technique, pilot-scale fermentations produced much less H_2S than did wild-type yeast.

Flavor stability in the finished beer is maintained by sufficient levels of sulfur dioxide, which acts both as an antioxidant and "catcher" of carbonyl compounds responsible for stale flavors. Many strategies have been used on various genes of the sulfate metabolic pathway to alter sulfur dioxide levels. Overexpression of certain genes such as the *MET3* and *MET14* would be expected to increase metabolic flux towards SO_2. Other strategies have also been aimed at disrupting the *MET2* and *MET10* genes, the latter approach being more successful than the former (see Ref. [78], and references therein).

3.3.1.2 Wine

Wine has been produced for thousands of years from grapes and their inhabitant microflora, provided either by the vineyard or the winery. The most predominant yeast species responsible for the wine-making process is *S. cerevisiae* or subspecies such as *S. bayanus* (referred to in the following text as wine yeast), which converts grape sugars into ethanol, carbon dioxide, and other characteristic metabolites. The objectives of recombinant technologies with wine yeast are the improvement of fermentation performance, processing efficiency, and flavor characteristics. Various attempts have recently been carefully reviewed [89].

Fermentation performance In order to extend the range of carbon and nitrogen assimilation beyond catabolite repression, experiments have been conducted based on extensive knowledge of the genes involved in the glycolytic and nitrogen pathways. Efficient sugar utilization was the aim of a study in which all glycolytic enzymes were overproduced, although ethanol formation could not be increased [90]. An alternative approach was to overexpress genes of the *HXT* permease gene family, thereby enabling high-affinity glucose uptake [89]. In addition, deregulation of the proline utilization pathway proved successful in overcoming nitrogen catabolite repression of poorly utilized nitrogen sources (e.g., proline, which is not taken up under anaerobic fermentative conditions) [91].

Processing efficiency Stages of refining and clarification to complete the processing of wine comprise removing excess levels of certain compounds. For example, polysaccharides have a detrimental influence on the clarification and stabilization of both must and wine. Therefore, both heterologous and endogenous genes have been overexpressed to enable enzymatic degradation of pectins, glucans, and hemicellulose (mainly xylans) (see Ref. [89] for an overview on heterologous expression). Recently, the overexpression of an endogenous *PGU1* gene in an oenological *S. cerevisiae* strain led to a significant reduction in filtration time of the wines produced. This recombinant strain secretes an active endopolygalacturonase, which serves as an alternative to supplementing pectolytic enzymes of fungal origin [92].

Flavor characteristics The improvements of wine flavor and other sensory qualities relate to the enhancement of terpenoids of grape origin, volatile esters and glycerol produced, and last – but not least – to the adjustment of wine acidity.

Terpenoids are desirable volatile compounds in wine, and secondary metabolites of grapes. A high percentage of these metabolites occurs as the precursor-forms, the nonvolatile *O*-glycosides. By utilizing heterologous expression of β-1,4-glucanase gene of *Trichoderma longibrachiatum* in wine yeast, aroma intensity was increased – presumably due to the hydrolysis of glycosylated flavor precursors [93].

Levels of esters formed during primary fermentation vary significantly. These compounds form characteristic fruity odors and require acetyltransferase activity for their synthesis. The *ATF1*-encoded alcohol acetyltransferase (AAT) is one of the best-studied in *S. cerevisiae*. Integration of the *ATF1* gene under the control of the constitutive *PGK1*-promoter into commercial wine yeast strains resulted in an improved flavor profile of the wine produced due to increased formation of desirable esters [94].

Glycerol, a nonvolatile compound, contributes to sensory characteristics such as sweet taste, smoothness, consistency, and overall body of the wine. Attempts at overproducing glycerol have been made in order to improve the wine's organoleptic quality. This is especially useful for white wine, which has lower amounts of glycerol than red wine. Remize and co-workers [95] were able to show that both overexpression of *GPD1* together with disruption of *ALD6* and *ALD7* led to a 2- to 3-fold higher glycerol production (at the expense of ethanol) and a significant decrease of acetate formation, when compared to the wild-type strain.

The predominant organic acids in wine are tartaric and malic acid, accounting for 90% of the titratable acidity of grapes, and lack of their adjustment can cause imbalances or even spoilage of the product. The ability of *S. cerevisiae* strains to assimilate L-malate acid varies widely, mainly due to the absence of an active malate transport system. In addition to *S. cerevisiae*'s inefficient metabolism of malate, there is only a low substrate specificity of malic enzyme. Only recently has it been shown that integration of *Schizosaccharomyces pombe* genes encoding for malate permease (*mae1*) and malic enzyme (*mae2*) into the genome of *S. cerevisiae* resulted in efficient degradation of L-malic acid in Chenin Blanc musts [96].

3.3.1.3 Sake

Sake is a typical Asian alcoholic beverage which is distilled from rice, that has been fermented with the sake yeast (mainly *S. cerevisiae*) and *A. oryzae*. Some attempts have been made to genetically engineer sake yeast strains to enhance flavor and to prevent the formation of toxic (often mutagenic) substances.

Flavor The rose-like aroma of sake is mainly due to phenylethyl alcohol and its acetate ester. Selected mutants with increased formation of these aroma compounds have been shown to exhibit a mutation in the *ARO4* gene which is involved in aromatic amino acid biosynthesis. The mutated *ARO4* gene was cloned into a centromere-based plasmid and used for transformation of sake yeast [97]. The respective sake yeast produced sake with increased phenylethyl alcohol production.

Toxic substances A recombinant sake yeast strain was constructed to prevent the occurrence of ethyl carbamate (ECA), a suspected carcinogen [98]. Urea is reported to be a main precursor of ECA in yeast cells, and is formed during the breakdown of arginine. A recombinant sake yeast strain was constructed by successive disruption of the two copies of the *CAR1* gene which encodes an arginase, in order to minimize urea production. This was found to be successful in eliminating urea accumulation in sake yeast; no ECA was detected in the resulting sake, even after storage for 5 months at 30°C.

3.3.1.4 Bread

Bread is made by fermenting dough (made mainly from wheat and other cereals) with the help of baker's yeast, and with *Lactobacillus* in the case of sourdough. The quality of baker's yeast relies on efficient respiratory metabolism during its production in order to yield (economically) a biomass from raw substrates such as molasses. During the dough-leavening process, the yeast also needs to produce considerable quantities of carbon dioxide via ethanolic fermentation of various sugars and by-products which contribute to the flavor and aroma of the product. Improvements to baker's yeast involve such aspects as fermentation characteristics in dough substrates, and the amount of yeast produced from a given amount of molasses [6, 71, 79, 80].

Fermentation in dough substrates A good fermentative capacity of baker's yeast depends on a high potential to utilize the maltose which results from hydrolysis of starch by the action of endogenous amylases present in the dough. The presence of glucose prevents maltose from being taken up into the cell and metabolized; this general regulatory mechanism is widely known as "catabolite repression". Glucose control also comprises repression of maltase (MalS) and maltose-permease (MalT) synthesis as well as inactivation of maltase enzyme (MalR), leading to a substantial lag in adaptation to the maltose shown by the yeast. Strategies for metabolic engineering to diminish the extent of glucose control imply the constitutive expression of the *MAL* genes together with disruption of regulatory genes such as *MIG1* [99]. This attempt could partly alleviate glucose control, and the specific growth rate was increased on both glucose and maltose, thus fortunately leading to a shorter process time for bread production. Similar results were obtained when concomitant deletion of the specific genes involved in the regulatory mechanism of glucose control (*MIG1*, *MIG2*) gave rise to alteration of sucrose metabolism [100].

Biomass yield
Baker's yeast is produced by aerobic fed-batch methods using molasses as carbon and nitrogen sources. Molasses contains mainly glucose, fructose, sucrose, and the trisaccharide raffinose, which is decomposed by invertase into fructose and melibiose. Only a few strains of *S. cerevisiae* are able to assimilate melibiose, and many strains are missing the genes for melibiase (α-galactosidase). Baker's yeast strains able to utilize melibiose by expressing the *MEL1* gene have been constructed by genetic engineering [84], thus increasing biomass yield. Another promising experimental strategy to alleviate glucose control in order to expand productivity has been shown by Blom et al. [101]. Overexpression of the transcription factor Hap4p partly relieves glucose repression of respiration; this results in a significant reduced ethanol production, and an improved growth rate with a 40% gain in biomass yield.

3.3.2
Other Industrial Yeast Strains

Most of the fermented foods of plant origin in the West are produced with industrial strains of *S. cerevisiae* (see Table 3.1). Other industrial yeast strains for the production of foodstuffs include e.g., *Schizosaccharomyces pombe* and *Zygosaccharomyces rouxii* for the fermentation of millet beer, palm juice gin, and soy sauce.

Schizosaccharomyces pombe is also known for fermenting rum and for the deacidification of wines, because of its ability to utilize malate. This yeast is also a candidate for biomass protein and expression of heterologous genes [13]. Its role in the production of traditionally fermented beverages such as pombe and ogogoro has been stated, but little is known about strain improvement either by the classical breeding approach or genetic modification. This is most likely due to the relatively limited use of *S. pombe* in biotechnology [24].

Zygosaccharomyces rouxii is an osmotolerant and halophilic yeast used to provide characteristic aromas to Japanese soy sauce and miso. Strain improvement for the production of soy sauce has been achieved only by simple selection for mutants [102]. An L-methionine resistant mutant strain was found to be partly deficient in conversion of methionine into S-adenosyl-methionine (SAM), and exhibited a 60-fold increase in methionol (3-methylthio-1-propanol) production, a characteristic flavor compound of soy sauce.

Other yeasts are used as an attractive host for the commercial production of heterologous protein (e.g., *Pichia pastoris*, *Hansenula polymorpha*, *Kluyveromyces lactis*) [106]. Despite their genetic modification for these purposes, those are of no relevance for the production of foodstuffs.

3.3.3
Industrial Filamentous Fungi

The production of foods of plant origin in the East is achieved largely with the help of filamentous fungi, with in particular *Aspergillus*, *Rhizopus*, and *Mucor* strains being involved in the fermentation of soy beans, rice, and peanut products (see Table 3.1). In Japan, these starter cultures are known as koji, a traditional inoculum. Strains are selected for their specific enzymes. For example, in soy sauce production the strains should contain a high proteolytic and amylolytic activity, whereas strains used for rice-wine production need a high amylolytic activity to convert starch into sugar [3].

Strain improvement for koji-molds should result in new strains with respect to the production of desirable enzymes, aroma compounds, organic acids, and pigment; improvements should also provide enhanced characteristics for growth and tolerance to both temperature and salt. The method of choice is genetic breeding, i.e., the protoplast fusion technique. Various filamentous fungi such as *Aspergillus*, *Mucor*, *Penicillium* and *Trichoderma* have become subject to intra- and interspecific protoplast fusion. The principles and applications for genetic recombination of industrial koji-molds have been extensively reviewed [103].

Methods for transforming filamentous fungi were outlined in section 3.2. The application of these techniques mainly involve heterologous expression of enzymes and other proteins or secondary metabolites of commercial value [104, 105].

3.4
Commercial use of Genetically Modified Organisms (GMO)

Despite extensive use of recombinant techniques in the production of food strains, these methods have made little impact on commercial applications. To date, only two GMO strains have become commercially available – a brewer's and a baker's yeast strain.

In 1991, a GMO baker's yeast was approved by the UK government [13]. The recombinant strain now expresses genes for maltose permease and maltase constitutively, due to altered promotion. During bread-making, carbon dioxide is produced faster than with conventional baker's yeast, and this ensures that the dough rises more rapidly.

In 1994, novel food approval was obtained for the commercial use of a genetically modified brewer's yeast strain, also in the UK [78]. The recombinant strain contains the *STA2* gene of *S. diastaticus*, encoding a glucoamylase for extracellular production, enabling a partial hydrolysis of maltodextrins. Beer produced with this novel bottom-fermenting yeast strain has a lower carbohydrate content, was bottled and labeled for research purposes only, and called Nutfield Lyte [1].

No approved GMO has yet been reported for a fungal starter culture.

3.5
The Future

As described above, considerable progress has been made in developing novel food strains of yeast and filamentous fungi with the help of metabolic engineering. Despite its GRAS (Generally Regarded As Safe) status, no genetically modified strains of *S. cerevisiae* are contemporarily being used in the production of beer, bread, or wine.

The benefits of application of recombinant DNA technology are yet to be elucidated for both the producer and the consumer. However, the resulting products should have a clear advantage over a conventionally produced commodity. Improvements in the nutritive value of yeasts could be an important objective for future markets. For instance, the enrichment of an industrial yeast with essential amino acids such as lysine, methionine, or threonine would improve products derived from cereals lacking these vital compounds [79].

Targets for modification must be carefully chosen with respect to both yeast physiology and commercial performance, and this calls for a "systems approach". Thus, it is "necessary to consider the complete metabolic network, or the complete set of signal transduction pathways that are involved in regulation of cellular function, and it is exactly this systems approach that distinguishes metabolic engineering from applied molecular biology." [71].

Other approaches relate to the safety of the product, simply to ensure that no recombinant microorganism used for the production of food contains resistance genes. In addition, endogenous genes should be used rather than heterologous genes. Guidelines for the approval of genetically modified (GM) products and the release of GMOs meet some of the proposals stated above. These include "a complete definition of the DNA sequence introduced, and the elimination of any sequence that is not indispensable for expression of the desired property; the absence of any selective advantage conferred on the transgenic organism that could allow it to become dominant in natural habitats; no danger to human health and/or the environment from the transferred DNA; and a clear advantage to both the producer and the consumer."

At present, however, regulatory authorities often appear more willing to approve the use of GMOs than the public is to use them. This is not so in the case of pharmaceuticals produced by recombinant technology, though there seems rather to be a psychological barrier against accepting food produced in the fashion described above. For this reason, providing the public with comprehensive information represents the major challenge for all scientists, producers and governments involved.

References

1 Bamforth, C. W. (1998) *Tap into the art and science of brewing*, Plenum Press, New York and London.
2 Sachse, G. E. (1996) Gentechnik in der Lebensmittelindustrie, in: *Gentechnik* (Gassen, H. G., Kemme, M., eds.), Fischer Taschenbuch Verlag, Frankfurt am Main, pp. 144–195.
3 Wolf, G. (1997) Traditional fermented food, in: *Fungal Biotechnology* (Anke, T., ed.), Chapman & Hall, Weinheim, pp. 3–13.
4 Tornai-Lehoczki, J., Dlauchy, D. (2000) Delimination of brewing yeast strains using different molecular techniques. *Int. J. Food Microbiol.* **62**, 37–45.
5 Van der Aa Kuhle, A., Jesperen, L., Glover, R. L., Diawara, B., Jakobsen, M. (2001) Identification and characterization of *Saccharomyces cerevisiae* strains isolated from West African sorghum beer. *Yeast* **18**, 1069–1079.
6 Attfield, P. V. (1997) Stress tolerance: the key to effective strains of industrial baker's yeast. *Nature Biotechnol.* **15**, 1351–1357.
7 Nakazato, A., Kadokura, T., Amano, M., Harayama, T., Murakami, Y., Takeda, M., Ohkuma, M., Kudo, T., Kaneko, T. (1998) Comparison of the structural characteristics of chromosome VI in *Saccharomyces* sensu stricto: the divergence, species-dependent features and uniqueness of sake yeasts. *Yeast* **14**, 723–731.
8 Krämer, J. (1992) *Lebensmittel-Mikrobiologie* 2. Auflage, Verlag Eugen Ulmer, Stuttgart.
9 Zulu, R. M., Dillon, V. M., Owens, J. D. (1997) Munkoyo beverage, a traditional Zambian fermented maize gruel using *Rhynchosia* root as amylase source. *Int. J. Food Microbiol.* **34**, 249–258.
10 Guerra, J. B., Araujo, R. A., Pataro, C., Franco, G. R., Moreira, E. S., Mendonca-Hagler, L. C., Rosa, C. A. (2001) Genetic diversity of *Saccharomyces cerevisiae* strains during the 24 h fermentative cycle for the production of the artisanal Brazilian cachaca. *Lett. Appl. Microbiol.* **33**, 106–111.
11 Schwan R. F., Mendonca, A. T., da Silva, J. J., Rodrigues, V., Wheals, A. E. (2001) Microbiology and physiology of Cachaca (Aguardente) fermentations. *Antonie Van Leeuwenhoek* **79**, 89–96.
12 Mortimer, R., Polsinelli, M. (1999) On the origins of wine yeast. *Res. Microbiol.* **150**, 199–204.
13 Walker, G. M. (1998) *Yeast. Physiology and Biotechnology*. John Wiley & Sons, Chichester.
14 Chen, C., Liu, B. Y. (2000) Changes in major components of tea fungus metabolites during prolonged fermentation. *J. Appl. Microbiol.* **89**, 834–839.
15 Greenwalt, C. J., Steinkraus, K. H., Ledford, R. A. (2000) Kombucha, the fermented tea: microbiology, composition, and claimed health effects. *J. Food Prot.* **63**, 976–981.
16 Hammes, W. P., Knauf, H. J. (1994) Starters in the processing of meat products. *Meat Sci.* **36**, 155–168.
17 Eliskases-Lechner, F., Ginzinger, W. (1995) The yeast flora of surface ri-

pened cheeses. *Milchwissenschaft* **50**, 458–462.

18 Hong, S. G., Chun, J., Bae, K. S. (**2000**) *Metschnikowa koreensis* sp. nov., a new yeast species isolated from sweet potato flower in Korea. 10th International Symposium on Yeasts (ISY 2000), Papendal, Arnhem, The Netherlands, Abstract Book (van Dijken, J. P. and Scheffers, W. A., eds.), Delft University Press, p. 292.

19 Buzzini, P., Martini, A. (**2000**) Biodiversity of killer activity in yeasts isolated from the Brazilian rain forest. *Can. J. Microbiol.* **46**, 607–611.

20 Zacchi, L., Vaughan-Martini, A. (**2000**) Insects as a source of yeasts with potential applications in agriculture and biotechnology. 10th International Symposium on Yeasts (ISY 2000), Papendal, Arnhem, The Netherlands, Abstract Book (van Dijken, J. P. and Scheffers, W. A., eds.), Delft University Press, p. 427.

21 Goffeau, A., Coster, F., Del Bino, S., et al. (**1997**) The yeast genome directory. *Nature* **387**, S1–S105.

22 Cherry, J. M., Adler, C., Ball, C., Chervitz, S. A., Dwight, S. S., Hester, E. T., Jia, Y., Juvik, G., Roe, T., Schroeder, M., Weng, S., Botstein, D. (**1998**) SGD: *Saccharomyces* Genome Database. *Nucleic Acids Res.* **26**, 73–79.

23 Romanos, M. A., Scorer, C. A., Clare, J. J. (**1992**) Foreign gene expression in yeast: a review. *Yeast* **8**, 423–488.

24 Sudbery, P. E. (**1993**) Genetic engineering of yeast, in: *Biotechnology.* Second Edition, Vol. 2 (Rehm, H.-J., Reed, G., Pühler, A., Stadler, P., eds.), VCH Weinheim, pp. 507–528.

25 Hadfield, C. (**1994**) Construction of cloning and expression vectors, in: *Molecular Genetics of Yeast. A practical approach* (Johnston, J. R., ed.), Oxford University Press, pp. 17–48.

26 Archer, D. B., MacKenzie, D. A., Jeenes, D. J. (**2001**) Genetic engineering: yeasts and filamentous fungi, in: *Basic Biotechnology.* Second Edition (Ratledge, C., Kristiansen B., eds.), Cambridge University Press, pp. 95–126.

27 Gietz, R.D, Woods, R. A. (**2001**) Genetic transformation of yeast. *BioTechniques* **30**, 816–831.

28 Hinnen, A., Hicks, J. B., Fink, G. R. (**1978**) Transformation of yeast. *Proc. Natl. Acad. Sci. USA* **75**, 1929–1933.

29 Hutchison, H. T., Hartwell, L. H. (**1967**) Macromolecule synthesis in yeast spheroplasts. *J. Bacteriol.* **94**, 1697–1705.

30 Kao, K. N., Michayluk, M. R. (**1974**) A method for high-frequency intergeneric fusion of plant protoplasts. *Planta* **115**, 355–367.

31 Ito, H., Fukuda, Y., Murata, K., Kimura, A. (**1983**) Transformation of intact yeast cells treated with alkali cations. *J. Bacteriol.* **153**, 163–168.

32 Klebe, R. J., Harriss, J. V., Sharp, Z. D., Douglas, M. G. (**1983**) A general method for polyethylene-glycol-induced genetic transformation of bacteria and yeast. *Gene* **25**, 333–341.

33 Yamakawa, M., Hishinuma, F., Gunge, N. (**1985**) Intact cell transformation of *Saccharomyces cerevisiae* by polyethylene glycol. *Agric. Biol. Chem.* **49**, 869–871.

34 Karube, I., Tamiya, E., Matsuoka, J. (**1985**) Transformation of *Saccharomyces cerevisiae* spheroplasts by high electric pulse. *FEBS Lett.* **182**, 90–94.

35 Hashimoto, H., Morikawa, H., Yamada, K., Kimura, A. (**1985**) A novel method for transformation of intact yeast cells by electroinjection of plasmid DNA, *Appl. Microbiol. Biotechnol.* **21**, 336–339.

36 Becker, D. M., Guarente, L. (**1991**) High-efficiency transformation of yeast by electroporation. *Methods Enzymol.* **194**, 182–187.

37 Grey, M., Brendel, M. (**1992**) A ten-minute protocol for transforming *Saccharomyces cerevisiae* by electroporation. *Curr. Genet.* **22**, 335–336.

38 Manivasakam, P., Schiestl, R. H. (**1993**) High efficiency transformation of *Saccharomyces cerevisiae* by electroporation. *Nucleic Acids Res.* **21**, 4414–4415.

39 Costanzo, M. C., Fox, T. D. (**1988**) Transformation of yeast by agitation with glass beads. *Genetics* **120**, 667–670.

40 Klein, T. M., Wolf, E. D., Wu, R., Sanford, J. C. (1987) High-velocity microprojection for delivering nucleic acids into living cells. *Nature* **327**, 70–73.

41 Johnston, S. A., Anziano, P. Q., Shark, K., Sanford, J. C., Butow, R. A. (1988) Mitochondrial transformation in yeast by bombardment with microprojectiles. *Science* **240**, 1538–1541.

42 Xiao, W., Rank, G. H. (1989) The construction of recombinant industrial yeasts free of bacterial sequences by directed gene replacement into a nonessential region of the genome. *Gene* **76**, 99–107.

43 Beggs, J. D. (1978) Transformation of yeast by a replicating hybrid plasmid. *Nature* **275**, 104–109.

44 Güldener, U., Heck, S., Fiedler, T., Beinhauer, J., Hegemann, H. (1996) A new efficient gene disruption cassette for repeated use in budding yeast. *Nucleic Acids Res.* **24**, 2519–2524.

45 Wach, A., Brachat, A., Pöhlmann, R., Philippsen, P. (1994) New heterologous modules for classical or PCR-based gene disruptions in *Saccharomyces cerevisiae*. *Yeast* **10**, 1793–1808.

46 Puig, S., Ramon, D., Perez-Ortin, J. E. (1998) Optimized method to obtain stable food-safe recombinant wine yeast strains. *J. Agric. Food Chem.* **46**, 1689–1693.

47 Mauersberger, S., Ohkuma, M, Schunck, W.-H., Takagi, M. (1996) *Candida maltosa*, in: *Nonconventional yeasts in biotechnology. A handbook* (Wolf, K., ed.), Springer-Verlag, Berlin and Heidelberg, pp. 411–580.

48 Barth, G., Gaillardin, C. (1996) *Yarrowia lipolytica*, in: *Nonconventional yeasts in biotechnology. A handbook* (Wolf, K., ed.), Springer-Verlag, Berlin, Heidelberg, pp. 313–388.

49 Gellissen, G. (2000) Heterologous protein production in methylotrophic yeasts. *Appl. Microbiol. Biotechnol.* **54**, 741–750.

50 Romanos, M. (1995) Advances in the use of *Pichia pastoris* for high level gene expression. *Curr. Opin. Biotechnol.* **6**, 527–533.

51 Dohmen, R. J., Hollenberg, C. P. (1996) *Schwanniomyces occidentalis*, in: *Nonconventional yeasts in biotechnology. A handbook* (Wolf, K., ed.), Springer-Verlag, Berlin, Heidelberg, pp. 117–137.

52 Cregg, J. M., Cereghino, J. L., Shi, J., Higgins, D. R. (2000) Recombinant protein expression in *Pichia pastoris*. *Mol. Biotechnol.* **16**, 23–52.

53 Van den Hondel, C. A. M. J. J., Punt, P. J. (1992) Gene transfer systems and vector development for filamentous fungi, in: *Applied Molecular Genetics of Fungi* (Peberdy, J. F., Caten, C. E., Ogden, J. E., Bennett, J. W., eds.), Cambridge University Press, Cambridge, pp. 1–29.

54 Turner, G. (1993) Genetic engineering of filamentous fungi, in: *Biotechnology. Second Edition*, Vol. 2 (Rehm, H.-J., Reed, G., Pühler, A., Stadler, P., eds.), VCH Weinheim, pp. 529–543.

55 Jarai, G. (1997) Heterologous gene expression in filamentous fungi, in: *Fungal Biotechnology* (Anke, T., ed.), Chapman & Hall, Weinheim, pp. 251–264.

56 Tudzynski, P., Tudzynski, B. (1997) Fungal genetics: novel techniques and regulatory circuits, in: *Fungal Biotechnology* (Anke, T., ed.), Chapman & Hall, Weinheim, pp. 229–249.

57 Case, M. E., Schweizer, M., Kushner S. R., Giles N. H. (1979) Efficient transformation of *Neurospora crassa* by utilizing hybrid plasmid DNA. *Proc. Natl. Acad. Sci. USA* **76**, 5259–5263.

58 Fincham, J. R. S. (1989) Transformation in fungi. *Microbiol. Rev.* **53**, 147–170.

59 Goldman, G. H., Van Montagu, M., Herrera-Estrella, A. (1990) Transformation of *Trichoderma harzianum* by high-voltage electric pulse. *Curr. Genet.* **17**, 169–174.

60 Dhawale, S. S., Paietta, J. V., Marzluf, G. A. (1984) A new rapid and efficient transformation procedure for *Neurospora*. *Curr. Genet.* **8**, 77–79.

61 Armaleo, D., Ye, G.-N., Klein, T. M., Shark, K. B., Sanford, J. C., Johnston, S. A. (1990) Biolistic nuclear transformation of *Saccharomyces cerevisiae* and other fungi. *Curr. Genet.* **17**, 97–103.

62 De Groot, M. J. A., Biundock, P., Hooykaas, P. J. J., Beijersbergen,

A. G. M. (1998) *Agrobacterium tumefaciens*-mediated transformation of filamentous fungi, *Nature Biotechnol.* **16**, 839–842.

63 Dunn-Coleman, N., Wang, H. (1998) *Agrobacterium* T-DNA: a silver bullet for filamentous fungi? *Nature Biotechnol.* **16**, 817–818.

64 Tilburn, J., Scazzocchio, C., Taylor, G. G., Zabicky-Zissman, J. H., Lockington, R. A., Davies, R. W. (1983) Transformation by integration in *Aspergillus nidulans*. *Gene* **26**, 205–221.

65 Kelley, J. M., Hynes, M. J. (1985) Transformation of *Aspergillus niger* by the *amdS* gene of *Aspergillus nidulans*. *EMBO J.* **4**, 475–479.

66 Gems, D., Johnstone, E. L., Cluterbuck, A. J. (1991) An autonomously replicating plasmid transforms *Aspergillus nidulans* at high frequency. *Gene* **98**, 61–67.

67 Barreau, C., Iskandar, M., Turcq, B., Javerzat, J. P. (1998) Use of a linear plasmid containing telomeres as an efficient vector for direct cloning in the filamentous fungus *Podospora anserina*. *Fung. Genet. Biol.* **25**, 22–30.

68 Gouka, R. J., Gerk, C., Hooykaas, P. J. J., Bundock, P., Musters, W., Verrips, C. T., de Groot, M. J. A. (1999) Transformation of *Aspergillus awamori* by *Agrobacterium tumefaciens*-mediated homologous recombination. *Nature Biotechnol.* **17**, 598–601.

69 Bailey, J. E. (1991) Toward a science of metabolic engineering. *Science* **252**, 1668–1674.

70 Cameron, D. C., Tong, I.-T. (1993) Cellular and metabolic engineering. *Appl. Biochem. Biotechnol.* **38**, 105–140.

71 Nielsen, J. (2001) Metabolic engineering. *Appl. Microbiol. Biotechnol.* **55**, 263–283.

72 Mast-Gerlach, E., Stahl, U. (1997) Superoxiddismutase aus Hefe als Antioxidans. *Branntweinwirtschaft* **6**, 90–94.

73 Donalies, U., Stahl, U. (2001) Phase-specific gene expression in *Saccharomyces cerevisiae*; using maltose as a carbon source under oxygen limiting conditions. *Curr. Genet.* **39**, 150–155.

74 Donalies, U., Stahl, U. (2002) Increasing sulphite formation in *Saccharomyces cerevisiae* by overexpression of *MET14* and *SSU1*. *Yeast* **19**, 475–484.

75 Nevoigt, E., Pilger, R., Mast-Gerlach, E., Schmidt, U., Freihammer, S., Eschenbrenner, M., Garbe, L., Veen, M., Lang, C., Stahl, U. (2002) Genetic engineering of brewing yeast to reduce the content of ethanol in yeast. *FEMS Yeast Res.* **2**, 225–232.

76 Barnett, J. A. (1992) The taxonomy of the genus *Saccharomyces* Meyen *ex* Rees: a short review for non-taxonomists. *Yeast* **8**, 1–23.

77 Hansen, J., Kielland-Brandt, M. C. (1996) Modification of biochemical pathways in industrial yeasts. *J. Biotechnol.* **49**, 1–12.

78 Hammond, J. R. M. (1995) Genetically-modified brewing yeasts for the 21st century. Progress to date. *Yeast* **11**, 1613–1627.

79 Benitez, T. Gasent-Ramirez, J. M., Castrejon, F., Codon, A. C. (1996) Development of new strains for the food industry. *Biotechnol. Prog.* **12**, 149–163.

80 Ostergaard, S., Olsson, L., Nielsen, J. (2000) Metabolic engineering of *Saccharomyces cerevisiae*. *Microbiol. Mol. Biol. Rev.* **64**, 34–50.

81 Hollenberg, C. P., Strasser, A. W. M. (1990) Improvement of baker's and brewer's yeast by gene technology. *Food Biotechnol.* **4**, 527–534.

82 Berghof, K., Stahl, U. (1991) Improving the filterability of beer by the use of β-glucanase active brewers' yeast. *BioEngineering* **7**, 27–32.

83 Volschenk, H., Viljoen, M., Grobler, J., Petzold, B., Bauer, F., Subden, R. E., Young, R. A., Lonvaud, A., Denayrolles, M., van Vuuren, H. J. J. (1997) Engineering pathways for malate degradation in *Saccharomyces cerevisiae*. *Nature Biotechnol.* **15**, 253–257.

84 Ostergaard, S., Roca, C., Ronnow, B., Nielsen, J., Olsson, L. (2000) Physiological studies in aerobic batch cultivations of *Saccharomyces cerevisiae* strains harbouring the *MEL1* gene. *Biotechnol. Bioeng.* **68**, 252–259.

85 Watari, J., Numora, M., Sahara, H., Koshino, S., Keranen, S. (**1994**) Construction of flocculent brewer's yeast by chromosomal integration of the yeast flocculation gene *FLO1*. *J. Inst. Brew.* **100**, 73–77.

86 Verstrepen, K., Bauer, F., Michiels, C., Derdelinckx, G., Delvaux, F., Pretorius, I. S. (**2000**) Controlled expression of *FLO1* in *Saccharomyces cerevisiae*, EBC Monograph 28: Symposium Yeast Physiology, Fachverlag Hans Carl, Nürnberg, pp. 30–42.

87 Mithieux, S. M., Weiss, A. S. (**1995**) Tandem integration of multiple *ILV5* copies and elevated transcription in polyploid yeast. *Yeast* **11**, 311–316.

88 Omura, F., Shibano, Y., Fukui, N., Nakatani, K. (**1995**) Reduction of hydrogen sulfide production in brewing yeast by constitutive expression of *MET25* gene. *J. Am. Soc. Brew. Chem.* **53**, 58–62.

89 Pretorius, I. S. (**2000**) Tailoring wine yeast for the new millennium: novel approaches to the ancient art of winemaking. *Yeast* **16**, 675–729.

90 Schaaff, I., Heinisch, J., Zimmermann, F. K. (**1989**) Overproduction of glycolytic enzymes in yeast. *Yeast* **5**, 285–290.

91 Salmon, J.-M., Barre, P. (**1998**) Improvement of nitrogen assimilation and fermentation kinetics under enological conditions by derepression of alternative nitrogen-assimilatory pathways in an industrial *Saccharomyces cerevisiae* strain. *Appl. Environ. Microbiol.* **64**, 3831–3837.

92 Vilanova, M., Blanco, P., Cortes, S., Castro, M., Villa, T. G., Sieiro, C. (**2000**) Use of a *PGU1* recombinant *Saccharomyces cerevisiae* strain in oenological fermentations. *J. Appl. Microbiol.* **89**, 876–883.

93 Perez-Gonzalez, J. A., Gonzalez, R., Querol, A., Sendra, J., Ramon, D. (**1993**) Construction of a recombinant wine yeast strain expressing β-(1,4)-endoglucanase and its use in microvinification processes. *Appl. Environ. Microbiol.* **59**, 2801–2806.

94 Lilly, M., Lambrechts, M. G., Pretorius, I. S. (**2000**) Effect of increased yeast alcohol acetyltransferase activity on flavour profiles of wine and distillates. *Appl. Environ. Microbiol.* **66**, 744–753.

95 Remize, F., Roustan, J. L., Sablayrolles, J. M., Barre, P., Dequin, S. (**1999**) Glycerol overproduction by engineered *Saccharomyces cerevisiae* wine yeast strains leads to substantial changes in by-product formation and to a stimulation of fermentation rate in stationary phase. *Appl. Environ. Microbiol.* **65**, 143–149.

96 Volschenk, H., Viljoen-Blom, M., Subden, R. E., van Vuuren, H. J. (**2001**) Malo-ethanolic fermentation in grape must by recombinant strains of *Saccharomyces cerevisiae*. *Yeast* **18**, 963–970.

97 Fukuda, K., Watanabe, M., Asano, K., Ouchi, K., Takasawa, S. (**1992**) Molecular breeding of a sake yeast with a mutated *ARO4* gene which causes both resistance to o-fluoro-DL-phenylalanine and increased production of β-phenylethyl alcohol. *J. Ferment. Bioeng.* **73**, 366–369.

98 Kitamoto, K., Oda, K., Gomi, K., Takahashi, K. (**1991**) Genetic engineering of a sake yeast producing no urea by successive disruption of arginase gene. *Appl. Environ. Microbiol.* **57**, 301–306.

99 Klein, C. J. L., Olsson, L., Ronnow, B., Mikkelsen, J. D., Nielsen, J. (**1997**) Glucose and maltose metabolism in *MIG1*-disrupted and *MAL*-constitutive strains of *Saccharomyces cerevisiae*. *Food Technol. Biotechnol.* **35**, 287–292.

100 Klein, C. J. L., Rasmussen, J., Ronnow, B., Olsson, L., Nielsen, J. (**1999**) Investigation of the impact of *MIG1* and *MIG2* on the physiology of *Saccharomyces cerevisiae*. *J. Biotechnol.* **68**, 197–212.

101 Blom, J., De Mattos, M. J. T., Grivell, L. A. (**2000**) Redirection of the respirofermentative flux distribution in *Saccharomyces cerevisiae* by overexpression of the transcription factor Hap4p. *Appl. Environ. Microbiol.* **66**, 1970–1973.

102 Aoki, T., Uchida, K. (**1991**) Enhanced formation of 3-(methylthio)-1-propanol in a salt-tolerant yeast, *Zygosaccharomyces rouxii*, due to deficiency of

S-adenosyl-methionine synthase. *Agric. Biol. Chem.* **55**, 2113–2116.

103 Ogawa, K. (**1994**) Approaches to hybridization and applied genetics by protoplast fusion: koji-moulds and mushrooms, in: *Recombinant microbes for industrial and agricultural applications* (Murooka, Y. and Imanaka, T., eds.), Marcel Dekker Inc., New York, pp. 581–603.

104 Gouka, R. J., Punt, P. J., van den Hondel, C. A. (**1997**) Efficient production of secreted proteins by *Aspergillus*: progress, limitations and prospects. *Appl. Microbiol. Biotechnol.* **47**, 1–11.

105 Van den Hombergh, J. P., van den Vondervoort, P. J., Fraissinet-Tachet, L., Visser, J. (**1997**) *Aspergillus* as a host for heterologous protein production: the problem of proteases. *Trends Biotechnol.* **15**, 256–263.

106 Faber, K. N., Harder, W., Ab, G., Veenhuis, M. (**1995**) Review – methylotrophic yeasts as factories for the production of foreign proteins. *Yeast* **11**, 1331–1344.

4
Production of Food Additives using Filamentous Fungi

Carsten M. Hjort

4.1
Filamentous Fungi in Food Production

Fungi are eukaryotic microorganisms that have been used either as foods or for the manufacture of food for more than a thousand years. The fungal kingdom consists of yeasts that are unicellular organisms, and also of filamentous fungi that are multicellular organisms with the cells organized in chains known as hyphae. The hyphae can be branched to greater or lesser degrees. Some fungi are dimorphic, which means that they have both unicellular and filamentous growth stages.

Yeasts have been used extensively for food production. The yeast *Saccharomyces cerevisiae* (baker's yeast) is used in baking, in brewing and in winemaking. In all of these applications the ability to ferment glucose to ethanol and carbon dioxide is the key feature.

Following the advent of recombinant production technology, a variety of yeasts have been used for the production of enzymes and metabolites, and recombinant expression systems suitable for large-scale production have now been developed for *S. cerevisiae*, *Schizosaccharomyces pombe*, *Pichia pastoris*, *Pichia methanolica*, *Hansenula polymorpha*, *Kluyveromyces lactis*, *Yarrowia lipolytica*, and other species [1, 2].

In the production of food enzymes however, filamentous fungi are much more broadly employed than yeast systems, and have been used in this role for a very long time. In some cases, the fungus is the food itself (an example is mushroom, *Agaricus bisporus*), but more recently the ascomycete *Fusarium venenatum* was developed as a single cell protein food source marketed under the tradename Quorn™ by the company Marlow Foods [3]. In a few cases, the fungus is actually an ingredient of the food, as in the case of cheese, where various species of *Penicillium* (e. g., *P. roqueforti* and *P. carmemberti*) form part of the cheese product. The oldest examples of using filamentous fungi in food production are in the fermentation of food. In Japan, the filamentous fungi *Aspergillus oryzae* (Fig. 4.1), *Aspergillus sake* and related species have been used for fermenting sake, shoyu, and

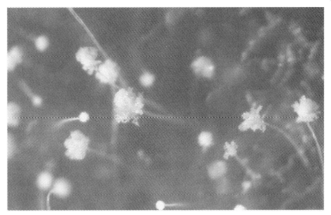

Figure 4.1 *Aspergillus oryzae* agar culture. A strain of *A. oryzae* was grown on an agar plate; various stages of sporulating phialides can be seen in the photomicrograph.

miso. In these processes the fungus ferments sugar to alcohol, but an equally important point is that the fungus secretes enzymes such as proteases and amylases that modify the raw material into the desired product.

4.1.1
Industrial Applications

The potential for the food industry of the enzyme complex produced by *A. oryzae* was acknowledged late in the nineteenth century when, in 1894, the Japanese-American enzyme pioneer Jockichi Takamine first manufactured an enzyme complex from *A. oryzae* that was sold under the tradename Takadiastase^{TM} [4]. Later, during the 1950s the submerged fermentation industry was developed in the United States, with *A. oryzae* being used to produce an array of enzyme products in submerged fermentation, but still with the emphasis on proteases and amylases. Other *Aspergillus* species such as the black aspergilli, *Aspergillus niger*, *Aspergillus awamori*, *Aspergillus foetidus*, *Aspergillus aculeatus* and *Aspergillus japonicus*, were also used mainly to produce the enzyme glucoamylase [5].

Glucoamylase formed the basis for the enzyme revolution in the starch industry. Traditionally, starch was acid-hydrolyzed to glucose using hydrochloric acid, followed by neutralization. During this process several by-products were formed, and heavy salt formation resulted from the neutralization step. By introducing enzymatic hydrolysis, the by-product formation was avoided and so a product of a better quality could be manufactured at a lower price.

A. niger was also used in the production of primary metabolites in bulk amounts, for example citric acid [5]. Hence, fermentation process development was carried out both for enzyme and metabolite production. The production strains were also dramatically improved by mutagenic manipulation of the strains using either chemicals or radiation, followed by yield screening.

Even though major improvements have been achieved using classical strain improvement and fermentation optimization, recombinant DNA technology has revolutionized microbial enzyme production, and for three main reasons:

1. It is possible to produce enzymes isolated from virtually any organism, in good yields, in a highly developed and safe expression system. The accumulated toxicity and safety studies for products created in these systems provides excellent documentation for their safety. In this way, several different enzyme products can be generated using the same equipment, and essentially the same process.
2. The host organism (the organism into which the foreign DNA is transformed) can be modified to suit the quality specifications of the enzyme product. These modifications include removal of unwanted side activities that might result in a product purity that cannot be obtained using nonrecombinant methods. The fermentation yields of the enzyme can often also be increased several fold compared with yields obtained from a donor organism.
3. Using recombinant DNA technology, it is possible to produce genetically engineered enzymes. Such enzymes may have one or more amino acids substituted for other amino acids, or they may be hybrids between two or more enzyme genes. By using genetic engineering, it is possible to obtain enzymes with substantially improved properties.

At the dawn of the age of recombinant technology production, one of the first – and one of the most critical – choices that had to be made was that of expression systems. The extensive experience acquired with *Aspergillus* sp. made them clear candidates as host organisms: first, they were known to be able to produce huge amounts of extracellular protein; second, they were well-suited for production in submerged cultures in stainless steel tanks; and third, mutants that were even better adapted to this production environment had been selected. Due to the long-term use of *A. oryzae* [6] and *A. niger* [7] as production organisms for food enzymes, materials produced by these organisms have for several years been recognized by the FDA as GRAS (Generally Regarded As Safe).

Although *Aspergillus* species such as *A. oryzae* [8], *A. niger* and *A. awamori* [9] are dominant in enzyme production, other fungal systems have also been successfully developed for this purpose. *Trichoderma reesei* (also known as *Trichoderma longibrachiatum*) is a wood-degrading fungus that first attracted attention due to its ability to produce huge amounts of cellulases and hemicellulases. *T. reesei* proved to be suitable for submerged fermentation and was subsequently developed as an expression system [10].

More recently, a new fungal expression system – the *Fusarium venenatum* expression system – has been developed [11]. This fungus is used for single cell protein production (see above) and is well-suited for submerged fermentation, largely because the fermentation broth is easily aerated as a consequence of its rheological properties. The long-term use of *F. venenatum* as a food has led to the establishment of a history of safe use for this fungus.

Clearly, recombinant DNA technology has not only been used to improve enzyme production systems. Rather, by manipulating the metabolism of the fungus it has been possible to increase the yields of certain desirable metabolites (e.g., citric acid [12]), to eliminate unwanted metabolites (e.g., oxalic acid [13]), or to enable the production of metabolites that are "foreign" to the fungus (e.g., astaxanthin [14]). The manipulation of metabolism in this way is referred to as "metabolic engineering".

4.2
Additives for the Food Industry

Nowadays, the food industry uses additives of microbial origin for many different purposes, with enzymes being the best established examples of materials that are produced using genetically modified microorganisms (GMM) and used in food production.

In fact, enzymes have been used by the food industry for thousands of years. For example, chymosin isolated from calves has been used in cheese-making, amylases produced by barley have been used in brewing, and amylases and pectinases produced by grapes in wine-making. However, the deliberate use of microbial enzymes is a more recent development, and enzymes produced by GMM for the food industry were not introduced until the late 1980s. Depending on the country in which the enzymes are used, they are classified as either "food additives" or as "food processing aids".

Some examples of food industries that currently use enzymes on a regular basis include:

- Starch industry: this uses (by volume) the largest amounts of enzymes. Alpha amylases and amyloglycosidase are used for degrading starch into glucose, and part of the glucose syrup which is formed is further processed. Glucose isomerase is used to convert glucose into fructose for the production of high-fructose corn syrups.
- Baking industry: here, amylases and xylanases are used to improve the quality of bread by giving it more volume, a better crumb structure, and a longer shelf-life. More recently, lipases have been introduced to baking to provide *in situ* formation of emulsifiers.
- Diary industry: here, chymosin extracted from calves for cheese production has been supplemented with proteases of microbial origin. Chymosin produced by recombinant *A. niger* has also become a relatively important product. Other enzymes than rennet proteases are also used in the diary industry; for example, the microbial lactase which is used to hydrolyze lactose can benefit people suffering from lactose intolerance.
- Brewing industry: microbial amylases are used in the brewing industry to achieve a more robust and efficient mashing process. Enzymes have also been developed to remove the bad-tasting metabolite diacetyl from beer. However, as beer is seen

by many as a very traditional product, many countries have strict regulations on the type of additives which may be used in brewing.
- Wine industry: this is another example of a very conservative industry where the penetration of microbial enzyme use has been slow, and this is especially true for enzymes produced by recombinant microorganisms. Pectinases are used to increase juice yield during mashing, while β-glucanases are used to remove haze resulting from *Botrytis* growth on the grapes. A more curious use of enzymes in this industry is that of laccases to preserve cork stoppers, thereby preventing cork off-flavor in the wine.

4.3
Design of GMM for Production of Food Additives and Processing Aids

As enzymes are important examples of food additives and food processing aids produced using GMM, a description of the development of host strains and expression vectors for enzyme production is provided in the following section, together with a brief description of the tools used to produce metabolites.

As mentioned earlier, *A. oryzae* is one of the most important filamentous fungi for the recombinant production of enzymes. This fungus was selected as the host strain due to its history of safe use in products with GRAS status, to previous experience with the organism in production processes, and finally to the huge protein production potential of this microorganism.

The production of especially amylases and proteases by *A. oryzae* was a major drawback, however. Two primary requirements in enzyme production are product purity and product stability. When producing a particular enzyme, the presence of amylases is undesirable, not only because of the contaminating protein but also because amylase activity may cause unwanted side reactions. Likewise, any host strain proteases present may also cause contamination and negatively affect the stability of the enzyme product. Hence, these unwanted enzyme activities must – if possible – be totally removed, and the most important method to achieve this is gene disruption [15]. Gene disruption is also used to improve *A. niger, T. reesei* and *F. venenatum* expression systems.

4.3.1
Gene Disruption

The basic steps of gene disruption are:

1. The gene to be disrupted is cloned as a genomic clone together with 1–2 kilobases (kb) of upstream and downstream sequence (i.e., the 1000–2000 base pairs (bp) preceding the coding sequence of the gene and the 1000–2000 bp preceding the coding sequence).
2. A part (or all) of the coding sequence is replaced with a selectable marker to form the gene disruption plasmid.

3. The gene disruption plasmid is linearized (typically by digesting it with a restriction enzyme) and subsequently transformed into the fungus.
4. Transformants are recovered using the selection system of the disruption plasmid, after which the transformants are screened for the desired genotype.

The gene disruption event is outlined in Fig. 4.2.

The *pyrG* (orotidine-5'-phosphate decarboxylase) marker shown in the example is an auxotrophic marker that is widely used in *A. oryzae* and in other *Aspergillus* sp. for gene disruptions [16]. It is a biosynthetic gene in the pyrimidine pathway, and it is necessary that the strain to be transformed is *pyrG* negative. The advantage of this marker is that it is bi-directional; that is, selection for both presence and absence of the marker is possible. Selection for the presence of *pyrG* is carried out simply by transforming a *pyrG* negative strain and selecting on a minimal medium for pyrimidine prototrophy (only cells having an intact *pyrG* gene can grow on such plates). Selection for absence of the *pyrG* gene is carried out by plating on a minimal medium containing uridine and the compound 5-fluoro-orotic acid. The uridine addition will enable cells to grow without a functional *pyrG* gene, while the 5-flouro-orotic acid enables counter-selection for the

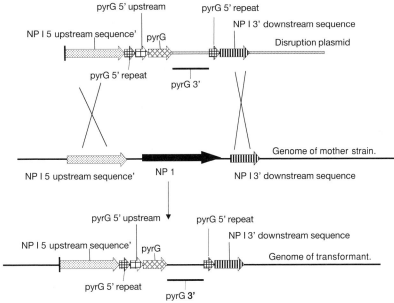

Figure 4.2 Gene disruption in filamentous fungi. The principle is illustrated by disruption of the gene NP I. A disruption plasmid was made by replacing the coding part of the NP I gene with the selectable marker *pyrG* (orotidine-5'-phosphate decarboxylase). The details of the components involved are described in the main text. In this example, a repeating sequence flanking the *pyrG* selection cassette has been included (*pyrG* 5' repeat). Recombination between the repeats results in the loss of the *pyrG* gene, and so these repeats greatly facilitate the selection of a *pyrG* negative strain with a predictable genotype for subsequent gene disruptions.

pyrG gene. This compound is converted into a toxic product by cells harboring a functional *pyrG* gene.

The bi-directional *pyrG* selection system is a very convenient tool for sequential disruption of several genes. A disruption plasmid is constructed for each of the genes that should be disrupted, and a *pyrG* negative mutant of the selected host strain is isolated using 5-fluoro-orotic acid. The *pyrG* mutant is then transformed with the first selection plasmid using selection for pyrimidine prototrophy. Transformants with the correct genotype, typically confirmed by Southern blot analysis [17], are then counter-selected for *pyrG*. The resulting strain is then ready for a new round of gene disruption using the next gene-disruption plasmid.

Other bi-directional markers are known, for example the *niaD* (nitrate reductase) system [18], but the *pyrG* system is the most applied system.

It is clear from Fig. 4.2 that the gene disruption event leads to major changes in the DNA sequence of the affected locus. These changes are irreversible, which means they will be present in all strains later in the pedigree of that particular host strain. They will thus also be present in transformants transformed with a particular product gene that is in the final GMM used for production. The gene disruption locus is thus an obvious target for analysis for the presence of DNA from the final GMM. In this way, recombinant host strain DNA may be detected, but if this host strain has been used for more products, it cannot reveal the specific product of the GMM.

By using gene disruptions, several amylase, glycoamylase, and protease genes have been disrupted in both *A. oryzae*, *A. niger* and *A. awamori*. A very similar technology has been used to disrupt primarily cellulase genes in *T. reesei*, metabolic pathways in *F. venenatum* [19], and in *A. niger* [13]. Rather than disrupt all of the individual genes, it is also possible to disrupt global activators that regulate the expression of entire classes of enzymes. For example, in *A. niger* a general protease regulator activates the expression of a range of extracellular proteases. Simply by disrupting this single gene, all of these proteases are silenced [20].

4.3.2
Expression Vectors

Development of the host strain is important in maintaining the efficiency of the expression system as well as the quality of the final product. Moreover, host strain improvement generates sequence tags in the final GMM suitable for analysis.

The expression vector is the other major factor in the final GMM, and is usually specific for the product to be expressed. It is therefore a natural target for analysis when analyzing for a specific product. In general, expression vectors comprise the following elements:

- The gene encoding the product to be generated.
- A promoter to drive expression of the product gene; this can be a promoter developed for the expression system (i.e., a generic promoter), or it can be specific to the product gene, typically the product gene's own promoter

- A translational terminator; as for the promoter, this can be a terminator generic for the expression system or it can be a specific terminator, typically the terminator of the product gene.
- A selection marker for selection in the expression host.
- Additional vector elements used to build the vector, for example an *Escherichia coli* selection marker and an *E. coli* origin of replication.

The expression vector is transformed into the fungal host cell, and then integrated into the chromosome of the host cell typically in more copies by tandem integration into one locus [21, 22]. An ideal integration of two copies is shown schematically in Fig. 4.3.

Typically, expression vectors are constructed in *E. coli*, and so elements necessary for this are often present in the final expression vector. The most widely used selection marker in *E. coli* is the β-lactamase gene which gives resistance to penicillins such as ampicillin. Public concern about the spread of antibiotic resistance markers has drawn attention to this part of the expression construct, even though antibiotic resistance markers from GMM used in contained production cannot be refound in nature, even when the biomass is spread to surrounding fields [23]. Hence, antibiotic markers have been replaced by other markers (e.g., auxotrophic markers) by most enzyme producers, and the *E. coli* portions of the expression vector are often completely removed before the construct is transformed into the host strain.

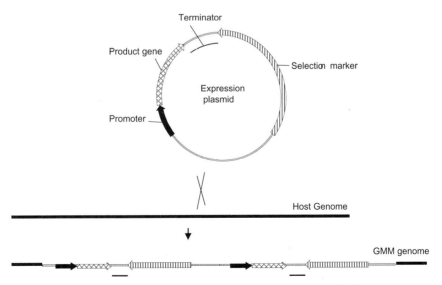

Figure 4.3 Integration of an expression plasmid into the genome. Tandem integration of an expression plasmid featuring a promoter, a product gene, a terminator and a selection marker into the genome of a filamentous fungus is shown schematically. The integration shown is head-to-tail in one locus, with no further recombination.

The choice of selection marker to select the expression vector in the fungus is very important, as different markers result in different ranges of copy numbers of integrated expression vector copies. The *pyrG* gene is sometimes used as a selection expression vector, and this is typically the case when the strategy is to insert the heterologous gene into a particular locus (perhaps the amyloglycosidase locus of *A. awamori*). The pyrG marker can also be used when the strategy is to remove the marker by recombination in a step following the transformation step, very much as described for gene disruption [24]. The drawback of these strategies is that typically only one copy of the expression vector is integrated into the genome, and this leads to low production yields.

The amdS (acetamidase) gene of *Aspergillus nidulans* [25] is often used as selection marker in *Aspergillus* sp., and even in other fungi. The selection principle is that the acetamide gene enables transformed cells to hydrolyze acetamide and, upon hydrolysis, to use the formed ammonium as sole source of nitrogen or the formed acetate as sole source of carbon. The advantage of this marker is that in both *A. oryzae* and *A. niger* high copy numbers are obtained, and thus high fermentation yields in production.

The choice of promoter is crucial for the yields, and this is one of the elements that has been most extensively studied and developed. For expression in *Aspergillus* sp., the amylase promoters were the natural choice as starting points for promoter development. These promoters are very powerful, and the ability to regulate them is very convenient in a production set-up as they are induced by starch, maltodextrins, and maltose. However, one of their drawbacks is that they are carbon catabolite-repressed, and this places a restraint on production. In order to avoid low productivity, the feed rate in fed-batch or in continuous fermentation must be carefully controlled. The *Aspergillus* amylase promoters are regulated very much in the same way, and they have been shown to be activated by the same activator [26].

The TAKA amylase is a very well-expressed amylase in *A. oryzae*, and hence the TAKA amylase promoter is one of the most widely used *Aspergillus* amylase promoters [8].

The neutral amylase promoter from *A. niger* is highly homologous to the TAKA promoter from *A. oryzae*, and is likewise broadly used [27]. These two promoters share sufficient sequence similarity to establish analytical methods that will be able to detect both.

The amyloglycosidase promoter from either *A. niger* or *A. awamori* is a promoter being in the same range of promoter strength. The sequence is quite distant from the sequence of the TAKA and the neutral amylase promoters, but it is regulated by the same activator. The amyloglycosidase promoter is also broadly used [9].

The choice of terminator is less critical than the choice of promoter. In *Aspergillus*, the amyloglycosidase terminator from *A. niger* is widely used [8].

The expression vectors for *T. reesei* are somewhat different than the *Aspergillus* expression vectors. The most frequently used promoter is the cellobiohydrolase I promoter, or variants of this [28]. This promoter is induced by cellulose-like sophorose, and also by more readily available carbon sources. This promoter is also carbon catabolite-repressed, but promoter variants where the creA sites mediating this repression have been removed have been described [29]. Several different selection markers are used in *Trichorderma*; these are often dominant selection markers against drugs, for example the hygromycin or bleomycin resistance markers.

Various yeast systems have also been developed for the production of enzymes, including *Kluyveromyces lactis*, *Hansenula polymorpha* and *P. pastoris* [30]. One notable difference between the yeast systems and the filamentous fungal systems is that the yeasts have self-replicating plasmids; that is, the transforming DNA is not necessarily integrated into the genome of the host strain. Genetic stability using self-replicating plasmids is significantly lower than the stability of strains where the DNA has been chromosomally integrated, and so integrating expression vectors are also preferred in yeasts. Host strain development is carried out in very much the same way as for the filamentous fungi. In particular, proteases are removed by gene disruption. The methylotrophic yeasts *H. polymorpha* and *P. pastoris* produce large amounts of alcohol utilization genes when induced by methanol. All or some of these genes have been disrupted in some host strains, or the expression plasmid is integrated into one of these genes when making the final GMM, thereby disrupting the gene.

The promoters used to drive expression in the methylotrophic yeasts are typically promoters from the alcohol utilization genes such as alcohol oxidase 1 or formate dehydrogenase promoter in *P. pastoris*, or the methanol oxidase 1 or formate dehydrogenase promoter in *H. polymorpha*. In *K. lactis* the β-galactosidase promoter (*LAC4*) has been identified as a strong promoter. For all of the yeast systems several different selection markers including both dominant and auxotrophic markers have been developed. The *URA3* marker, the yeast homologue of *pyrG* is very often used.

The genetic manipulations for metabolites are very much of the same type as those described for enzyme production. The main tools in metabolic engineering are gene disruption, gene replacement, overexpression of genes (e. g., by increasing the copy number), or the introduction of new genes.

Gene replacement is a variant of gene disruption. The result of gene replacement is that a target gene is replaced with another gene, typically in a two-step strategy very similar to the gene disruption strategy previously described.

4.4
Industrial Enzyme Production Processes

Almost all industrial enzyme products are formulations of enzymes that are secreted from the fungus during fermentation. Following fermentation, the fungal biomass is removed and the enzyme recovered from the broth. As the GMM DNA is present in the biomass, it is removed together with the biomass. Hence, only GMM DNA that is released into the broth as a result of cell lysis, and which is neither removed nor degraded during the subsequent recovery and formulation processes, are present in the final product. This means that the utility of analysis on the final product for GMM based on the recombinant DNA will be totally dependent on the production process. A brief introduction to industrial fermentation processes is provided in the following section.

Traditionally, enzyme production using filamentous fungi has been based on both surface fermentations and on submerged fermentation processes. Today, enzymes are produced almost exclusively by the latter approach.

The production of industrial enzymes by fermentation processes using recombinant fungi starts with the inoculation of a vial of the organism into a small flask containing an agar medium. The flask is then placed in an incubator that provides the optimal temperature for culture sporulation. The spores are then transferred to a seed fermenter (this is a small fermenter in which the biomass for the main fermentation is generated). Seed fermentation allows the cells to adapt to the environment and nutrients that they will encounter later on.

Following seed fermentation, the cells are transferred to the main fermenter, where temperature, pH and dissolved oxygen are carefully controlled to optimize enzyme production. The fermentation process can either be run as a batch process, a fed-batch process, or a continuous fermentation. In the batch process, all media components are added from the start of the fermentation, while in the fed-batch system the fungus is fed with additional medium during the fermentation. In a continuous fermentation, a steady state is reached by supplying fresh medium and harvest from the tank simultaneously. The degree of cell lysis during fermentation is mainly dependent on the organism that is fermented, the duration of the fermentation process, and the design of the fermenter (i.e., the shear stress that the biomass will encounter). A typical fermentation process is outlined in Fig. 4.4. When the main fermentation is complete, the mixture of cells, nutrients and enzymes – referred to as the "broth" – is ready for filtration, recovery and purification – this is collectively referred to as "downstream processing".

The next stage is to separate the broth containing the enzyme from the biomass. This is achieved by various chemical treatments of the fermentation broth to ensure efficient separation, followed by removal of the biomass using either centrifugation or filtration.

Following separation, the enzyme is concentrated by means of semi-permeable membranes or evaporation. In the case of products with high purity demands, the downstream process often requires special steps to remove unwanted impurities. This is often done by selective precipitation or adsorption of the impurities, or by crystallization by which very pure enzyme products can be obtained. In rare cases, a costly column chromatography step may be applied. The recovery process is outlined in Fig. 4.5.

The final step in the process is formulation of the enzyme product. The enzymes can be formulated either as liquid products or as granulates, depending on their application. The critical issues of the formulation are to secure stability of the enzyme product, release of the enzyme in the application, and to prevent enzyme dust formation that can cause allergy.

4 Production of Food Additives using Filamentous Fungi | 97

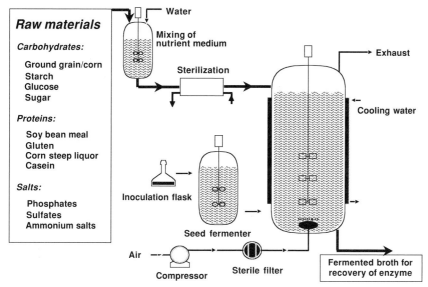

Figure 4.4 Schematic outline of a typical fermentation process flow. Spores from an inoculation flask are used to inoculate a seed fermenter, and this is in turn used to inoculate the main fermenter. The fermentation broth is aerated by stirring with large propellers. Additional medium components can be prepared and added during fermentation.

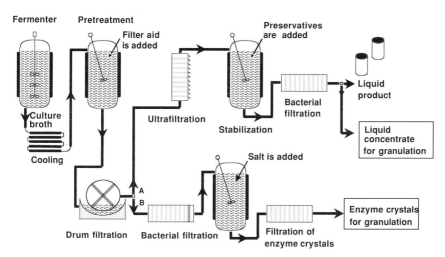

Figure 4.5 Outline of a recovery process. Typical process flows are demonstrated. The biomass is separated from the broth by filtration. The broth is then concentrated and sterile filtered, or alternatively the enzyme is recovered by crystallization.

References

1 G. Gellissen, K. Melber, Z. A. Janowicz, U. M. Dahlems, U. Weydemann, M. Piontek, A. W. Strasser, C. P. Hollenberg. *Antonie Van Leeuwenhoek* **1992**, 62, 79–93.

2 C. K. Raymond, T. Bukowski, S. D. Holderman, A. F. Ching, E. Vanaja, M. R. Stamm. *Yeast* **1998**, 14, 11–23.

3 A. P. J. Trinci. *Mycol. Res.* **1992**, 96, 1–13.

4 J. W. Bennet. *SIM News* **2001**, 51, 65–71.

5 J. W. Bennet, M. A. Klich. *Aspergillus, Biology and industrial applications*. Butterworth-Heinemann, Stoneham, USA, **1992**.

6 P. Barbesgaard, H. P. Heldt-Hansen, B. Diderichsen. *Appl. Microbiol. Biotechnol.* **1992**, 36, 569–572.

7 S. D. Martinelli, J. R. Kinghorn. *Aspergillus: 50 years on, Progress in industrial microbiology*, Volume 29, Elsevier, Amsterdam, **1994**.

8 T. Christensen, H. Woeldike, E. Boel, S. B. Mortensen, K. Hjortshoej, L. Thim, M. T. Hansen. *Biotechnology* **1988**, 6, 1419–1422.

9 R. M. Berka, K. H. Kodama, M. W. Rey, L. J. Wilson, M. Ward. *Biochem. Soc. Trans.* **1991**, 19, 681–685.

10 S. Keränen, M. Pentillä. *Curr. Opin. Biotechnol.* **1995**, 6, 534–537.

11 J. C. Royer, D. L. Moyer, S. G. Reiwitch, M. S. Madden, E. B. Jensen, S. H. Brown, C. C. Yonker, J. A. Johnston, E. J. Golightly, W. T. Yoder. *Biotechnology* **1995**, 13, 1479–1483.

12 B. Christensen, J. Nielsen. *Adv. Biochem. Eng. Biotechnol.* **2000**, 66, 209–231.

13 H. Pedersen, B. Christensen, C. Hjort, J. Nielsen. *Metab. Eng.* **2000**, 2, 34–41.

14 T. Hoshino, K. Ojima, Y. Setoguchi, Hoffmann La Roche & Co ag, Roche Vitamins Inc., European Patent EP-0108210, **1998**.

15 J. Lehmbeck, Novo Nordisk A/S, European Patent EP-0956338-A1, **1996**.

16 D. J. Ballance, G. Turner, *Gene*, **1985**, 36 (3), 321–331.

17 J. Sambrook, E. F. Fritsch, T. Maniatis, *Molecular cloning, a laboratory manual*, second edition, Cold Spring Harbor Laboratory Press **1989**.

18 S. E. Unkles, E. I. Campbell, D. Carrez, C. Grieve, R. Contreras, W. Fierss, C. A. M. J. J. van den Hondel, J. R. Kinghorn. *Mol. Gen. Genet.* **1989**, 218, 99–104.

19 J. C. Royer, L. M. Christianson, W. T. Yoder, G. A. Gambetta, A. V. Klotz, C. L. Morris, H. Brody, S. Otani. *Fungal Genet. Biol.* **1999**, 28, 68–78.

20 C. Hjort, A. M. J. J. van den Hondel, P. Punt, F. H. J. Schuren, Novo Nordisk A/S Patent WO 0020596, **1998**.

21 J. R. Fincham. *Microbiol. Rev.* **1989**, 53, 148–170.

22 B. Ruiz-Díez. *J. Appl. Microbiol.* **2002**, 92, 189–195.

23 J. T. Andersen, T. Schäefer, P. L. Joergensen, S. Moeller. *Res. Microbiol.* **2001**, 152, 823–833.

24 G. C. M. Selten, R. F. M. van Gorcom, B. W. Swinkels, Gist Brocades NV, European Patent EP 0635574A1.

25 C. M. Corrick, A. P. Twomey, M. J. Hynes, *Gene*, **1987**, 53 (1), 63–71.

26 K. L. Petersen, J. Lehmbeck, T. Christensen. *Mol. Gen. Genet.* **1999**, 262, 668–676.
27 E. Boel, T. Christensen, H. F. Woeldike, Novo Nordisk A/S, European Patent EP0238023, **1987**.
28 J. M. Uusitalo, K. M. Nevalainen, A. M. Harkki, J. K. Knowles, M. E. Penttilä. *J. Biotech.* **1991**, 17, 35–49.
29 M. Ilmén, M. L. Onnela, S. Klemsdal, S. Keränen, M. Penttilä. *Mol. Gen. Genet.* **1996**, 253, 303–314.
30 G. Gellissen, C. P. Hollenberg. *Gene* **1997**, 190, 87–97.

5
Perspectives of Genetic Engineering of Bacteria used in Food Fermentations

Arnold Geis

5.1
Introduction

Fermentation is a very old technology used for food production in many parts of the world. Indigenous fermented foods such as fermented milk products as cheeses, bread, fish, meat and numerous vegetables and fruits have been prepared for thousands of years and are often strongly linked to tradition and culture. Production of fermented foods involves the use of microorganisms (and/or enzymes) which alters properties of the raw material as taste, texture, digestibility, nutritional value, and shelf-life. Fermentation is a relatively efficient, low-energy preservation process and is therefore an appropriate technique for safe food production in developing countries with either no or limited access to sophisticated techniques for food preservation. In these countries, a great variety of raw materials, mainly of plant origin, have been fermented. Fermentations were performed mostly as spontaneous processes under nonaseptical conditions. The specific environment then gradually select for those microorganisms required for the desired product. The underlying microbial and/or enzymatic processes are in most cases poorly understood, which makes the process difficult to control. Any refinement and upscaling of these fermentation processes needs a better understanding of the microorganisms involved and their metabolic capabilities. Such knowledge would allow the selection or development of more productive bacterial cultures and a better control and manipulation of the culture conditions. This, together with an improvement of fermentation technology, would have a great impact on nutrition and food security in developing countries, while reducing post-harvest losses.

In developed countries, many of which are located in the northern hemisphere and have a moderate climate, milk – and to a lesser extent meat, fish, and vegetables – form the raw materials for the production of fermented foods. During the long history of fermentation – especially of milk – a sophisticated technology has been developed, and today well-characterized fermentation cultures allows the controlled, large-scale production of many safe, tasty products with high nutritional values. For example, during 2001 a total of about 13×10^6 tons of cheese was consumed in the USA and the EU.

In addition to efforts to improve the production of traditional fermented milk products, an increasing amount of scientific efforts has been undertaken during recent years to develop new products with additional beneficial effects on the health of the consumer (these are termed 'functional foods'). Products fermented with or containing added probiotic bacteria are thought to relieve lactose intolerance, reduce the duration of rotavirus-induced diarrhea, and also to alleviate allergic reactions. Other proposed beneficial health effects of probiotics await confirmation. Likewise, the mechanisms behind these health-promoting effects have still to be elucidated by *in vitro* and *in vivo* studies.

In addition to yeasts and filamentous fungi (see Chapter 3), lactic acid bacteria (LAB) play the predominant role in food fermentation, causing characteristic changes in and prolonging the shelf-life of the fermented food. LAB that grow during fermentations either originate from the raw material (plant material, meat) or are added as starter cultures (milk, meat). LAB used in food fermentations are generally regarded as save or even advantageous for human health (probiotics).

5.2
Lactic Acid Bacteria

The LAB form a heterogeneous group of Gram-positive bacteria, but they are physiologically related by their ability to ferment carbohydrates to lactic acid as predominant metabolic end-product. The homofermentative LAB convert fermentable carbohydrates almost quantitatively into lactic acid, whereas the heterofermentative LAB produce lactic acid and other compounds such as acetic acid, CO_2 and ethanol as primary metabolic end products. Some LAB also produce flavor compounds that are essential to the final taste of the fermented product.

Species from six genera of lactic acid bacteria have been used in the industrial production of food, feed, and alcoholic beverages (Table 5.1).

5.2.1
Lactococcus lactis subsp. *lactis* and subsp. *cremoris*

Lactococci are mesophilic, homofermentative LAB which are used extensively in milk fermentation to produce a great number of different products. Until now, *Lactococcus lactis* is the best-characterized LAB with regard to its physiology and molecular genetics. The determination of the complete nucleotide sequence of *L. lactis* IL1403 [1a, b] and some other strains which soon will be publicized [2] make these bacteria amenable to transcriptome and proteome analysis, and will promote comparative genetic analysis among the lactococci. During the past two decades, a plethora of methods and molecular devices has been developed to perform any kind of genetic engineering of these bacteria (see Chapter 12).

Table 5.1. Lactic acid bacteria (LAB) used for food fermentation.

Fermented foods	Lactic acid bacteria*
Dairy products (butter and buttermilk, yogurt and cheeses)	L. lactis subsp. lactis L. lactis subsp. lactis var. diacetylactis L. lactis subsp. cremoris Lb. delbrueckii subsp. bulgaricus Lb. casei Lb. helveticus Lb. acidophilus Ln. mesenteroides subsp. cremoris Ln. lactis Streptococcus thermophilus
Meat and fish products (fermented sausages, many indigenous products)	Lb. curvatus Lb. sake Lb. plantarum P. acidilactici P. pentosaceus various undefined LAB
Plant products (sauerkraut, olives, cassava products, soy sauce, and many indigenous products)	Lb. plantarum Lb. ssp. Ln. mesenteroides P. pentosaceus various undefined LAB
Bakery products (sourdough products, cracker)	Lb. sanfrancisco Lb. plantarum Lb. fermentum Lb. reuteri, Lb. amylolyticus Lb. ssp.

* Abbreviations: *L.*: Lactococcus; *Lb.*: Lactobacillus; *Ln.*: Leuconostoc; *P.*: Pediococcus

5.2.2
Lactobacillus spp.

The genus *Lactobacillus* comprises a large number of relatively diverse species. Lactobacilli are widespread in nature, and many have been used in food fermentation processes, including milk, meat, and plant material fermentations. In addition, a few species of *Lactobacillus* are used as probiotic microorganisms in functional foods (see Table 5.1). Lactobacilli were divided approximately into three groups that reflect mainly their abilities to ferment various hexoses, pentoses, and disaccharides to lactic acid or lactic acid and further metabolites [3, 4]. On the basis of genome sequence data which soon will be available for several different species [2], the biochemical and genetic analysis of these bacteria, which remains rudimentary for many of the species, will undergo rapid development to allow better and extended exploitation of this important bacterial group, both for traditional and innovative purposes.

5.2.3
Streptococcus thermophilus

Streptococcus thermophilus is the only streptococcal species that is used, along with *Lactobacillus* spp., as starter culture in food technology for the manufacture of yogurt, mozzarella and Swiss-type cheeses. It is distinguished from the other streptococci by its ability to grow at elevated temperatures (up to 52 °C), and its limited capability to ferment sugars. The increasing manufacture of the above-mentioned products demands deeper understanding of the performance and production requirements of such starter cultures. The number of industrial strains which comply with these demands is limited. Several traits such as phage resistance and exopolymer synthesis have been the target for a strain improvement program [5]. The availability of the complete genome sequences of two *S. thermophilus* strains presently in the annotation process [2] will greatly improve our knowledge of the metabolism and molecular genetics of this species, and also facilitate genetic engineering of new strains and the control of the current large-scale fermentation processes.

5.2.4
Leuconostoc spp.

Bacteria of the genus *Leuconostoc* have been found in a great number of natural and man-made habitats. Numerous strains have been isolated from grass, herbage, and silage [6, 7]. These heterofermentative bacteria play an essential role in the fermentation of vegetables such as cabbage and cucumbers, where they initiate spontaneous lactic acid fermentation [8, 9]. The ability of some species (especially *Ln. mesenteroides* subsp. *cremoris*) to ferment the citric acid of milk to the flavor compound diacetyl has led to their use in dairy starter cultures for this purpose [10]. The ability of *Ln. mesenteroides* subsp. *mesenteroides* to produce dextrans and levans has been exploited for industrial production of this substances for use in the biochemical and pharmaceutical industry [11]. The genome sequence of *Ln. mesenteroides* subsp. *mesenteroides* will be completely available in the near future [2].

5.2.5
Pediococcus spp.

Members of the genus *Pediococcus* are homofermentative, acidophilic bacteria that divide alternately in two perpendicular directions to form tetrads [12]. Several species, as *P. pentosaceus* and *P. acidilactici* are used as starter cultures for sausage making, vegetables and soy milk fermentation and for silage inoculation [12, 13] and are also found on ripening cheeses as constituents of the nonstarter lactic acid microflora [14]. *P. pentosaceus* strains may contain up to five plasmids linked to the abilities to ferment raffinose, melibiose, and sucrose and to produce bacteriocins [15, 16]. Recent advances in genomic sequence analysis [2] will make this

5.2.6
Oenococcus spp.

Oenococcus oeni (formerly *Leuconostoc oenos*) is a highly acid- and alcohol-tolerant lactic acid bacterium found naturally in fruit mashes and related habitats [17]. The most studied aspect of *O. oeni* is its ability to convert malate to lactate. This malolactic conversion is employed in wineries to reduce the acidity and improve the stability and quality of wines [18, 19]. The genome sequence analysis of two commercially used *O. oeni* strains is currently in progress [2]. Knowledge of the *O. oeni* genome will be valuable for the development of new tools to control malolactic fermentation in wine.

5.3
Perspective and Aims

5.3.1
Bioconservation

Preservation of food by drying, salting, and fermentation are traditional methods of food preservation. The industrialization of food processing had increased the need and use of chemical additives to prolong shelf-life and to prevent or reduce deleterious effects due to recontamination by spoilage microorganisms. The growing demand of the consumer for more natural food, and an awareness of the health risks posed by some chemical food preservatives, has led to a demand for alternatives.

The preservation of food by antagonistic microorganisms has been known for many decades [20], but as yet has not been applied to food preservation on a larger scale. The term 'bioconservation' was coined to distinguish this type of preservation from the chemical preservation of foods. Bioconservation may consist of adding bacterial strains that produce antagonistic compounds or by adding the purified antagonistic substances. Some LAB produce, in addition to their main metabolic products, a variety of inhibitory substances such as diacetyl, CO_2, hydrogen peroxide, and bacteriocins.

Bacteriocins – proteinaceous compounds that inhibit the growth of closely related bacteria – have been produced by a variety of lactococcal strains. In a survey of 280 strains isolated from dairy environments, 5% were found to produce such substances. On the basis of biochemical and physical properties, host range and cross-reactivity, eight bacteriocin types were predicted [21, 22]. In recent years, a number of these bacteriocins have been characterized using both genetic and biochemical methods and assigned to two main classes: lantibiotics-class I; and small, heat stable, nonlantibiotics-class II [23]. Most of the bacteriocins of LAB show a narrow host range and are effective only against closely related bacteria, whereas others are also active against a broad spectrum of Gram-positive bacteria, including food pathogens (e. g., *Listeria* spp., *Clostridia* spp.). Bacteriocins of the latter group have the greatest potential for use as food preservatives.

Nisin, a lantibiotic, is the best known bacteriocin and the only one approved for biopreservation in food. It is produced by some *L. lactis* strains and shows a broad antibacterial spectrum. Nisin is used as a protective agent in a variety of food products worldwide (for a review, see [24]). Nisin is most effective in an acidic environment however, and consequently its use is limited.

Lacticin 3147 is a two-component bacteriocin that is produced by *Lactococcus lactis* DPC 3147 [25] and inhibits the growth of a variety of Gram-positive bacteria, including food-pathogens such as *Listeria monocytogenes*, *Clostridia*, *Staphylococci* and *Streptococci*. It can be used as food protective when delivered either as a food ingredient or through the action of a bacteriocin-producing starter. Lacticin 3147 is encoded on a conjugative plasmid and can easily be transferred into industrially used starter strains by conjugation. The effectiveness of this bacteriocin has been proven in cheese manufacture and ripening and even in mastitis control. Due to its broad pH range, Lacticin 3147 may have an advantage over nisin as food preservative [26, 27].

A steadily increasing number of bacteriocins produced by different genera of LAB are currently under investigation by various genetic and biochemical means. Some of these compounds show promise for potential use as food preservatives. Many of these bacteriocins can be expressed in heterologous hosts, and this may extend their application in food preservation [28]. Better knowledge of the genetics and biochemistry of these substances may, in future, allow the design of tailor-made bacteriocins by site-directed mutagenesis with improved properties of stability, pH range, and activity spectrum. Placing a bacteriocin gene/operon under the control of growth-phase-, pH-, or salt-dependent promoters might allow secretion of the protective substance on completion of the fermentation process, without causing any detrimental effects on the starter and/or the ripening culture. The construction of multi-bacteriocinic strains using recombinant gene technological methods may further extend the usefulness of these compounds in food preservation.

5.3.2
Bacteriophage Resistance

Bacteriophage-inhibition during fermentation, especially in the dairy industry and in other submerged fermentations (e. g., acetic acid) is a ever-present danger that may cause significant economic losses. Thus, bacteriophage resistance of fermentation cultures is a property of great importance which has attracted extensive scientific interest. Most of these investigations have been – and still are – concentrated mainly on lactococci and very recently on other LABs. In general, bacteria possess a variety of phage-defense systems that interfere with all steps of phage development – from phage adsorption, DNA injection and replication, expression of the phage genes, and assembly and maturation of the phage particle. Research into phage-resistance mechanisms of dairy starter strains, and especially of lactococci, has been summarized recently in a number of comprehensive reviews [29–33]. In lactococci, the phage-defense systems are predominantly encoded on plasmids, many of which are self-transmissible, and this offers the opportunity to transfer phage-resistance genes in other, industrially important strains by conjugation, which is a naturally occurring (food grade) gene transfer mechanism. Conjugation has been employed successfully to create starters that perform reliable in dairy fermentations. The genetic linkage for phage-defense systems with bacteriocin production and resistance/immunity genes as selection marker on lactococcal plasmids was used for strain construction. Three different, potent phage-resistance mechanisms encoded on different plasmids were introduced into a recipient strain by sequential transfer. The resulting strain was highly phage-resistant and successfully used for cheese manufacture [34]. The use of this method for strain construction is limited by the lack of dominant selection markers and often by instability of the transferred plasmids, though this may be due to incompatibility of the transferred plasmid with the indigenous plasmids of the recipient. In addition, many phage-resistance systems reside on nonconjugative plasmids.

More than 50 different phage-defense systems including abortive infection (Abi)-, restriction-modification (R/M) systems and some others, all of which act on other stages of phage propagation, have been identified in lactococci. Moreover, the number of these systems identified in other LABs has steadily increased. The current approach to perform a genome sequence program on LABs will further extend our knowledge about phage-resistance in all members of this bacterial group.

The tools of recombinant DNA technology and the growing number of suitable gene cloning systems (see Chapter 12) will allow improvements to be made in the phage-resistance of industrially used starter strains. In principle, any combination of phage-defense systems acting on different stages of the phage development and/or on different phages could be transferred in many industrial strains by the use of electro-transformation or more sophisticated conjugation methods [35]. So far, the use of recombinant DNA technology to improve starter bacteria has been limited by legislative constraints, though the development of food-grade gene delivery systems (see Chapter 12) may overcome these limitations. Vectors with bacteriocin-resistance/immunity genes as selection markers are available, as well as other systems based on complementation [36]. A two-component food-grade cloning system was developed recently, which consists of a vector entirely made up of L. lactis DNA without any selection marker, and a replication-deficient plasmid encoding a selectable marker gene that depends on the presence of the vector-plasmid for maintenance [37]. After cloning of the desired gene into the vector, both plasmids were transferred in the host strain by co-electroporation. Under selective conditions, only cells can grow which contain both plasmids. Due to intrinsic incompatibility the plasmid carrying the selection marker is easily lost from the cells during growth under nonselective conditions. In this way, a phage-defense system (AbiQ) was stably introduced into two industrial strains, resulting in an effective resistance phenotype against phages.

An efficient recombinant anti-phage system, designated PER (phage encoded resistance) was first described by Hill et al. [38]. It is based on the observation that the lytic cycle of a phage can be interrupted by presenting the origin of replication (*ori*) of the phage in multiple copies in the host cell. After infection, the phage replication factors are titrated by the multi-copy *ori* present on a plasmid-vector. This results in induction of the plasmid rather than phage–DNA replication. By cloning multiple copies of the phage *ori* an increased efficiency of the PER phenotype was obtained [39, 40].

Antisense-RNA technology also has been used to construct phage-resistant strains. Transcription of the noncoding strand of phage genes results in non-translatable RNA hybrids, which are rapidly degraded. The reduction of transcripts for essential phage components severely disturbs the phage propagation [41]. Several antisense RNA cassettes have been developed and used in combination with inducible promoters, explosive expression vectors and phage-triggered suicide traps [42–44]. The suicide system makes use of phage-inducible promoters and a gene that is lethal to the host cell. The introduction of these constructs in *L. lactis* and *S. thermophilus* strains resulted in significant protection against phage attack.

These approaches, in combination with the food-grade, two-component gene delivery system, will open new opportunities to construct highly phage-resistant, food-grade starter strains.

5.3.3
Exopolysaccharides

Polysaccharides, which form a very diverse group of polymers, have been applied to many purposes in industry. High molecular-weight exopolysaccharides (EPS) are important ingredients in many food products as thickeners, stabilizers, emulsifiers, fat-replacers or gelling agents. Most of these polysaccharides have been derived from plant material, but alternative sources for these compounds include bacteria. For food manufacture, EPS produced by lactic acid bacteria are of major importance for the texture and organoleptic properties of many fermented milk products. The EPS of LABs form a very diverse group with regard to their chemical composition, size, molecular structure and physico-chemical properties [45]. The genetic determinants for EPS synthesis in *Lactococcus lactis* [46], *Streptococcus thermophilus* [47–49], and *Lactobacillus* spp. [50] have been sequenced and analyzed and reveal a common operon structure which suggests a common biosynthesis mechanism (for a review, see [51]). The gene sequence in all EPS-biosynthesis clusters is as follows: regulation, chain length determination, biosynthesis of the repeating unit, polymerization, and export [52].

Structure–function relationships of EPS and interaction of specific EPS with different food components are poorly understood, but are the ongoing aim of extensive scientific efforts. A better knowledge of these relationships, combined with a steadily increasing supply of genetic and biochemical data, will doubtless open the opportunity to construct recombinant LABs that produce tailor-made EPS for specific food products. The genetic modification of *eps* genes could result in EPS with different repeating units or with different chain lengths and, consequently, specific desired properties. Most LAB strains produce (if any) only low amounts of EPS and cannot be used for EPS-linked purposes. However, investigations to improve EPS yield have shown that the production of EPS is influenced by bacterial growth phase, carbon- and nitrogen source, and pH and temperature. EPS overexpression was demonstrated in *L. lactis* by placing part of the EPS-operon under the control of an inducible promoter [53]. In *S. thermophilus*, ESP production was increased by altering expression of the enzymes involved in central carbohydrate metabolism [54]. The glycosyltransferases could be the target in engineering approaches. Introduction of heterologous glycosyltransferase genes or replacing existing genes with different sugar specificity may allow the synthesis of EPS with different sugar composition. A better knowledge of the polymerization and export processes might allow optimization of EPS production with altered repeating units or changing the rheological properties of native EPS by engineering of the molecular mass [51].

5.3.4
Proteolysis

Due to amino acid auxotrophy and lack of sufficient amounts of free amino acids in milk, lactic acid bacteria require a proteolytic system for growth in milk. Proteolysis of these bacteria has been studied in great detail using a variety of genetic, biochemical and ultrastructural methods, and from these it has emerged that the proteolytic system of dairy lactococci and lactobacilli are remarkable similar both in their components and mode of action. The proteolytic system consists of a cell-wall-bound serine-protease (PrtP), three peptide transporters (Opp, DtpT, DtpP) and a variety of peptidases (Pep). The genes for the majority, if not all, of the enzymes necessary for casein degradation and transport of the degradation products were cloned, sequenced and analyzed in detail. From these data the following pictures emerges. Caseins of the milk were partially degraded by the cell-wall-bound protease into a large number of oligopeptides, some of which (≤ 10 amino acid residues) were taken up by the oligo- and the di-/tripeptide transport systems and subsequently hydrolyzed to amino acids by a plethora of peptidases. Mutants missing transporter or/and peptidase genes have been constructed by targeting deletion or disruption of the corresponding genes. Those missing the Opp system but still having di-/tripeptide transport activity are unable to grow in milk. Mutants with an increasing number of peptidase mutations showed decreasing growth rates in milk – a five-fold peptidase mutant grew 10 times more slowly than the wild-type. In lactococci, the genes for peptide transport systems and the peptidase were located on the chromosome, and those for the proteases exclusively on plasmids [56–59].

The degradation of casein is a crucial process in the formation of texture and flavor in many fermented dairy products, especially cheeses. With the genetic and biochemical tools presently available, it is now possible to manipulate the pathways of protein and peptide degradation in LABs by quantitative and qualitative means. *L. lactis* strains that express heterologous proteases, as neutral proteinase from *Bacillus subtilis* [60] and peptidase from LABs and other genera [61, 62] have been constructed, and some were tested in cheese manufacture, ripening and flavor development.

The expected wealth of genetic information arising within the LAB genome sequencing project [2], in combination with the development of sophisticated genomic/proteomic tools, will result in a better understanding of the proteolytic/catabolic pathways in LABs and allow the construction of strains with controlled synthesis of amino acids catabolites. This may include reduction of biogenic amines, as well as controlled production of specific flavor compounds in fermented foods [55].

Milk proteins are precursors of many different biologically active peptides [63] that can be released in active form during intestinal digestion or during fermentation. Fermented milks containing angiotensin-I-converting-enzyme (ACE)-inhibitory peptides have been produced with selected LAB strains [63]. ACE inhibition results in an antihypertensive effect, but may also have influence on other regulatory systems [65, 66]. Genetic engineering of the proteolytic system of LAB may open the opportunity to overproduce these valuable peptides.

5.4
Metabolic Engineering of Lactic Acid Bacteria

Genetic engineering means the direct improvement of biochemical properties of cells through the modification of specific reactions, or by the introduction of new reactions using recombinant DNA technology. To develop strategies for metabolic engineering, a thorough understanding of the metabolic network and metabolic fluxes is necessary. LAB possess a relatively simple and (in some species) well-known metabolism, where sugars are fermented mainly to lactic acid. *Lactococcus lactis* is the most extensively studied LAB and, therefore, the first one employed in metabolic engineering efforts. A *L. lactis* strain has been constructed for the efficient conversion of sugar into diacetyl by α-acetolactate decarboxylase inactivation and simultaneous NADH-oxidase overexpression. About 80% of the carbon flux was found to be re-routed via α-acetolactate to the production of diacetyl [67]. Diacetyl is essential as an aroma compound in many fermented dairy products, such as butter and fresh cheeses. Overexpression of serine hydroxymethyltransferase in *S. thermophilus* resulted in an increase of acetaldehyde (a major flavor compound in yogurt) and also folic acid formation [68].

The introduction and efficient expression of a heterologous *alaD*-gene, encoding alanine dehydrogenase, in lactate dehydrogenase-deficient *L. lactis* cells resulted, in the presence of ammonium, in complete conversion of pyruvate to alanine. Inactivation of the endogenous alanine racemase (*alr*) gene led to stereo-specific L-alanine production as the only end-product of fermentation [69].

An international, collaborative research project funded by the European Commission deals with the formation of health-promoting components in food as a result of bacterial activities. These metabolic engineering activities focus on the production of nonfermentable (low-calorie) sugars, removal of galactose and/or lactose from dairy products, removal of raffinose from soy products, oligosaccharide and vitamin production (www.nutricells.com) [70].

5.5
Stress Responses in Lactic Acid Bacteria

Lactic acid bacteria used in fermentation processes are at different stages of their life cycle (starter culture handling, fermentation and subsequent storage), subjected to a wide variety of harsh conditions. On order to survive the physical (high and low temperatures) and chemical (high acidity, nutritional starvation, oxidative and osmotic stress) changes in their environment, LAB have developed numerous stress-resistance mechanisms (for recent reviews, see [71, 72]. Bacterial stress responses rely on complex regulatory circuits, which act together to improve stress tolerance. The bacterial heat shock response is characterized by an elevated expression of a number heat shock proteins such as chaperons, small heat shock proteins and specific proteases. Induction of the heat shock response by a short heat shock at sublethal temperatures increased the viability of LAB under subsequent stress conditions [73, 74].

A variety of different systems that may contribute to acid tolerance have been identified in LAB, including different ATPases, the arginine deiminase pathway, ureases, decarboxylation reactions coupled to electrogenic transporters and other less well-characterized systems [72, 75].

Low-temperature adaptation and cryoprotection were found to be connected with the synthesis of cold-shock proteins induced at temperatures far below the optimal growth temperature in LAB [76–78].

The presence of oxygen is a factor that greatly affects the outcome of a fermentation process. In the presence of oxygen, hydrogen peroxide is formed which, if not degraded by peroxidases and/or catalases, may accumulate and lead to deleterious effects and even cell death. After cloning and expression of the catalase gene of *Lb. sake* in a catalase-deficient *Lb. curvatus* strain which was frequently used for meat fermentation, no H_2O_2 accumulation was observed and the recombinant strain remained viable in the stationary phase [79].

At present, the stress response of LAB is a field of extensive research, and a better understanding of the molecular basis of the different stress response mechanism and their interaction will undoubtedly allow the development of new strategies to screen and select for stress-tolerant or -sensitive strains. Alternately, strains used in many fermentation processes may be directly manipulated by the use of recombinant gene technology.

5.6
Methods

5.6.1
Transformation and Vector Systems

During the past fifteen years or so, sophisticated genetic tools – namely efficient transformation and specialized vector systems – have been developed which allow the analysis of gene cloning, expression, and regulation of homologous and heterologous genes. The transformation of LAB is routinely performed using electroporation [80–83]. Several protocols exist which allow efficient transformation even of strains which were resistant to the formerly used method of polyethylene glycol (PEG)-induced protoplast transformation and are also useable for many industrial strains. In addition, conjugal gene transfer systems allow efficient exchange of plasmids among lactic acid bacteria.

A large number of vector systems now allow the cloning and expression of genes from different sources, random and targeted inactivation of genes, selection for promoter-, terminator-, and signal-sequences and anchoring of proteins to the surface of lactococci [36, 42, 84–91]. Only one of the most sophisticated expression vector systems which allows very efficient control of gene expression in *Lactococcus* and other lactic acid bacteria will be described [92].

This system makes use of the autoregulatory properties of the nisin gene cluster. Nisin – at concentrations far below its minimal inhibitory concentration – acts as an inducer from outside via a two-component signal transduction system – a histidine protein kinase (NisK), and a response repressor (NisR). The two genes, *nisK* and *nisR*, under the control of a constitutive promoter are delivered by recombinant plasmids or integrated into the chromosome of the expression strain. The gene to be expressed is fused to the *nisA* promoter, which is part of the expression vectors. Vectors for transcriptional and translational fusions have been constructed; as a consequence, homologous as well as heterologous proteins can be produced up to 47% of the total cell protein with increasing amounts of the inducer [92].

A variety of "food-grade" vectors are now available, which consist only of DNA derived from LAB and are devoid of antibiotic resistance genes as selectable markers (see Chapter 12).

In recent years, several systems have been developed to anchor and to display heterologous (poly-)peptides at the surface of lactococci [93, 94]. This allowed changes to be made of the outside composition of the cells, which may help in an understanding of the mechanisms of protein targeting on its own. Changes in the surface composition may also influence interactions between the bacteria and its environment, and allow potentially important biotechnological applications such as immobilization of enzymes at the bacterial surface, fixing of cells to special carrier surfaces, and the display of entire peptide libraries. *Lactococcus lactis*, by presenting the correct antigens, may in future be used as a live bacterial vaccine delivery system [95–97].

By fusion of genes for heterologous polypeptides to genes of proteins known to be fixed to the cell envelope and presented at the cell surface, a variety of peptides and proteins could be anchored and at least partially presented at the cell surface of lactococci. These include enzymes such as β-lactamase [98], α-amylase [98], and nucleases [88], as well as epitopes of the human cytomegalovirus [99], of the human immunodeficiency virus [94], the *Plasmodium falciparum* merocoite stage surface antigen [94] and the *Clostridium tetani* toxin C-fragment [95, 100].

5.7 Conclusions

In the industrial scale production of fermented dairy products, a good understanding of the microbial processes has emerged during the past few decades. Some of the microorganisms involved in milk fermentation, and especially *Lactococcus lactis*, have been well-characterized such that sophisticated molecular and genetic tools are now available for the manipulation of these bacteria in many respects. Starter strains have been developed and will continue to be constructed to allow better control of the traditional fermentation processes, and also to develop new products with higher nutritional values and health-promoting properties. Genetically engineered LAB may also be used for the production of food-related products such as neutraceuticals, as well as other food additives such as flavor compounds and enzymes. Today, extensive research effort is being committed to the use of LAB as both production and delivery systems for pharmaceutically active substances and vaccines.

Despite the importance of fermented foods for nutritional purposes in less-developed regions of the world, little is known about the microbial and enzymatic processes in many fermentations, and this indicates that there is a clear and urgent need for further research in this field. The selection and development of optimized starter cultures which can be easily propagated, stored at ambient temperatures, and which are resistant to different stress factors encountered during fermentation and storage is only one aspect for improvement among the many fermentation processes. At present, the genomes of about thirty strains among all the biotechnologically important LAB have been determined. Hence, with the vast amount of genetic data available, a better understanding should soon be available of the biochemical potential, and the similarities and differences of these bacteria. Undoubtedly, this will result in a wealth of new screening, selection and genetic engineering strategies applicable to fermentation processes in the production of food.

References

1a Bolotin, A., Mauger, S., Malarme, K., Ehrlich, S. D., and Sorokin, A. **1999**. Low redundancy sequencing of the entire *Lactococcus lactis* IL 1403 genome. *Antonie van Leeuwenhoek* **76**:27–76.

1b Bolotin, A., Winker, P., Mauger, S., Jaillon, S., Malarme, K., Weissenbach, J., Ehrlich, S. D., and Sorokin, A. **2001**. The complete genome sequence of the lactic acid bacteria *Lactococcus lactis* subsp. *lactis* IL 1403. *Genome Res.* **11**:128–132.

2 Klaenhammer, T. R., Altermann, E., Arigoni, F., Bolotin, A., Breidt, F., Broadbend, J., Cano, R., Chaillou, S., Deutscher, J., Gasson, M., van der Guchte, M., Guzzo, J., Hartke, A., Hawkins, T., Hols, P., Hutkins, R., Kleerebezem, M., Kok, J., Kuipers, O., Lubbers, M., Maguin, E., McKay, L., Mills, D., Nauta, A., Overbeek, R., Pel, H., Pridmore, D., Saier, M., van Sinderen, D., Sorokin, A., Steele, J., O'Sullivan, D., de Vos, W., Weimer, B., Zagorec, M. and Siezen, R. **2002**. Discovering lactic acid bacteria by genomics: *Antonie van Leeuwenhoek* **82**:29–58.

3 Hammes, W. P. and Vogel, R. F. **1995**. The genus *Lactobacillus*, pp. 19–54. In, B. J. B. Wood and W. H. Holzapfel (eds). *The Genera of Lactic Acid Bacteria*. Chapman & Hall, London.

4 Axelsson, L. **1998**. Lactic acid bacteria: classification and physiology, pp. 1–72. In, S. Salminen and A. Von Wright (eds). *Lactic acid bacteria: Microbiology and Functional Aspects*, 2nd ed., Marcel Dekker, Inc, New York.

5 Delcour, J., Ferain, T., and Hols, P. **2000**. Advances in the genetics of thermophilic lactic acid bacteria. *Curr. Opin. Biotechnol.* **11**:497–504.

6 Garvie, E. **1960**. The genus *Leuconostoc* and its nomenclature. *J. Dairy Science* **27**:301–306.

7 Whittenbury, R. **1966**. A study of the genus *Leuconostoc*. *Archiv für Mikrobiologie* **53**:317–327.

8 Peterson, C. S. **1960**. Sauerkraut, In C. O. Chichester, E. M. Mrak, E. M. Steward (eds.) *Advances in Food Research*, Vol. 10. Academic Press, London.

9 Daeschel, M. A., Andersson, R. E., and Fleming, H. P. **1987**. Microbial ecology of fermenting plant material. *FEMS Microbiol. Rev.* **46**:357–367.

10 Sandine, W. E. and Elliker, P. R. **1970**. Microbially induced flavors and fermented foods – Flavor in fermented dairy products. *J. Agric. Food Chemistry* **18**:557–567.

11 Sutherland, I. W. **1996**. Extracellular polysaccharides, p. 613–657. In: H.-J. Rehm, G. Reed, A. Pühler & P. Stadler (eds.) *Biotechnology*, 2nd ed., Vol. 6: Products of Primary Metabolism. VCH, New York.

12 Simpson, W. J. and Taguchi, H. **1995**. The genus *Pediococcus*, with notes on the genera *Tetragenococcus* and *Aerococcus*, pp. 125–172. In, B. J. B. Wood and W. H. Holzapfel (eds). *The Genera of Lactic Acid Bacteria*. Chapman & Hall, London

13 Hammes, W. P., Bantleon, A., and Min, S. **1990**. Lactic acid bacteria in meat

fermentation. *FEMS Microbiol. Rev.* **87**:165–173

14 Fox, P. F., Lucey, J. A., and Cogan, T. M. **1990**. Glycolysis and related reactions during cheese manufacture and ripening. *Crit. Rev. Food Sci. Nutr.* **29**:237–253.

15 Daeschel, M. A. and Klaenhammer, T. R. **1985**. Association of a 13.6-megadalton plasmid in *Pediococcus pentosaceus* with bacteriocin activity. *Appl. Environ. Microbiol.* **50**:1528–1541.

16 Gonzales, C. F. and McKay, L. L. **1986**. Evidence for plasmid linkage of raffinose utilization and associated α-galactosidase and sucrose hydrolase activity in *Pediococcus pentosaceus*. *Appl. Environ. Microbiol.* **51**:105–109

17 Dicks, L. M., Dellaglio, F., and Collins M. D. **1995**. Proposal to reclassify *Leuconostoc oenos* as *Oenococcus oeni* [corrig.] gen. nov., comb. nov. *International Journal of Systematic Bacteriology* **45**:395–397.

18 Kunkie, R. E. **1991**. Some role of malolactic fermentation in wine making. *FEMS Microbiol. Rev.* **88**:55–72.

19 Versari, A., Parpinello, G. P. and Cattaneo, M. **1999**. *Leuconostoc oenos* and malolactic fermentation in wine: A review. *Journal of Industrial Microbiology & Biotechnology* **23**:447–455.

20 Hurst, A. **1973**. Microbial antagonism in foods. *Can. Inst. Food Sci. Technol. J.* **6**:80–90.

21 Geis, A., Singh, J., and Teuber, M. **1983**. Potential of lactic streptococci to produce bacteriocin. *Appl. Environ. Microbiol.* **45**:205–211.

22 Klaenhammer, T. R. **1993**. Genetics of bacteriocins produced by lactic acid bacteria. *FEMS Microbiol. Rev.* **12**:39-86.

23 Nes, I. F., Diep, D. B., Håvarstein. L. S., Brurberg, M. B., Eijsink, V., and Holo, H., **1996**. Biosynthesis of bacteriocins in lactic acid bacteria. *Antonie van Leeuwenhoek* **70**:17–32.

24 Vandenbergh, P. A. **1993**. Lactic acid bacteria, their metabolic products and interference with microbial growth. *FEMS Microbiol. Rev.* **12**:221–238.

25 Ryan, M. P., Rea, M. C., Hill, C. and Ross, R. P. **1996**. An application in cheddar cheese manufacture for a strain of *Lactococcus lactis* producing a novel broad-spectrum bacteriocin, lacticin 3147. *Appl. Environ. Microbiol.* **62**:612–619.

26 Ross, R. P., Galvin, M., McAuliffe, O., Morgan, S. M., Ryan, M. P., Twomey, D. P., Meaney, W. J. and Hill, C. **1999**. Developing applications for lactococcal bacteriocins. *Antonie van Leeuwenhoek* **76**:337–346.

27 Ross, R. P., Morgan, S. and Hill, C. **2002**. Preservation and fermentation: past, presence and future. *Int. J. Food Microbiol.* **79**:3–16.

28 Rodriguez, J. M., Martinez, M. I., Horn, N. and Dodd, H. M. **2002**. Heterologous production of bacteriocins by lactic acid bacteria. *Int. J. Food Microbiol.* **80**:101–116.

29 Klaenhammer, T. R. **1987**. Plasmid-directed mechanisms for bacteriophage defense in lactic streptococci. *FEMS Microbiology Reviews* **46**:313–325.

30 Hill, C. **1993**. Bacteriophage and bacteriophage resistance in lactic acid bacteria. *FEMS Microbiol. Lett.* **12**:132–142.

31 Josephsen, J. and Neve, H. **1998**. Bacteriophages and lactic acid bacteria, p. 385–436. In: S. Salminen & A. van Wright, (eds.). Lactic acid bacteria: Microbiology and functional aspects, 2nd ed. Marcel Dekker Inc., New York.

32 Forde, A. and Fitzgerald, G. F. **1999**. Bacteriophage defence systems in lactic acid bacteria. *Antonie van Leeuwenhoek* **76**:89–113.

33 Coffey, A. and Ross, R. P. **2002**. Bacteriophage-resistance systems in dairy starter strains: molecular analysis to application. *Antoine van Leeuwenhoek* **82**:303-321.

34 O'Sullivan, D., Coffey, A., Fitzgerald, G. F., Hill, C., and Ross, R. P. **1998**. Design of a phage-insensitive lactococcal dairy starter strain via sequential transfer of naturally occurring conjugative plasmids. *Appl. Environ. Microbiol.* **64**:4618–4622.

35 Hickey, R. M., Twomey, D. P., Ross, P., and Hill, C. **2001**. Exploitation of plas-

mid pMRC01 to direct transfer of mobilizable plasmids into commercial lactococcal starter strains. *Appl. Environ. Microbiol.* **67**:2853–2858.

36 Sørensen, K. J., R. Larsen, A. Kibenich, M. P. Junge, and E. Johansen. **2000**. A food-grade cloning system for industrial strains of *Lactococcus lactis*. *Appl. Environ. Microbiol.* **66**: 1253–1258.

37 Émond, É., Lavallée, R., Drolet, G., Moineau, S., and Lapointe, G. **2001**. Molecular characterization of a theta replication plasmid and its use for development of a two-component food-grade cloning system for *Lactococcus lactis*. *Appl. Environ. Microbiol.* **67**:1700–1709.

38 Hill, C., Miller, L. A., and Klaenhammer, T. R. **1990**. Cloning, expression, and sequence determination of a bacteriophage fragment encoding bacteriophage resistance in *Lactococcus lactis*. *J. Bacteriol.* **172**:6419–6426.

39 McGrath, S., Seegers, J. F., Fitzgerald, G. F. and van Sinderen, D. **1999**. Molecular characterization of a phage-encoded resistance system in *Lactococcus lactis*. *Appl. Environ. Microbiol.* **65**:1891–1899.

40 McGrath, S., Fitzgerald, G. F. and van Sinderen, D. **2001**. Improvement and optimization of two engineered phage resistance mechanisms in *Lactococcus lactis*. *Appl. Environ. Microbiol.* **67**:608–616.

41 Kim, S. G., and Batt, C. A. **1991**. Antisense mRNA-mediated bacteriophage resistance in *Lactococcus lactis* ssp. *lactis*. *Appl. Environ. Microbiol.* **57**:1109–1113.

42 O'Sullivan, D. J., S. A. Walker, S. G. West, and T. R. Klaenhammer. **1996**. Development of an expression strategy using a lytic phage to trigger explosive plasmid amplification and gene expression. *Bio/Technology* **14**: 82–87.

43 Djordjevic, G. M., O'Sullivan, D. J., Walker, S. A., Conkling, M. A., and Klaenhammer, T. R. **1997**. Bacteriophage-triggered defence systems: phage adaptation and design improvements. *Appl. Environ. Microbiol.* **63**:4370–4376.

44 Sturino, J. M. and Klaenhammer, T. R. **2002**. Expression of antisense RNA targeted against *Streptococcus thermophilus* bacteriophages. *Appl. Environ. Microbiol.* **68**:588–596.

45 De Vuyst, L. and Degeest, B.**1999**. Heteropolysaccharides from lactic acid bacteria. *FEMS Microbiol. Rev.* **23**:153–177.

46 van Kranenburg, R., Marugg, J. D., van Swam, I. I., Willem, N. J. and de Vos, W. M. **1997**. Molecular characterization of the plasmid encoded *eps* gene cluster essential for exopolysaccharide biosynthesis in *Lactococcus lactis*. *Mol. Microbiol.* **24**:387–397.

47 Stingele, F., Neeser, J. R. and Mollet, B. **1996**. Identification and characterization of the *eps* (Exopolysaccharide gene cluster from *Streptococcus thermophilus* Sfi6. *J. Bacteriol.* **178**:1680–1690.

48 Almiron-Roig, E., Mulholland, E., Gasson, M. J., and Griffin, A. M. **2000**. The complete *cps* gene cluster from *Streptococcus thermophilus* NCFB 2393 involved in the biosynthesis of a new exopolysaccharide. *Microbiology*, **146**:2793–2802.

49 Germond, J. E., Delley, M., DÀmico, N. and Vincent, S. J. **2001**. Heterologous expression and characterization of the exopolysaccharide from *Streptococcus thermophilus* Sfi39. *Eur. J. Biochem.* **268**:5149–5156.

50 Lamothe, T., Jolly, L., Mollet, B. and Stingele, F. **2002**. Genetic and biochemical characterization of exopolysaccharide biosynthesis by *Lactobacillus delbrueckii* subsp. *bulgaricus*. *Arch. Microbiol.* **178**:218–228.

51 Jolly, L. and Stingele, F. **2001**. Molecular organization and functionality of exopolysaccharide gene cluster in lactic acid bacteria. *Int. Dairy J.* **11**:733–745.

52 Jolly, L., Vincent, S. J. F., Duboc, P. and Neeser, J.-R. **2002**. Exploiting exopolysaccharides from lactic acid bacteria. *Antonie van Leeuwenhoek* **82**:367–374.

53 van Kranenburg, R., Vos, H. R., van Swam, I. I., Kleerebezem, M. and de Vos, W. M. **1999**. Functional analysis of glycosyltransferase genes from *Lactococcus lactis* and other gram-positive

cocci: Complementation, expression, and diversity. *J. Bacteriol.* 181:6347–6353.

54 Levander, F., Svensson, M. and Rådström, P. **2002**. Enhanced exopolysaccharide production by metabolic engineering of *Streptococcus thermophilus*. *Appl. Environ. Microbiol.* 68:784–790.

55 Christensen, J. E., Dudley, E. G., Pederson, J. A. and Steele, J. L.. **1999**. Peptidases and amino acid catabolism in lactic acid bacteria. *Antonie van Leeuwenhoek* **76**: 217–246.

56 Kok, J., and W. M. de Vos. **1994**. The proteolytic system of lactic acid bacteria. pp. 169–210. In: Gasson, M. J., and W. M. de Vos (eds.) *Genetic and biotechnology of lactic acid bacteria.* Blackie and Professional, London.

57 Kunji, E. R. S., J. Mierau, A. Hagting, B. Poolman, and W. N. Konings. **1996**. The proteolytic system of lactic acid bacteria. *Antonie van Leeuwenhoek* **70**: 87–221.

58 Mierau, I., E. R. S. Kunji, G. Venema, and J. Kok. **1997**. Casein and peptide degradation in lactic acid bacteria. *Biotechnol. Genet. Engin. Rev.* 14:279–301.

59 Siezen, R. J. **1999**. Multi-domain, cell-envelope proteinases of lactic acid bacteria. *Antonie van Leeuwenhoek* **76**: 139–155.

60 van der Guchte, M., Kodde, J., van der Vossen, J. M. B. M., Kok, J. and Venema, G. **1990**. Heterologous gene expression in *Lactococcus lactis* subsp. *lactis*: Synthesis, secretion, and processing of *Bacillus subtilis* neutral protease. *Appl. Environ. Microbiol.* 56:2606–2611.

61 Leenhouts, K., Bolhuis, A., Venema, G. and Kok, J. **1998**. Construction of a food-grade multiple-copy integration system for *Lactococcus lactis*. *Appl. Environ. Microbiol.* 49: 417–423.

62 Joutsjoki, V., Luoma, S., Tamminen, M., Johansen, E. and Palva, A. **2002**. Recombinant Lactococcus starters as a potential source of additional peptidolytic activity in cheese ripening. *Appl. Microbiol.* 92:1159–1166.

63 Meisel, H. and Bockelmann, W. **1999**. Bioactive peptides encrypted in milk proteins: proteolytic activation and tropho-functional properties. *Antonie van Leeuwenhoek* 76:207–215.

64 Gobbetti, M., Ferranti, P., Smacchi, E., Goffredi, F. and Addeo, F. **2000**. Production of angiotensin-I-converting-enzyme-inhibitory peptides in fermented milks started by *Lactobacillus delbrueckii* subsp. *bulgaricus* and *Lactococcus lactis* subsp. *cremoris* FT4. *Appl. Environ. Microbiol.* 66:3898–3904.

65 Meisel, H. **1993**. Casokinins as inhibitors of angiotensin-converting-enzyme. pp.153–159. In G. Sawatzki and B. Renner (eds.). *New perspectives in infant nutrition.* Thieme, Stuttgart, Germany.

66 Takano, T. **2002**. Anti-hypertensive activity of fermented dairy products containing biogenic peptides. *Antonie van Leeuwenhoek* 82:333–340.

67 Hugenholtz, J., Kleerebezem, M., Starrenburg, M., Delcour, J., de Vos, W. and Hols, P. **2000**. *Lactococcus lactis* as a cell factory for high-level diacetyl production. *Appl. Environ. Microbiol.* 66:4112–4114.

68 Chaves, A. C. S. D., Fernandez, M., Lerayer, A. L. S., Mierau, I., Kleerebezem, M. and Hugenholtz, J. **2002**. Metabolic engineering of acetaldehyde production by *Streptococcus thermophilus*. *Appl. Environ. Microbiol.* 68:5656–5662.

69 Hols, P., Kleerebezem, M., Schank, A. N., Ferain, T., Hugenholtz, M., Delcour, J. and de Vos, W. **1999**. Conversion of *Lactococcus lactis* from homolactic to homoalanine fermentation through metabolic engineering. *Nature Biotechnol.* 17:588–592.

70 Hugenholtz, J., Sybesma, W., Groot, M. N., Wisselink, W., Ladero, V., Burgess, K., van Sinderen, D., Piard, J.-C., Eggink, G., Smid, E. J., Savoy, G., Sesma, F., Jansen, T., Hols, P. and Kleerebezem, M. **2002**. Metabolic engineering of lactic acid bacteria for the production of nutraceuticals. *Antonie van Leeuwenhoek* 82:217–235.

71 Duwat, P., Cesselin, S. S. and Gruss, A. **2000**. *Lactococcus lactis*, a bacterial

72 van der Guchte, M., Serror, P., Chervaux, C., Smokvina, T., Ehrlich, S. D. and Maguin, E. 2002. Stress response in lactic acid bacteria. *Antonie van Leeuwenhoek.* **82**: 187–216.

73 Guzzo, J., Calvin, J. and Divies, C. 1994. Induction of stress proteins in *Leuconostoc oenos* to perform direct inoculation of wine. *Biotechnol. Lett.* **16**:1189–1194.

74 Gouesbet, G., Jan, G. and Boyaval, P. 2002. Two-dimensional electrophoresis study of *Lactobacillus delbrueckii* subsp. *bulgaricus* thermotolerance. *Appl. Environ. Microbiol.* **68**:1055–1063.

75 Rallu, F., Gruss, A. Ehrlich, S. D. and Maguin, E. 2000. Acid- and multistress-resistant mutants of *Lactococcus lactis*: identification of intracellular stress signals. *Mol. Microbiol.* **35**: 517–528.

76 Mayo, B., Derzelle, S., Fernández, M., Léonard, C., Ferain, T., Hols, P., Suárez, J. E. and Delcour, J. 1997. Cloning and characterization of *cspL* and *cspP*, two cold-inducible genes from *Lactobacillus plantarum*. *J. Bacteriol.* **179**:3039–3042.

77 Wouters, J. A., Rombouts, F. M., de Vos, W. M., Kuipers, O. P. and Abee, T. 1999. Cold shock proteins and low-temperature response of *Streptococcus thermophilus* CNRZ302. *Appl. Environ. Microbiol.* **65**:4436–4442.

78 Wouters, J. A., Frenkiel, H., de Vos, W. M., Kuipers, O. and Abee, T. 2001. Cold shock proteins of *Lactococcus lactis* MG1363 are involved in cryoprotection and in the production of cold-induced proteins. *Appl. Environ. Microbiol.* **67**:5171–5178.

79 Hertel, C., Schmidt, G., Fischer, M., Oellers, K. and Hammes, W. 1998. Oxygen-dependent regulation of the expression of the catalase gene *katA* of *Lactobacillus sakei* LTH677. *Appl. Environ. Microbiol.* **64**:1359–1365.

80 Harlander, S. K. 1987. Transformation of *Streptococcus lactis* by electroporation, p. 229–233. In: J. J. Ferretti and R. Curtiss III (eds), *Streptococcal genetics*. American Society for Microbiology, Washington, D. C.

81 Holo, H. and I. F. Nes. 1989. High-frequency transformation by electroporation of *Lactococcus lactis* subsp. *cremoris* grown with glycine in osmotically stabilized media. *Appl. Environ. Microbiol.* **55**:3119–3123.

82 Powell, I. B., M. G. Achen, A. J. Hillier, and B. E. Davidson. 1988. A simple and rapid method for genetic transformation of lactic streptococci by electroporation. *Appl. Environ. Microbiol.* **54**:655–660.

83 Serror, P., Sasaki, T., Ehrlich, S. D. and Maguin, E. 2002. Electrotransformation of *Lactobacillus delbrueckii* subsp. *bulgaricus* and *Lactobacillus delbrueckii* subsp. *lactis* with various plasmids. *Appl. Environ. Microbiol.* **68**:46–52.

84 De Vos, W. M. 1986. Gene cloning in lactic streptococci. *Neth. Milk Dairy J.* **40**:141–154.

85 Kok, J., van der Vossen, J. M. B. M. and Venema, G. 1984. Construction of plasmid cloning vectors for lactic streptococci which also replicate in *Bacillus subtilis* and *Escherichia coli*. *Appl. Environ. Microbiol.* **48**:726–731.

86 Leenhouts, K., Bolhuis, A., Boot, J., Deutz, I., Toonen, M., Venema, G., Kok, J. and Ledeboer, A. 1998. Cloning, expression, and chromosomal stabilization of the *Propionibacterium shermanii* proline iminopeptidase gene (*pip*) for food-grade application in *Lactococcus lactis*. *Appl. Environ. Microbiol.* **64**:4736–4742.

87 Maguin, E., H. Prevost, S. D. Ehrlich, and A. Gruss. 1996. Efficient insertional mutagenesis in lactococci and other Gram-positive bacteria. *J. Bacteriol.* **178**:931–935.

88 Poquet, I., S. D. Ehrlich, and A. Gruss. 1998. An export-specific reporter designed for Gram-positive bacteria: application to *Lactococcus lactis*. *J. Bacteriol.* **180**:1904–1912.

89 Simon, D., and A. Chopin. 1988. Construction of a vector plasmid family and its use for molecular cloning in *Streptococcus lactis*. *Biochimie* **70**:559–566.

90 van der Guchte, M., J. M. B. M. van der Vossen, J. Kok, and G. Venema. **1989**. Construction of a lactococcal expression vector: expression of hen egg white lysozyme in *Lactococcus lactis*. *Appl. Environ. Microbiol.* **55**:224–228.

91 van der Vossen, J. M. B. M., J. Kok, and G. Venema. **1985**. Construction of cloning-, promoter and terminator screening shuttle vectors for *Bacillus subtilis* and *Streptococcus lactis*. *Appl. Environ. Microbiol.* **50**:540–542.

92 De Ruyter, P. G. G. A., O. P. Kuiper, and W. M. de Vos. **1996**. Controlled gene expression for *Lactococcus lactis* with the food-grade inducer Nisin. *Appl. Environ. Microbiol.* **62**:3662–3667.

93 Piard, J. C., I. Hauteford, V. A. Fischetti, S. D. Ehrlich, M. Fons, and A. Gruss. **1997**. Cell wall anchoring of the *Streptococcus pyogenes* M6 protein in various lactic acid bacteria. *J. Bacteriol.* **179**:3068–3072.

94 Leenhouts, K., Buist, G. and Kok, J.. **1999**. Anchoring of proteins to lactic acid bacteria. *Antonie van Leeuwenhoek* **76**:367–376.

95 Norton, P. M., R. W. F. LePage, and J. M. Wells. **1995**. Progress in the development of *Lactococcus lactis* as a recombinant mucosal vaccine delivery system. *Folia Microbiol.* **40**:225–230.

96 Robinson, K., L. M. Chamberlain, K. M. Schofield, J. K. Wells, and R. W. F. LePage. **1997**. Oral vaccination of mice against tetanus with recombinant *Lactococcus lactis*. *Nature Biotechnol.* **15**:653–657.

97 Steidler L., J. Viaene, W. Fiers, and E. Remaut. **1998**. Functional display of a heterologous protein on the surface of *Lactococcus lactis* by means of the cell wall anchor of *Staphylococcus aureus* protein A. *Appl. Environ. Microbiol.* **65**:342–345.

98 Buist, G. **1997**. AcmA of *Lactococcus lactis*, a cell-binding major autolysin. Ph. D. thesis, University of Groningen, Haren, The Netherlands.

99 Franke, C. M. **1998**. Topology of a type I secretion system for bacteriocins of *Lactococcus lactis*. Ph. D. thesis, University of Groningen, Haren, The Netherlands.

100 Norton, P. M., H. W. G. Brown, J. M. Wells, A. M. Macpherson, P. W. Wilson, and R. W. F. LePage. **1996**. Factors affecting the immunogenicity of tetanus toxin fragment C expressed in *Lactococcus lactis*. *FEMS Immunol. Med. Microbiol.* **14**:167–177.

Part II
Legislation in Europe

6
The Legal Situation for Genetically Engineered Food in Europe

Rudolf Streinz / Lars O. Fuchs

6.1
Introduction

6.1.1
The Need for Regulation

In 1985, in the context of its so-called "new strategy", the Commission mentioned both "certain bio-technological techniques" and "other techniques and treatments" as some of those areas of law which required harmonization. Thereby, it intended to secure the functioning of the common market by standardizing the relevant national rules which could be changed with regard to art. 30 TEC if they protected public health, but nevertheless might create conditions of unfair competition (cf. recitals 1 and 2 NFR). The Netherlands and Denmark, for example, had statutory provisions containing rules concerning "novel foods", in particular genetically modified foodstuffs. In the UK, too, there existed a voluntary agreement between the food industry and the appropriate authorities, which could be declared binding in accordance with the Food Safety Act 1990.

6.1.2
The History of the Novel Food Regulation

Although the draft proposal for a corresponding regulation (OJ (EC) 1992, C 190/3) which the EC-Commission (henceforth: Commission) presented in 1992, after years of discussions, was only the eleventh version, it was widely criticized – especially by the European Parliament. The criticism focused in particular on the rules concerning the scope of application and the labeling. Unfortunately, the Commission considered this only partially when it presented its amended proposal for a Regulation (EC) of the European Parliament and of the Council concerning novel foods and food ingredients (OJ (EC) 1994, C 16/10). Therefore, the following legislative procedures proved to be awkward: only a compromise put forward by the French Presidency enabled the Council to eventually formulate a common position which was adopted on 23 October 1995 by a qualified majority against the votes of Denmark, Germany, Austria and Sweden (OJ (EC) 1995, C 320/1). These four Member States regarded the rules on labeling as insufficient. This criticism was reflected during the joint legislative procedure in the EP (art. 251 TEC): Despite general acceptance of large parts of the proposal, the EP requested a number of amendments, especially regarding the question of labeling. These were in turn rejected by the Commission. Finally, a compromise proposal which originated in the Conciliation Committee on the 27 November 1996 under the Irish presidency was accepted by both the Council and the EP. The Novel Food Regulation (Regulation (EC) No 258/97 of the EP and of the Council of 27 January 1997 concerning novel foods and novel food ingredients, OJ (EC) 1997, 43/1), which entered into force on the 15 May 1997, supersedes within its scope of applicability the relevant national rules as pre-eminent and directly applicable law.

This chronological and content-related development of the history of the Novel Food Regulation already shows both the need for such rules and the explosive force inherent in this area of law and the controversial positions on the dealings with novel foods and food ingredients.

6.2
Status Quo

6.2.1
The Novel Food Regulation

6.2.1.1 Introduction
Beyond the legal definition of novel foods and food ingredients – which is used in the regulation to define its scope of application – the Novel Food Regulation contains rules concerning the safety requirements for these foodstuffs and food ingredients, rules on their placing on the market and on their labeling and provisions concerning the relation of the Novel Food Regulation and other legislation.

Since the Novel Food Regulation does not contain any rules on its implementation, these rules are part of the laws of the Member States, in Germany in the *Statutory Instrument for the Implementation of Community Legislation concerning Novel Foods and Food Ingredients and the Labeling of Products made from Genetically Modified Soya Beans and Genetically Modified Maize as well as the Labeling of Foodstuffs which have been Produced Without Means of Genetic Engineering* (Novel Foods and Food Ingredients Instrument – NFI), as re-published on 14 February 2000 (Bundesgesetzblatt, Federal Law Gazette 2000, part. I, p. 123; cf. *infra* chapter 2, section 6).

Since the Novel Food Regulation could not include in its scope of application other products made from maize or soya beans, in relation to which the use of methods of genetic engineering was very advanced and which therefore could not be regarded as "novel" (cf. art. 2 para. 2 NFR and *infra* section 6.2.1.2), these required specific rules on labeling. These were contained in the so-called "Supplementary Regulation" (EC) No 1813/97 of 19 September 1997 concerning the compulsory indication on the labeling of certain foodstuffs produced from genetically modified organisms of particulars other than those provided for in Directive 79/112/EEC (OJ (EC) 1997, L 257/7), which only extends the rules on labeling as contained in the Novel Food Regulation to these products. The "Supplementary Regulation" itself therefore required implementation. The Council Regulation (EC) No 1139/98 of 26 May 1998 concerning the compulsory indication of the labeling of certain foodstuffs produced from genetically modified organisms of particulars other than those provided for in Directive 79/112/EEC (OJ (EC) 1998, L 159/4), made to achieve this purpose, is though less of an implementation regulation but rather replaces the "Supplementary Regulation" (cf. art. 3 of Regulation (EC) No. 1139/98). It is therefore generally and also here referred to as "Replacement Regulation" (cf. *infra* section 6.2.4).

Despite the long time spent on the Novel Food Regulation, a good number of its rules are unclear and still meet with criticism. Even though a new version taking account of this is in the process of preparation (cf. *infra* section 6.3), this chapter first presents the contents of the regulation currently in force.

6.2.1.2 Scope of application

According to art. 1 para. 2 NFR, the Novel Food Regulation applies to foods and food ingredients not hitherto used for human consumption to a significant degree within the Community:

- containing or consisting of genetically modified organisms within the meaning of Directive 90/220/EEC (art. 1 para. 2 lit. a NFR),
- produced from, but do not contain, genetically modified organisms (art. 1 para. 2 lit. b NFR),
- with a new or intentionally modified primary molecular structure (art. 1 para. 2 lit. c NFR),
- consisting of or have been isolated from micro-organisms, fungi or algae (art. 1 para. 2 lit. d NFR),
- consisting of or isolated from plants and food ingredients isolated from animals, except for foods and food ingredients obtained by traditional propagating or breeding practices and having a history of safe food use (art. 1 para. 2 lit. e NFR),
- to which has been applied a production process not currently used, where that process gives rise to significant changes in the composition or structure of the foods or food ingredients which affect their nutritional value, metabolism or level of undesirable substances (art. 1 para. 2 lit. f NFR).

Already the pre-requisite "not hitherto [...] to a significant degree", which is intended to define the novelty of a relevant product, shows a major flaw of the Novel Food Regulation: the use of many indeterminate legal terms.

With regard to the novelty the Commission holds the opinion, that the industry is under a duty to present the relevant information and proof, thereby shifting the assessment risk – which is already high due to the imprecise definition – effectively to the person placing the product on the market. He (or she) is obliged to assess autonomously, whether his/her product qualifies as "novel food" and to which type of procedure it is subject (either the notification procedure according to art. 3 para. 4 read in conjunction with art. 5 NFR, cf. *infra* section 6.2.1.4(b), or the more complex, two-level authorization procedure according to art. 6 and 7 NFR, cf. *infra* section 6.2.1.4(a)). In addition to the cost of the assessment, this causes further economic consequences, for example with respect to delays for the entry into the market or the competitive advantages of competitors who assess their respective products differently [1].

The definition contained in art. 1 para. 2 NFR covers not exclusively, but certainly also genetically modified foodstuffs and food ingredients, that is the genetically modified "FlavrSavr"-Tomato as well as the genetically not modified fat replacement "Olestra". Food additives (art. 2 para. 1 lit. a NFR), flavorings for use in foodstuffs (art. 2 para. 1 lit. b NFR) and extraction solvents used in the production of foodstuffs (art. 2 para. 1 lit. c NFR) on the other hand are explicitly excluded from the scope of the Novel Food Regulation. These are covered by the rules of the relevant specialized regulations, while the Commission is under a duty to ensure that the safety levels laid down in those directives correspond to the safety level of the Novel Food Regulation, which might be relevant for example with regard to the labeling of such products (cf. *infra* section 6.2.1.6)

6.2.1.3 Requirements for novel foods and food ingredients

The novel foods and food ingredients covered by the Novel Food Regulation must not – in accord with the general principles of food law – present a danger for the consumer or mislead him or her. Likewise, they must not differ from foods or food ingredients which they are intended to replace to such an extent that their normal consumption would be nutritionally disadvantageous for the consumer.

6.2.1.4 Procedures

Food law is based on the so-called "abuse principle" (Mißbrauchsprinzip), according to which foodstuffs may be placed on the market autonomously, that is without previous authorization, if they meet the statutory requirements (under German law especially those of the protection of health, art. 8 Lebensmittel- und Bedarfsgegenständegesetz (LMBG – Food Act), and of the prohibition of deception, art. 17 LMBG). In this respect, a general authorization to place foodstuffs on the market, coupled with an corresponding power of prohibition reserved for the authorities, has been put in place. This principle gives way to the so-called prohibition principle (Verbotsprinzip) whenever specific needs of the preventive protection of health require this. The prohibition principle therefore provides for a general prohibition with reserved powers of authorization, forbidding everything which has not been explicitly permitted in the course of an authorization procedure.

The Novel Food Regulation itself distinguishes between two different procedures with respect to placing novel foods and food ingredients on the market: According to art. 3 para. 2 NFR, novel products are generally subject to the so-called authorization procedure provided for in art. 4, 6, 7 and 8 NFR. Differing from this rule, foodstuffs or food ingredients as defined in art. 1 para. 2 lit. b, lit. d and lit. e NFR, which, on the basis of the scientific evidence available and generally recognized or on the basis of an opinion delivered by one of the competent bodies referred to in art. 4 para. 3, are substantially equivalent to existing foods or food ingredients with regard to their composition, nutritional value, metabolism, intended use and the level of undesirable substances contained therein, are subject to the so-called notification procedure provided for in art. 5 NFR (see also the possibility of a definition using the procedures according to art. 13 NRF). The different procedures depend on their relevance for health as required by the principle of proportionality. But again, the indeterminate legal term "substantially equivalent" poses numerous problems of interpretation and is the target of severe criticism. At first glance, the legal definition of "no longer equivalent" contained in art. 8 para. 1 lit. a subpara. 2 NFR might seem to alleviate this problem, but this reference would overlook the shading of the terms "no longer equivalent" (art. 8 para. 1 lit. a to c and para. 2 NFR) and "substantially equivalent" (art. 3 para. 4 subpara. 1 NFR). Incidentally, art. 8 para. 1 lit. a subpara. 2 NFR restricts the definition to the term "no longer equivalent for the purpose of this article".

(a) **The authorization procedure**
According to art. 4 para. 1 sentence 1 NFR, the applicant shall submit his/her request for authorization to the Member State in which the product is to be placed on the market for the first time, at the same time forwarding a copy of the request to the Commission, art. 4 para. 1 sentence 2 NFR. The mandatory content of the request is regulated by art. 6 para. 1 NFR, which requires in particular proof that the product complies with the criteria laid down in art. 3 para. 1 NFR. Likewise, an appropriate proposal for the labeling of the product must be suggested. Finally, according to art. 6 para. 1 sentence 2 NFR the request shall be accompanied by a summary of the dossier. The Commission forwards to the Member States a copy of this summary (art. 6 para. 2 subpara. 2 NFR).

In the context of the initial assessment the member state, according to art. 6 para. 2 subpara. 1 NFR, notifies the Commission of the name of the competent food assessment body responsible for preparing the initial assessment report, or alternatively asks the Commission to propose a suitable body. The Commission for that purpose maintains a list of those food assessment bodies, of which the Member States have notified it according to art. 4 para. 3 NFR.

Within a period of three months from receipt of a request, according to art. 6 para. 3 NFR, the Commission must be informed by an initial assessment report whether or not the product requires additional assessment in accordance with art. 6. This report, too, is forwarded to the other Member States (art. 6 para. 4 subpara. 1 sentence 1 NFR), which may make comments or present a reasoned objection (art. 6 para. 4 subpara. 1 sentence 2 NFR), that in turn are forwarded by the

Commission to the other Member States (art. 6 para. 4 subpara. 2 NFR). Where such an objection is raised, art. 7 para. 1 NFR additionally requires an authorization decision to be taken in accordance with the procedure laid down in art. 13 NFR.

Following the procedure referred to in art. 6 para. 4 NFR, the Member State which has received the request shall inform the applicant without delay of its result. Where the additional assessment is not required, and no reasoned objection has been presented the applicant may place the product concerned on the market. Where however a decision of the Commission is necessary, this decision must define the scope of the authorization and must establish, where appropriate, the conditions of use of the product, its designation, its specification, and its specific labeling requirements, art. 7 para. 2 NFR. According to art. 7 para. 3 sentence 1 NFR, the Commission shall without delay inform the applicant of the decision taken; in addition the decision, according to art. 7 para. 3 sentence 2 NFR, has to be published in the Official Journal of the European Communities. Where a decision must be taken, the Commission is assisted by the Standing Committee for Foodstuffs, art. 13 para. 1 NFR. For details of this procedure which has been regulated in minute detail reference should be made to the wording of art. 13 NFR.

(b) **The notification procedure**

Those foodstuffs and food ingredients which are named in art. 3 para. 4 NFR however are subject to the simpler notification procedure provided for in art. 5 NFR. Here, the applicant is obliged to notify, giving the relevant details provided for in art. 3 para. 4 NFR, the Commission of the placing on the market when he/she does so, art. 5 subpara. 1 sentence 1 and 2 NFR. The Commission forwards to the Member States a copy of that notification (and, if requested, a copy of the relevant details) within sixty days. Each year, a summary of the notifications is published in the "C" series of the Official Journal of the European Communities, art. 5 subpara. 2 NFR. However, the rules on labeling contained in art. 8 NFR also apply to the products which are subject to the simpler notification procedure (cf. the following section).

6.2.1.5 Labeling

(a) **Conditions**

In addition to the standard requirements of Community law concerning the labeling, art. 8 NFR contains additional specific labeling requirements for novel foods and food ingredients. Art. 8 para. 1 lit. a to c and para. 2 NFR all refer to the condition that a novel food or food ingredient is no longer equivalent to an existing food or food ingredient. According to the legal definition contained in art. 8 para. 1 lit. a subpara. 2 NFR, a novel food or food ingredient is deemed to be no longer equivalent if scientific assessment, based upon an appropriate analysis of existing data, can demonstrate that the characteristics assessed are different in comparison with a conventional food or food ingredient, having regard to the accepted limits of natural variations for such characteristics.

The problems arising from this definition are obvious:

- First, the accepted limits of natural variations of conventional foodstuffs must be determined, since differences, which lie within these limits, are irrelevant.
- Differences found must then be scientifically assessed, especially with regard to their nutritional relevance. But such an assessment offers a certain scope for evaluation, allowing for different results.
- Finally, the scientific assessment must be based upon an appropriate analysis of existing data. It is unclear when such an analysis is appropriate and the question remains unanswered to what extent an improvement of existing methods of analysis must be taken into account in assessing novel food and food ingredients (on this cf. also art. 12 NFR and *infra* section 6.2.1.6(b)).

All these different examples of scope for interpretation make it very difficult to determine whether there is a duty of labeling, which leads to corresponding risks for the person placing the product on the market. There might be a certain "advantage" for those, who follow the authorization procedure (cf. *supra* section 6.2.1.1.4(a)): if an additional assessment according to art. 7 NFR is carried out, the resulting decision must establish, according to art. 7 para. 2 last indent NFR, specific labeling requirements as referred to in art. 8 NFR. However, already when requesting the authorization the person placing the product on the market is, according to art. 6 para. 1 sentence 1 NFR, under a duty to make an appropriate proposal for the presentation and labeling of the novel product.

(b) **The elements of the label**
Where an appropriate label is required it must show the following elements:

- A list of the characteristics or properties modified, together with the method by which that characteristic or property was obtained if the product is no longer equivalent to existing products (art. 8 para. 1 lit. a subpara. 3 NFR)
- A list of any characteristic or food property such as composition, nutritional value or nutritional effects, or intended use of the food, which renders a novel food or food ingredient no longer equivalent to an existing food or food ingredient (art. 8 para. 1 lit. a subpara. 1 NFR).
- The name of any material which is not present in an existing equivalent foodstuff and which may have implications for the health of certain sections of the population (for example persons suffering from an allergy) (art. 8 para. 1 lit. b NFR);
- The name of any material which is not present in an existing equivalent foodstuff and which gives rise to ethical concerns; (art. 8 para. 1 lit. c NFR)
- A list of any organism genetically modified by techniques of genetic modification, the nonexhaustive list of which is laid down in Annex I A, Part 1 of Directive 90/220/EEC (art. 8 para. 1 lit. d NFR).

In case that there is no existing equivalent product, according to art. 8 para. 2 NFR appropriate provisions shall be adopted where necessary in order to ensure that consumers are adequately informed of the nature of the product.

(c) **Examples**

Labeling is therefore required for example in the case of:

- vegetable oils, the fatty acid structure of which has been altered due to genetic engineering;
- rice, from which an allergenic protein has been removed to reduce allergies; and
- generally all products which contain a new protein.

Although the ethical concerns mentioned in art. 8 para. 1 lit. c NFR are rather difficult to define, they will certainly cover concerns, which are based on religious or other belief e.g. those of vegetarians (for example the transfer of pig genes to other animals or microorganisms or the transfer of animal genes to plants).

No labeling is however required for example in the case of:

- sugar which has been extracted from rhizomania-resistant sugar beet where the genetic modification of the raw fruit has no consequences in the final product;
- similar yeasts, which are used in beer brewing;
- products, which have been produced by means of genetic engineering but do not contain any genetically modified organisms themselves (for example, chymosin made from genetically modified microorganisms which is used in cheese production or the cheese so produced itself).

While according to art. 8 para. 1 lit. d NFR a genetically modified organism such as the "FlavrSavr"-Tomato must be labeled, this is questionable for purée made from it, as the puréeing might have destroyed the foreign genes which therefore might no longer be present. However, there might still be a duty to label it according to art. 8 para. 1 lit. a NFR if the purée produced in such a way is not "equivalent" to conventional tomato purée.

6.2.1.6 Other questions

(a) **Additives and flavorings**

Food additives are, according to art. 2 para. 1 lit. a NFR, excepted from the scope of application of the Novel Food Regulation, as are flavorings for use in foodstuffs according to art. 2 para. 1 lit. b NFR (cf. *supra* section 6.2.1.2).

Commission Regulation (EC) No 50/2000 on the labeling of foodstuffs and food ingredients containing additives and flavorings that have been genetically modified or have been produced from genetically modified organisms now contains specific rules on the labeling of these substances. Such regulation had become necessary both to comply with the consumer's interest in appropriate information (recital 4, 10 and 13 of Regulation (EC) No 50/2000) and to forestall Member States' provisions that are likely to create new obstacles to *infra*-Community trade (recital 5 and 6 of Regulation (EC) No 50/2000). Since separate measures imposing similar labeling requirements are envisaged for additives and flavorings, which are sold as such to the final consumer, these were expressly excluded from the scope of application of Regulation (EC) No 50/2000 (cf. also recital 9 of Regulation (EC) No 50/2000).

The conditions and the extent of the labeling of foodstuffs that contain the relevant additives and flavorings, are almost identical to those provided for in the relevant parts of the Novel Food Regulation (cf. *supra* sections 6.2.1.5(a) and 6.2.1.5(b)), art. 2 of Regulation (EC) No 50/2000 prescribes the information to be provided in case the characteristics named in it apply. Art. 4 of Regulation (EC) No 50/2000 explicitly regulates the manner of labeling and art. 3 of Regulation (EC) No 50/2000 defines the term "no longer equivalent". Differing from art. 8 para. 1 lit. a subpara. 2 NFR (cf. *supra* section 6.2.1.5(a)), art. 3 para. 2 of Regulation (EC) No 50/2000 states that the necessary differences in the characteristics between relevant additives or flavorings and conventional additives or flavorings are given, where the specified additives or flavorings contain protein and/or DNA resulting from genetic modification.

(b) **New insights**

According to art. 12 para. 1 sentence 1 NFR, the Member States may restrict or suspend the trade in and use of novel food or food ingredients complying with the Novel Food Regulation in their territory. The condition therefore is that the member state has, as a result of new insights, detailed grounds for considering that the use of the relevant product endangers human health or the environment. It must immediately inform the other Member States and the Commission thereof, giving the grounds for its decision, art. 12 para. 1 sentence 2 NFR.

The Commission, according to art. 12 para. 2 sentence 1 NFR is then obliged both to examine the grounds given by the member state within the Standing Committee for Foodstuffs and to take the appropriate measures in accordance with the procedure laid down in art. 13.

(c) **Sanctions permitted under the Novel Food Regulation**

The Novel Food Regulation itself neither contains rules for sanctions, nor does it provide for a duty of the Member States to establish such rules. Nevertheless, their general duty of loyalty arising from art. 10 TEC requires the Member States to take all necessary steps to ensure the authority and effectiveness of community law. In particular, they must sanction breaches of community law or national law in an effective, proportional and deterrent manner [2]. Under German law this is (mainly) done in the Novel Foods and Food Ingredients Instrument (cf. *infra* section 6.2.6).

6.2.2
Problems

6.2.2.1 **Negative labeling**

Recital 10 of the Novel Food Regulation expressly states that nothing shall prevent a supplier from informing the consumer on the labeling of a food or food ingredient that the product in question is not a novel food within the meaning of the Novel Food Regulation, or that the techniques indicated in art. 1 para. 2 NFR were not used in the production of that food or food ingredient. This so-called "negative labeling" (here, the term "negative" does not imply any negative evaluation, but refers only to a lack of novelty of a foodstuff or of a *novel* food ingredient or the fact that the techniques indicated in art. 1 para. 2 NFR have not been used) is not mentioned again in the Novel Food Regulation. With a view to European law it must be observed that such labeling – in accord with the general principles – must not breach primary community law, especially art. 28 TEC. While it was unclear for some time, which precise requirements such labeling was subject to, these have been provided for in German law by the Novel Foods and Food Ingredients Instrument (NFI) (cf. *infra* section 6.2.6.3).

6.2.2.2 Compliance with World Trade law

The Agreement Establishing the World Trade Organization (WTO) of 15 April 1994 has considerable influence on food law through the Agreement on the Application of Sanitary and Phytosanitary Measures (SPS-Agreement) and the Agreement on Technical Barriers to Trade (TBT-Agreement) which were concluded within its framework and through their combination with the rules of the Codex Alimentarius Commission. In the context of the Novel Food Regulation, problems arise in particular from the extent of its scope of application, from the requirements that transcend those of the Deliberate Release Directive and the large number of indeterminate legal terms. According to the USA, especially rules on labeling breach WTO rules since the sheer existence of protein and/or DNA resulting from genetic modification could not lead to a lack of "equivalence". In particular, the big differences in the approach to genetically modified organisms – an increasing part of genetically modified agrarian products and the enlargement of the corresponding area of cultivation on the side of the USA versus an authorization moratorium on the side of the EU – provoked tensions in the past. On the topic of the current proposals for regulation of the EC, see *infra* section 6.3.4.

6.2.2.3 Competent authorities in the member states

Competent authorities to receive requests according to art. 4 para. 1 NFR and food assessment bodies according to art. 4 para. 3 NFR in Germany are according to art. 1 para. 1 NFI (which was made by the Federal Minister for Health with the consent of the Federal Council [Bundesrat]) the Robert Koch-Institut, PO Box 650280, D-13303 Berlin, for novel foods and novel food ingredients referred to in art. 1 para. 2 lit. a NFR (s. 1 para. 1 no. 1 NFI) and the Bundesinstitut für gesundheitlichen Verbraucherschutz und Veterinärmedizin, PO Box 33013, 14191 Berlin, for foodstuffs and food ingredients referred to in art. 1 para. 2 lit. b to f NFR (s. 1 para. 1 no 2 NFI). The latter is according to art. 1 para. 2 NFI also competent to deliver the opinion as to the substantial equivalence of products referred to in art. 3 para. 4 NFR.

A table of the authorities competent to implement the Novel Food Regulation in the other Member States may be found in Table 6.1.

Table 6.1. List of national authorities responsible for the implementation of European Parliament and Council Regulation (EC) No. 258/97 concerning novel foods and novel food ingredients.

Country	Competent authority for the submission of applications, art. 4 para. 1 NFR	Food assessment bodies, art. 4 para. 3 NFR
Austria	Bundesministerium für Soziale Sicherheit und Generationen (BMSG) Sektion IX/B Radetzkystrasse 2 A-1030 WIEN	Bundesanstalt für Lebensmittel- untersuchung und -forschung Kinderspitalgasse 15 A-1090 WIEN
Belgium	Inspection Générale des Denrées Alimentaires Cité Administrative de l'Etat Quartier Esplanade – 11ème étage Boulevard Pachéo 19, boîte 5 B-1010 BRUXELLES	Conseil supérieur d'Hygiène Publique Cité Administrative de l'Etat Quartier Esplanade – 7ème étage Boulevard Pachéo 19, boîte 5 B-1010 BRUXELLES Hoge Gezondheidsraad RAC-Esplanadegebouw 7de verdieping Pachecolaan 19, bus 5 B-1010 BRUXELLES **For novel foods and novel food ingredients referred to in Article 1 (2), (a) and (b):** Conseil de Biosécurité Secrétariat Biosécurité et Biotechnologie Institut Scientifique de la santé publique – Louis Pasteur rue Juliette Wytsman, 14 B-1050 BRUXELLES
Denmark	Ministeriet for Fødevarer, Landbrug og Fiskeri Fødevaredirektoratet Mørkhøj Bygade 19 DK-2860 SØBORG	Fødevaredirektoratet Mørkhøj Bygade 19 DK-2860 SØBORG
Finland	Elintarvikevirasto Leena Mannonen PL 5 FIN-00531 HELSINKI	Uuselintarvikelautakunta Anna-Liisa Koskinen Kauppa- ja teollisuusministeriö PL 32 FIN-00023 VALTIONEUVOSTO

Table 6.1. (continued)

Country	Competent authority for the submission of applications, art. 4 para. 1 NFR	Food assessment bodies, art. 4 para. 3 NFR
France	DGCCRF – Bureau C2 Teledoc 051 59, boulevard Vincent Auriol F-75703 PARIS Cedex 13 Ministère de la Santé Veille Sanitaire – VS3 8 avenue de Ségur F-75350 PARIS Cedex 075 SP DGAL – Ministère de l'Agriculture et de la Pêche 251, rue de Vaugirard F-75732 PARIS cedex 15	AFSSA 23, avenue du Général de Gaulle, BP 19 F-94701 MAISONS-ALFORT Cédex AFSSA = Agence française de sécurité sanitaire des aliments
Germany	Bundesamt für Verbraucherschutz und Lebensmittelsicherheit Postfach 14 01 62 D-53056 BONN	Bundesamt für Verbraucherschutz und Lebensmittelsicherheit Postfach 14 01 62 D-53056 BONN **For novel foods and novel food ingredients referred to in Article 1 (2), (a):** Robert Koch-Institut Postfach 65 02 80 D-13303 BERLIN
Greece	Ministry of Finance General State Chemical-Laboratory a) Direction: Foodstuffs b) High Council of Chemistry Ave, Tsoha 16, GR-11521 ATHENS Ministry of Agriculture 1) General-Directorate for Plant Production 2) General-Directorate for Animal Production Acharnon 2 GR-Athens	Ministry of Finance General State Chemical-Laboratory a) Direction: Foodstuffs b) High Council of Chemistry Ave, Tsoha 16, GR-11521 ATHENS National Drug Organisation Ministry of Health a) Product Assessment Division b) Scientific Committee for Special Dietary Products and Food Supplements 284, Messogion Av., Holargos GR-15562 ATHENS Ministry of Agriculture Directorate for Plant Production Acharnon 2 GR-Athens

Table 6.1. (continued)

Country	Competent authority for the submission of applications, art. 4 para. 1 NFR	Food assessment bodies, art. 4 para. 3 NFR
Ireland	Food Safety Authority of Ireland Abbey Court Lower Abbey Street DUBLIN 1	Food Safety Authority of Ireland Abbey Court Lower Abbey Street DUBLIN 1
Italy	Ministero della Sanità Dipartmento Alimenti Nutrizione e Sanità Pubblica Veterinaria Paola Picotto UFF. IV Piazza Marconi 25 I-00144 ROMA	Ministero della Sanità Dipartmento Alimenti Nutrizione e Sanità Pubblica Veterinaria Piazza Marconi 25 I-00144 ROMA
Luxembourg	Ministère de la Santé Allée Marconi L-2120 LUXEMBOURG	Direction de la Santé Division de la Pharmacie et des Médicaments Allée Marconi L-2120 LUXEMBOURG Laboratoire National de Santé Division du Contrôle des Denrées Alimentaires 1A, Rue Auguste Lumière L-1950 LUXEMBOURG
Netherlands	Ministerie van Volksgezondheid Welzijn en Sport Directie Gezondheidsbeleid Postbus 20350 NL-2500 EJ DEN HAAG	The Health Council of the Netherlands Programme Director Dr. Jeanine AG van de Wiel PO Box 16052 NL-2500 BB DEN HAAG
Portugal	Direccao Geral de Fiscalizacao e Controlo de Qualidade Alimentar Av Conde Valbom 98 P-1064-824 LISBOA	Instituto de Biologica Experimental e Technologica Apartado 12 P-2780 OEIRAS Instituto Nacional de Engenharia e Technologica Industrial Estrada do Paco do Lumiar P-1649-001 LISBOA Instituto Nacional de Investigacao Agraria Rua Barata Salgueiro, 3 P-1200 LISBOA Laboratório Nacional de Investigacao Veterinaria Estrada de Benfica 701 P-1500 LISBOA

Table 6.1. (continued)

Country	Competent authority for the submission of applications, art. 4 para. 1 NFR	Food assessment bodies, art. 4 para. 3 NFR
Spain	Dirección General de Alimentación Ministerio de Agricultura, Pesca y Alimentacion Paseo Infanta Isabel, 1 E-28071 MADRID Dirección General de Salud Pública y Consumo Ministerio de Sanidad y Consumo Paseo del Prado No. 18–20 E-28071 MADRID	Dirección General de Alimentación Ministerio de Agricultura, Pesca y Alimentacion Paseo Infanta Isabel, 1 E-28071 MADRID Dirección General de Salud Pública y Consumo Ministerio de Sanidad y Consumo Paseo del Prado No. 18–20 E-28071 MADRID
Sweden	Statens livsmedelsverk Kåre Wahlberg/Reggie Vaz Box 622 S-75126 UPPSALA Ministry of Agriculture, Food and Fisheries National Food Administration (NFA) Monika Schere SE-10333 STOCKHOLM	Statens livsmedelsverk Kåre Wahlberg/Reggie Vaz Box 622 S-75126 UPPSALA
United Kingdom	Food Standards Agency Mr N. Tomlinson 235 Ergon House PO Box 31037 17 Smith Square LONDON SW1P 3WG (address to change in February 2001)	Food Standards Agency Mr N. Tomlinson 235 Ergon House PO Box 31037 17 Smith Square LONDON SW1P 3WG (address to change in February 2001) Food Standards Agency Scotland St Magnus House, 6th Floor 25 Guild Street ABERDEEN AB11 6NJ SCOTLAND Food Standards Agency Wales 1st Floor, Southgate House Wood Street CARDIFF CF10 1EW SOUTH WALES

(Source: European Commission, Health & Consumer Protection Directorate-General, Directorate D – Food Safety: production and distribution chain, D4 – Food law and biotechnology; partially modified)
[1] Rudolf Streinz, Anwendbarkeit der Novel Food-Verordnung und Definition von Novel Food, *Zeitschrift für das gesamte Lebensmittelrecht* **1998**, p. 19, p. 27 et seq.
[2] ECJ, Judgment of 8 June 1994, Case C-382/92 – Commission/United Kingdom – [1994] ECR I-2435/2475 citing further references.

6.2.4
Supplementary and Replacement Regulation

6.2.4.1 History
The so-called Supplementary Regulation and the so-called Replacement Regulation contain provisions for those products made from soya or maize that are not covered by the Novel Food Regulation due to their lack of novelty (cf. *supra* section 6.2.1.3). The Supplementary Regulation was repealed by art. 3 of the Replacement Regulation. Therefore the following explanations only deal with the latter.

6.2.4.2 Scope of application of the Replacement Regulation
According to art. 1 para. 1 Replacement Regulation it covers those foodstuffs and food ingredients which are to be delivered as such to the final consumer, produced, in whole or in part, from genetically modified soya beans or genetically modified maize. According to art. 1 para. 2 Replacement Regulation it does not apply to food additives, flavorings and extraction solvents which are also excepted from the scope of application of the Novel Food Regulation according to art. 2 NFR (cf. *supra* section 6.2.1.2 and especially 6.2.1.6(a)).

The duty to label relates to the existence of genetically modified proteins or DNA in foodstuffs or food ingredients (cf. recital 13 and art. 2 para. 2 lit. a Replacement Regulation). But it is impossible to completely rule out any accidental contamination of foodstuffs with genetically modified protein or DNA (cf. recitals 14 and 15 of the Replacement Regulation). It is possible, for example, that during a transport using a tanker which has previously been used to transport genetically modified organisms and has not been properly cleaned, a mixing of conventional foodstuffs or food ingredients and these remnants occurs. In these cases the threshold value in art. 2 para. 2 b subpara. 1 Replacement Regulation must be observed: Foodstuffs and food ingredients need not be labeled where material derived from the genetically modified organisms is present in their food ingredients or the food comprising a single ingredient in a proportion no higher than 1 % of the food ingredients individually considered or food comprising a single ingredient, provided this presence is adventitious. According to art. 2 para. 2 lit. b subpara. 2 Replacement Regulation the person responsible must be able to establish that the presence of this material is adventitious.

Art. 2 para. 2a Replacement Regulation provides that in order to facilitate the application of art. 2 para. 2 lit. a, a (non-exhaustive) list of products in which neither protein nor DNA resulting from the genetic modification is present, shall be drawn up. This is intended to allow products, which as a result of processing do not contain any genetically modified material any more, to be exempted from the labeling without the carrying out of tests in each individual case.

6.2.4.3 Requirements of labeling

The specific additional (cf. art. 2 para. 1, para. 2, para. 3 and para. 4 Replacement Regulation) elements of the labeling themselves have been regulated in art. 2 para. 3 Replacement Regulation. The wording to be used in each case is mandatorily prescribed in art. 2 para. 3 lit. a to lit. d Replacement Regulation. Deviations from it would not meet the labeling requirements – even if the content of the declaration would be identical. To avoid repetition, reference is made to the wording of art. 2 para. 3 Replacement Regulation (cf. also *infra* section 6.2.6.2)

6.2.5
Relation to *Council Directive No 90/220/EEC of 23 April 1990 on the deliberate release into the environment of genetically modified organisms* (OJ (EC) 1990 No L 117/15; from here on: Deliberate Release Directive)

Art. 9 NFR provides for the coordination with the Deliberate Release Directive. In particular, art. 9 para. 1 subpara. 2 NFR establishes additional formal requirements for a request for placing on the market according to art. 6 para. 1 NFR for those products which contain genetically modified organisms within the meaning of art. 2 paras. 1 and 2 Deliberate Release Directive (the correct reference would be art. 2 No. 1 and 2). It is further clarified in art. 9 para. 1 subpara. 2 NFR, that articles 11 to 18 of the Deliberate Release Directive shall not apply to foods or food ingredients which contain or consist of genetically modified organisms. These deal with the placing on the market of products that contain genetically modified organisms (essentially procedural provisions, labeling requirements, and so forth), whereby art. 10 para. 2 Deliberate Release Directive, too, emphasizes, that articles 11 to 18 shall not apply to any product covered by Community legislation which provides for a specific environmental risk assessment similar to that laid down in the Deliberate Release Directive. It is consistent for art. 9 para. 2 sentence 1 NFR to determine that in the case of foods or food ingredients containing or consisting of genetically modified organisms, the additional assessment shall respect the environmental safety requirements laid down by the Deliberate Release Directive.

6.2.6
**Supplementary National Provisions in German law:
The Novel Foods and Food Ingredients Instrument**

The *Statutory Instrument for the Implementation of community legislation concerning novel foods and food ingredients and the labeling of products made from genetically modified soya beans and genetically modified maize as well as the labeling of foodstuffs which have been produced without means of genetic engineering* (Novel Foods and Food Ingredients Instrument – NFI), as republished on 14 February 2000 (*Bundesgesetzblatt, Federal Law Gazette* 2000, part. I, p. 123), deals with novel foods within the meaning of the Novel Food Regulation only in articles 1 to 3. Otherwise, it contains rules concerning the labeling of products made from genetically modified soya beans and genetically modified maize (art. 4 NFI), rules concerning the labeling of foods made without using procedures of genetic engineering (articles 5 to 7 NFI) and rules on criminal offences and misdemeanors.

6.2.6.1 General rules for novel foods

Regarding provisions on competencies in art. 1 NFI, reference is made to section 6.2.3 to avoid repetitions.

Art. 2 NFI determines the procedural details of the authorization procedure (cf. section 6.2.1.4(a)), art. 3 NFI defines once more the conditions for the placing on the market of novel foods and food ingredients and repeats the reference to the specific labeling requirements contained in art. 8 para. 1 NFR (cf. *supra* section 6.2.11.5(b)).

6.2.6.2 Rules on labeling of genetically modified soya beans and genetically modified maize

As regards the area of regulation, art. 4 para. 1 NFI expressly refers to the scope of application of the Replacement Regulation (cf. *supra* section 6.2.4.2) and makes the marketability of genetically modified soya beans and genetically modified maize dependent on it being labeled in accordance with art. 2 para. 3 Replacement Regulation (cf. *supra* section 6.2.4.3). Furthermore, art. 4 para. 2 NFI regulates the compulsory modalities of the labeling requiring the data to be easily visible, clearly legible and indelible and prescribes the positioning of the label as a function of the way the product is delivered (cf. art. 4 para. 2 sentence 2 no. 1 to no. 5 NFI).

6.2.6.3 Availability of negative labeling of foodstuffs made without using procedures of genetic engineering

The provisions of the Novel Food Regulation do not contain a prohibition of negative labeling of products made without the use of genetic engineering, but states expressly in its recital 10 that such statements may be made (cf. on this and the concept of negative labeling; section 6.2.2.1).

(a) **Conditions of negative labeling**

The relevant foodstuff may neither consist of nor be produced from genetically modified organisms (art. 5 sentence 1 no. 1 NFI), nor be produced with the help of such material or of auxiliary material which in turn have been extracted from genetically modified organisms (art. 5 sentence 1 no. 2 NFI). If parts of a genetic modification enter into a food unintentionally and in inevitable traces, this does not prevent the negative labeling according to art. 5 sentence 2 NFI. To use the indeterminate legal term "inevitable traces" rather than defining a marginal value for the maximum contamination permitted is problematic, though: If we compare this to the regulation of the threshold value for the labeling according to art. 2 para. 2 lit. b subpara. 1 Replacement Regulation (cf. *supra* chapter 2, section 4.2), it is possible to imagine cases where, due to contamination above 1%, the rule of art. 2 para. 2 lit. b subpara. 1 Replacement Regulation does not apply any more, but still the relevant product may – if the necessary inevitability is given – be labeled "ohne Gentechnik" ("without genetic engineering") according to art. 5 sentence 2 NFI. How these valuation discrepancies, that seem simply inexplicable to the consumer, will be dealt with by the legislator or the courts (for example by presuming that only contaminations below 1% can be deemed inevitable, which would lead to an improvement in legal certainty but rather support the introduction of a precise marginal value in art. 5 sentence 2 NFI), remains to be seen.

If the food comes from an animal, this animal must not have been fed with feed, feed additives or drugs within the meaning of art. 2 Drug Act that have been produced with the help of genetic engineering. Here, those drugs are excluded which have been given for therapeutic or prophylactic purposes and which could not be replaced by an equivalent conventional drug, art. 4 sentence 3 NFI.

To understand art. 6 NFI it is necessary to realize that it contains an editorial error: The (actual) articles 5 to 7 NFI themselves were only introduced by the First Statutory Instrument for the Amendment of the Novel Foods and Food Ingredients Instrument (*Federal Law Gazette* 1998, Part I, p. 3167), originally as articles 4 to 6. The Second Statutory Instrument for the Amendment of the Novel Foods and Food Ingredients Instrument (*Federal Law Gazette* 1999, Part I, p. 1885) introduced a new art. 4 and renamed the then articles 4 to 6 as articles 5 to 6. It was forgotten, however, to replace the reference in (actual) art. 6 to art. 4 NFI by a reference to art. 5 NFI. But art. 6 NFI only makes sense, if the reference in it is understood to refer to the immediately preceding section, that is now art. 5 NFI. Having said that, art. 6 contains a number of additional requirements for labeling: art. 6 sentence 1 NFI demands proof of the requirements contained in art. 5 sentences 2 and 3 NFI, without which the corresponding labeling is unlawful, art. 6 sentence 2 NFI. Art. 6 sentence 3 NFI refers to the standard example ("especially") of a binding statement of producers or suppliers, rendering other forms of appropriate proof (at least theoretically) possible. Art. 7 NFI additionally increases these requirements by already allowing the prohibition of the labeling when the person responsible is unable to dispel reasonable doubts about whether the requirements are met.

(b) **The element of labeling**

Both the labeling of the relevant products and the advertisement for them (cf. art. 5 sentence 4 NFI, that makes the rules contained in art. 5 sentences 1 to 3 NFI applicable to advertisements) may – to protect the consumer against a plurality of terms, which may sound similar but differ as regards content – exclusively use the term "ohne Gentechnik" ("without genetic engineering"), art. 5 sentence 1 NFI. All other terms are unlawful.

6.2.6.4 Rules on criminal offences and misdemeanors

S. 8 NFI contains rules on criminal offences, art. 9 rules on misdemeanors, making it unlawful to breach the rules of the Novel Foods and Food Ingredients Instrument enumerated therein. The sanctions available are defined in art. 52 para. 1 no. 2 Food Act, art. 52 para. 1 no. 11 Food Act, art. 53 para. 1 Food Act or art. 54 para. 1 no. 2a Food Act, which are referred to by means of a blanket law (outline panel law).

6.3
Recent Development at the European Level

6.3.1
Introduction

In the Annex to its White Paper on Food Safety of 12 January 2000 [1], the Commission had already demanded in its catalogue of actions to:

- enact legal rules for the assessment, authorization and labeling of novel foods (White Paper on Food Safety, Annex: Action Plan on Food Safety, no. 6),
- make the procedures for authorization of novel foods more transparent (White Paper on Food Safety, Annex: Action Plan on Food Safety, no. 50),
- review Regulation (EC) No. 258/97 on novel foods and novel food ingredients (White Paper on Food Safety, Annex: Action Plan on Food Safety, no. 51),
- introduce a general requirement for a new safety evaluation (White Paper on Food Safety, Annex: Action Plan on Food Safety, no. 41) and
- complete and harmonize the provisions governing the labeling (White Paper on Food Safety, Annex: Action Plan on Food Safety, no. 52).

The Commission now pursues the implementation of these projects with its proposals for a "Regulation of the European Parliament and of the Council on genetically modified food and feed" [2] and a "Regulation of the European Parliament and of the Council concerning traceability and labeling of genetically modified organisms and traceability of food and feed products produced from genetically modified organisms and amending Directive 2001/18/EC" [3]. It is intended to establish a complete and solid framework of rules for the use of genetically modified organisms in Europe [4]. Consistently, the proposals affect numerous existing rules (that sometimes only recently entered into force). The following amendments shall be enacted:

- According to art. 37 CommProp GMFF the following are to be repealed:
  - *Council Regulation (EC) No 1139/98 of 26 May 1998 concerning the compulsory indication of the labeling of certain foodstuffs produced from genetically modified organisms of particulars other than those provided for in Directive 79/112/EEC, OJ (EC) 1998 L 159/4 of 3 June 1998,*

- *Commission Regulation (EC) No 49/2000 of 10 January 2000 amending Council Regulation (EC) No 1139/98 concerning the compulsory indication on the labeling of certain foodstuffs produced from genetically modified organisms of particulars other than those provided for in Directive 79/112/EEC*, OJ (EC) L 6/11 of 11 January 2000 and
- *Commission Regulation (EC) No 50/2000 of 10 January 2000 on the labeling of foodstuffs and food ingredients containing additives and flavorings that have been genetically modified or have been produced from genetically modified organisms*, OJ (EC) L 6/15 of 11 January 2000.

- The following shall be amended:
  - according to art. 38 CommProp GMFF *Regulation (EC) No 258/97 of the European Parliament and of the Council of 27 January 1997 concerning novel foods and novel food ingredients (Novel Food Regulation)*, OJ (EC) 43/1 of 14 February 1997,
  - according to art. 39 CommProp GMFF *Council Directive 82/471/EEC of 30 June 1982 concerning certain products used in animal nutrition*, OJ (EC) L 213/8 of 21 July 1982,
  - according to art. 40 CommProp GMFF *Council Directive 70/457/EEC of 29 September 1970 on the common catalogue of varieties of agricultural plant species*, OJ (EC) L 225/1 of 12 October 1970,
  - according to art. 41 CommProp GMFF *Council Directive 70/458/EEC of 29 September 1970 on the marketing of vegetable seed*, OJ (EC) L 225/7 of 12 October 1970,
  - according to art. 42 CommProp GMFF *Directive 2001/18/EC of the European Parliament and of the Council of 12 March 2001 on the deliberate release into the environment of genetically modified organisms and repealing Council Directive 90/220/EEC*, OJ (EC) L 106/1 of 17 April 2001.

Both Commission proposals do not stop at consolidating the existing rules, but contain substantial innovations which are considered in the following part.

The Commission further intends to present two supporting proposals concerning genetically modified seeds.

6.3.2
The Commission Proposal for a "Regulation of the European Parliament and of the Council on Genetically Modified Food and Feed"

In the course of four sections, this Commission proposal deals with "Objective and Definitions" (art. 1 and 2 CommProp GMFF), "Genetically Modified Food" (with the sections "Authorisation and Monitoring" and "Labelling"; art. 3 to 15 CommProp GMFF), "Genetically Modified Feed" (with the same divisions; art. 16 to 28 CommProp GMFF) and "Common Provisions" (art. 29 to 47 CommProp GMFF).

6.3.2.1 Objective and definitions

Art. 1 lit. a CommProp GMFF emphasizes as the first objective of the Regulation to provide the basis for the assurance of a high level of protection in every respect in relation to genetically modified food and feed (cf. also recitals 2 and 3 of the Commission Proposal). This shall be achieved on the one hand by Community procedures for the authorization and supervision (art. 1 lit. b CommProp GMFF; cf. also recitals 4 to 8) and on the other hand by appropriate rules on labeling (art. 1 lit. c CommProp GMFF; cf. also recitals 16 to 23).

If these objectives are generally comprehensible, the definitions contained in art. 2 CommProp GMFF are rather difficult to understand: references to *Regulation (EC) No 178/2002 of the European Parliament and of the Council of 28 January 2002 laying down the general principles and requirements of food law, establishing the European Food Safety Authority and laying down procedures in matters of food safety* (OJ (EC) 2002, No. L31/1; from here on: Basis Regulation) may be helpful to establish the coherence of food law rules; however, if this is done the way art. 2 CommProp GMFF employs, this makes it more difficult for the lawyer to maintain the overview. Out of thirteen definitions in art. 2 CommProp GMFF four refer to the Basis Regulation and six refer to Directive 2001/18/EC. Only three terms, viz. "genetically modified food or feed" (art. 2 No. 3 CommProp GMFF), "produced from genetically modified organisms" (art. 2 No. 6 CommProp GMFF), and "control sample" (art. 2 No. 7 CommProp GMFF), have been defined in the Commission Proposal itself [6]. Typically enough, art. 2 No. 6 CommProp GMFF defines part of the definition given in art. 2 No. 3 CommProp GMFF. To avoid misunderstandings at an early stage, the Commission explained in recital 15 CommProp GMFF as an example the difference between food and feed produced "from" a GMO and that produced "with" a GMO. The latter – for example cheese produced with the help of a genetically modified enzyme or products obtained from working animals, that were fed on genetically modified feed or treated with genetically modified drugs – are therefore not covered by the Commission Proposal.

6.3.2.2 Requirements of genetically modified food and feed

Sections 2 and 3 of the Commission Proposal are (apart from their scope of application) virtually identical: The rules on authorization and monitoring as well as labeling of genetically modified feed impose therefore essentially the same requirements as those for genetically modified food. This shows clearly the integral approach of the Commission, to fashion a high level of protection for the whole food chain from the producer to the consumer. Nevertheless, it is surprising that the Commission Proposal also applies to feed that is not for the use of working animals. Art. 16 CommProp GMFF does not contain a more precise term, the definition in art. 2 No. 1 CommProp GMFF only refers to the Basis Regulation. This in turn only mentions "intended to be used for oral feeding to animals" without any appropriate specification. This example shows the deficiencies of the reference technique, too.

According to art. 4 para. 1 CommProp GMFF genetically modified foods must not present a risk for human health or the environment, mislead the consumer, or differ from the food which they replace to such an extent that its normal consumption would be nutritionally disadvantageous for the consumer (the corresponding rules for feed are contained in art. 17 para. 1 CommProp GMFF; all rules on feeds corresponding to those on foods may be found by adding 13 to the number of the relevant article on foods). To guarantee this, genetically modified foods are subject to the prohibition principle (Verbotsprinzip), that is their marketability depends according to art. 4 para. 6 CommProp GMFF (for the time being) on the relevant authorization. The necessary authorization procedure is regulated in detail in art. 6 to 8 CommProp GMFF. A number of special features must be mentioned: The applicant must be resident in the Community. In the application for authorization, which according to art. 6 para. 1 CommProp GMFF must be submitted to the future European Food Authority, [7] the applicant must comment on Annex II CPBS (art. 6 para. 3 lit. c CommProp GMFF), provide information about ethical or religious concerns (art. 6 para. 3 lit. g CommProp GMFF) and describe a suitable method for detection (art. 6 para. 3 lit. i CommProp GMFF). To shorten the lengthy procedures, that occurred in the context of the Novel Food Regulation in the past, art. 7 para. 1 CommProp GMFF imposes a duty on the EFSA to give an opinion within six months of the receipt of a valid application. Given the complexity of the area and the duty to inform the Member States, the Commission and the public according to art. 7 para. 3 lit. b and c CommProp GMFF it must be feared that the exception of exceeding this time limit in "exceptionally complex cases", will turn out to be the standard situation. The relevant statement will not only be forwarded to the Commission but also, having regard to the requirements of confidentiality provided for in art. 31 CommProp GMFF, be made available to the public. Despite the name authority, which is misleading in so far, it is not the EFSA, but the Commission that will make the authorization decision according to art. 8 CommProp GMFF. The opinion of the EFSA is not decisive for the decision of the Commission, as is clear from the requirement laid down in art. 8 para. 1 sentence 1 CommProp GMFF "taking into account Community law and other legitimate factors relevant to the matter under consideration" that leaves a wide margin of interpretation.

Once given, the authorization – which can contain conditions or restrictions according to art. 10 para. 1 CommProp GMFF – is valid for ten years to start with according to art. 8 para. 5 sentence 1 CommProp GMFF, but can be renewed on an application according to art. 12 CommProp GMFF by the holder, which according to art. 12 para. 1 sentence 1 CommProp GMFF must be at the latest one year before the expiry date of the authorization.

According to art. 4 para. 2 CommProp GMFF, the authorization as such makes the relevant product marketable; according to art. 8 para. 7 CommProp GMFF, this, however, does not relieve the food business of its liability under private or criminal law in any way.

6.2.2.3 Labeling of genetically modified food and feed

The labeling of food produced from genetically modified organisms, which until now was covered by the Novel Food Regulation, shall now be subject to the rules on labeling as provided for in art. 13 et seq. CommProp GMFF. It complements the general rules of the Labeling Directive (*Directive 2000/13/EC of the European Parliament and of the Council of 20 March 2000 on the approximation of the laws of the Member States relating to the labeling, presentation and advertising of foodstuffs*, OJ (EC) 2000, No. L 109/29 of 6. 5. 2000). The transformation into German Law is to be found in the *Statutory Instrument on the Labeling of Foodstuffs* (LFI) as re-promulgated on 15 December 1999 (*Bundesgesetzblatt, Federal Law Gazette* 1999, Part I, p. 2464), as amended by art. 3 of the *Statutory Instrument for the Amendment of the Statutory Instrument on Spirits and of other Statutory Instruments in the Field of Food Law* of 8 December 2000 (*Bundesgesetzblatt, Federal Law Gazette* 2000, Part I, p. 1689). Both position and wording of the necessary labeling are mandatorily prescribed; to avoid repetitions reference is made to the wording of art. 14 CommProp GMFF. The most important novelty is to be found in the area of labeling provisions: the system is changed from a product- or proof-related to a procedure-related labeling: whereas until now only products which contained traces of DNA or proteins of genetically modified organisms had to be labeled, this is now true also for those foodstuffs for which such contents cannot be detected but which were nevertheless produced from genetically modified organisms. Highly refined maize or soya oil may serve as examples. A table showing the current and the intended labeling can be found in Annex I of the press release of the Commission mentioned in Ref. [4].

This labeling requirement finally realizes a demand which has been put forward by consumer protection groups for a long time. Nevertheless, the labeling requirements are not all-encompassing: According to art. 13 para. 2 sentence 1 CommProp GMFF they do not apply to foods containing material which contains, consists of, or is produced from genetically modified organisms in a proportion no higher than thresholds established by a scientific committee or the EFSA, provided that this presence is adventitious or technically unavoidable.

The corresponding provision of art. 5 para. 1 sentence 1 CommProp GMFF, however, must be distinguished: in accordance with it products which contain adventitious or technically unavoidable contaminations up to 1% (or up to a lower threshold to be established) are excluded from the scope of application of the Regulation. This is in recognition that such contaminations cannot be totally excluded during cultivation, harvest, transport, and processing, cf. recital 23 to 25 of the Commission Proposal. The addition, on purpose, of genetically modified organisms up to the threshold, however, is not covered by this provision, since art. 5 para. 2 Commission Proposal requires proof that the contamination was unavoidable despite taking appropriate steps to prevent it.

6.3.2.4 **General provisions**
Some of the general provisions should be discussed here.

The so-called "integral approach" ("one door – one key") of art. 29 para. 1 CommProp GMFF is designed to make sure that those products which might be used both as food and as feed need only one authorization procedure. If one product might be suitable as food but not as feed (or vice versa), the authorization must be rejected for both uses. It is not open to the applicant to restrict the assessment to one type of use, since the EFSA may consider whether an integral assessment should take place according to art. 29 para. 2 CommProp GMFF.

If it becomes clear that an authorized genetically modified food or feed endangers human health, animal health or the environment, then the procedure for emergency measures as provided for in art. 35 CommProp GMFF must be followed (as to the modification, suspension and revocation of authorizations cf. art. 11 and 24 CommProp GMFF). Again, the opinion of the EFSA shall be considered, but the Commission takes the decision on emergency measures [8]. It is worthwhile noting that the Commission Proposal does not enumerate the possible measures, but obviously leaves the choice of means to the Commission. That may be conducive to flexible action in a situation of crisis, but it comes at the expense of transparency and coherence. Given that emergency measures according to art. 35 para. 2 sentence 2 CommProp GMFF may remain in place until a final decision is taken in accordance with art. 11 or 24 CommProp GMFF, it seems inevitable that quarrels will occur which might have been avoided with the help of a catalogue of the measures available.

The Commission is also competent to fulfill the tasks resulting from the Cartagena-Protocol: According to art. 43 CommProp GMFF the Commission notifies the other parties of any authorization, renewal, modification, suspension or revocation of authorization of a genetically modified food or feed.

6.3.3
The Commission Proposal for a Regulation of the European Parliament and of the Council Concerning Traceability and Labeling of Genetically Modified Organisms and Traceability of Food and Feed Products Produced from Genetically Modified Organisms and Amending Directive 2001/18/EC

The traceability principle is intended to help the tracing of a product through the whole chain of production and sales back to its origin. This should on the one hand establish opportunities for effective monitoring and control (also of the necessary labeling), and on the other hand facilitate the possible withdrawal of a product from the market (cf. art. 1 and art. 3 no. 3 CommProp Traceability and its recitals 3 to 5 as well as the press release of the Commission *supra* [4]). The regulation shall apply to products consisting of, or containing GMOs (art. 2 para. 1 lit. a CommProp Traceability), foods and food ingredients, including food additives and flavorings, produced from GMOs (art. 2 para. 1 lit. b CommProp Traceability) and feed materials, compound feedstuffs and feed additives, produced from GMOs (art. 2 para. 1 lit. c CommProp Traceability). By the way of reference, the products subject to the corresponding exceptions concerning adventitious or technically unavoidable contaminations are excluded from the scope of application, art. 6 paras. 3 and 4 CommProp Traceability.

These are not only subject to the rules on labeling as provided for in art. 4 para. 1 to 3 CommProp Traceability, which however are without prejudice to other specific requirements in other Community legislation according to art. 4 para. 5 CommProp Traceability, but also to the traceability requirements: accordingly it must be possible, except where the product is delivered to the final consumer (art. 6 para. 2 CommProp Traceability), to identify, for a period of five years from each transaction, as to the person from whom and to whom the product has been made available (art. 5 para. 2 CommProp Traceability) with the help of systems to be established (art. 5 para. 2 CommProp Traceability).

6.3.4
Stage of Legislative Procedure

The Commission Proposal concerning Genetically Modified Food and Feed was forwarded to the European Parliament and to the Council on the 30 July 2001, the Committee of the Regions gave its opinion on 16 May 2002, and the Economic and Social Committee gave its opinion on 30 May 2002.

The Commission Proposal concerning Traceability was forwarded to the Council on 17 August 2001 and to the European Parliament on 20 August 2001. It was discussed in the Council on 29 October 2001, the Economic and Social Committee gave its opinion on 21 March 2002 and the Committee of the Regions gave its opinion on 16 May 2002. [9]

The first reading of each proposal has meanwhile taken place, with the European Parliament demanding numerous amendments which would mostly lead to stricter requirements. Concerning CommProp GMFF, the European Parliament adopted the resolution drafted by Karin Scheele, concerning CommProp Traceability, the European Parliament adopted a resolution on GMO labeling, where the rapporteur was Antonios Trakatellis. [10] Note that, at the final vote, the rapporteur voted against the report, arguing that the adopted amendments on traceability and labeling would prove "inapplicable in real life" [11].

Meanwhile, not only within the EU but also on the part of the USA there are doubts as to the conformity of the proposals to WTO rules [12]. The USA, in particular, criticize the labeling and traceability requirements as hardly workable, and to cause at least enormous and unnecessary costs. The objectives pursued could not be achieved, and it would rather be the case that international trade would suffer [13].

References

1 COM (1999) 719, on the internet at http://www.europa.eu.int/comm/dgs/health_consumer/library/pub/pub06_en.pdf. See on this topic Horst/Mrohs, Das Europäische Lebensmittelrecht am Scheideweg – Das Weißbuch der Kommission zur Lebensmittelsicherheit, ZLR 2000, p. 125.

2 COM (2001) 425 final. On the internet at http://www.europa.eu.int/eur-lex/de/com/dat/2001/de_501PC0425.html. From here on: CommProp GMFF.

3 COM (2001) 182 final. On the internet at http://www.europa.eu.int/eur-lex/de/com/dat/2001/de_501PC0182.html. From here on: CommProp Traceability.

4 Likewise the Commission in its press release of 25. 06. 2001 (IP/01/1095) under the headline: Commission improves rules on labeling and tracing of GMOs in Europe to enable freedom of choice and ensure environmental safety.

5 See on this the press release referred to in fn. 4 and "Commissioner supports a 'happy labeling day'", *EU Food Law Monthly*, No. 117, September 2001, p. 1 (3).

6 On the problems of such references with special emphasis on criminal responsibility under food law cf. Kühne, Rechtssicherheit und Kohärenz als Auftrag des Europäischen Lebensmittelstrafrechts. *Zeitschrift für das gesamte Lebensmittelrecht* 2001, p. 379.

7 From here on: EFSA. Cf. art. 22 et seq. Basis Regulation. Recent/current information on the development of the EFSA are published at http://www.efsa.eu.int.

8 On the concept of emergency measures and the corresponding procedures cf. also art. 53 and 54 Basis Regulation. In detail also Rudolf Streinz/Lars O. Fuchs, Lebensmittelsicherheit: Die Europäische Ebene – Behörden, Kompetenzen, Kontrollen. *Zeitschrift für das gesamte Lebensmittelrecht* 2002, p. 169, p. 201, et seq.

9 See also Danes to battle with GM dossiers, *EU Food Law Monthly*, No. 125, May 2002, 10–11.

10 Concerning CommProp GMFF see the report A5-0225/2002 final containing 47 amendments, Concerning CommProp Traceability, see the report A5-0229/2002 final.

11 See on the internet at http://www.db.europarl.eu.int/oeil/oeil_ViewDNL.ProcViewByNum?lang=2&procnum=COD/2001/0180.

12 GM proposals "breach WTO rules", *EU Food Law Monthly*, No. 121, January 2002, p. 5; cf. Tough European stance on GM Labelling increases the divide, *World Food Law Monthly*, No. 51/2002, p. 1, et seq.

13 On this see also Lars O. Fuchs/Christoph Herrmann, Die Regulierung genetisch veränderter Lebensmittel im Lichte aktueller Entwicklungen auf europäischer und internationaler Ebene. *Zeitschrift für das gesamte Lebensmittelrecht* 2001, p. 789, p. 803, et seq.

Part III
Methods of Detection

7
Detection of Genetic Modifications: Some Basic Considerations

Knut J. Heller

7.1
The Conversion of Genetic Information from DNA to Phenotypes

The genetic information of all organisms, the latter of which may consist of just one or of many cells, is stored within DNA in the form of a specific sequence of just four different nucleobases. This information becomes available for functional traits by two consecutive fundamental biological processes: transcription and translation (Fig. 7.1). In the first step of "transcription," RNA is formed. This single-stranded molecule is, with two exceptions, a one-to-one copy of one strand of the DNA. The first exception is that the backbone of RNA contains ribose instead of deoxy-ribose as in DNA. And the second exception is that, wherever DNA contains the nucleobase thymine in its sequence, RNA contains uracil instead. Three different forms of RNA are synthesized: transfer RNA (tRNA); ribosomal RNA (rRNA); and messenger RNA (mRNA) (Fig. 7.2). All three types of RNA are needed for protein synthesis, which is the second fundamental biological step, that of "translation." Only the mRNA, which is used as a template for protein synthesis, determines directly the amino acid sequence of proteins, while tRNA and rRNA are just helper molecules needed to constitute a functional protein synthesis machinery. The proteins serve different functions inside or outside the cell: as structural elements, regulators, transporters, and enzymes. Especially the latter two are involved in the synthesis of other structural components of the cell, the lipids and polysaccharides. Through the action of structural elements and of enzymes, the cytoplasm of a cell shows a composition of soluble, insoluble and high molecular-weight, low molecular-weight substances, which reflects the interaction between environmental conditions (chemical and physical factors) and endogenous cellular factors (growth phase, stage of differentiation, etc.). The structural and functional elements contribute to the measurable properties of a cell, which are called phenotypes. A phenotype, therefore, is a direct reflection of a genotype.

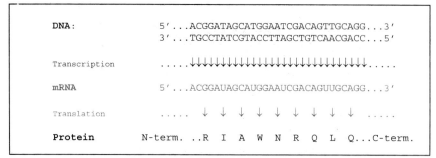

Figure 7.1. The flux of genetic information from DNA to protein. Deoxyribonucleic acid (DNA) consists of the four nucleobases adenine (A), guanine (G), cytosine (C) and thymine (T), which form two anti-parallel, complementary strands, the polarity of which is determined by their direction of biosynthesis and is indicated by their 5'- and 3'-ends. The backbone of the strands contains deoxy-ribose. Messenger-ribonucleic acid (mRNA) is synthesized using one DNA strand – in this case the lower one – as a template. RNA consists of one strand formed of the four nucleobases adenine (A), guanine (G), cytosine (C), and uracil (U). The polarity of RNA is determined by its direction of biosynthesis and is indicated by its 5'- and 3'-ends. The backbone of the strand contains ribose. Protein is synthesized using mRNA as a template. A frame of three nucleotides codes for the amino acids [e. g. arginine (R), isoleucine (I), alanine (A), tryptophan (W), asparagine (N), glutamine (Q), leucine (L)], which form one polypeptide strand. The polarity of the polypeptide is determined by its direction of biosynthesis, and this is indicated by its N-terminal and C-terminal ends.

7.2
DNA, Protein and Phenotypes as Targets for Detection Assays

With regard to genetically modified organisms – be they modified by natural mutation or by genetic engineering – information concerning the genotype of the organisms can be obtained on all levels of the process of conversion of genetic information into structural and functional traits: on the level of DNA, on the level of RNA, on the level of protein, on the level of cellular non-nucleic acid, nonprotein substances, and on the level of phenotypes (see also Chapter 13). However, the conclusions concerning the genetic modification that can be drawn from the detection on the different levels, may vary considerably. Three examples will be used here to illustrate this.

1. Cells of a bacterial strain, which normally is unable to degrade starch and use the sugar moieties for growth because of lack of suitable enzymes, show the phenotype "growth on starch as sole carbon source." The conclusion is very precise: the cells must have obtained at least a gene encoding a secreted starch-degrading enzyme. This already proves that a new gene has been introduced into the cells. In addition, the gene, the mRNA, the degrading enzyme, and the degradation products of the starch all may serve as substances on which a detection assay may be built.

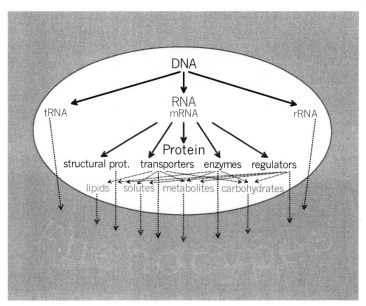

Figure 7.2. Conversion of genetic information from DNA to phenotypes. DNA is transcribed into RNA, consisting of the three species transfer RNA (tRNA), messenger RNA (mRNA), and ribosomal RNA (rRNA). mRNA is translated into protein, which is partitioned within the cell into structural proteins, transporters, enzymes, and regulatory proteins. The latter three proteins are involved in the synthesis of lipids and carbohydrates and in the production of metabolites. The cellular concentration of further solutes is adjusted mainly by transporters and regulators. The physical and functional elements of a cell determine its phenotypes. Solid arrows indicate transformation processes in which the upper molecule serves as a template for the synthesis of the lower one. Thus, the sequence of the upper molecule can, in principle, be deduced from the sequence of the lower molecule. The dotted arrows indicate transformation processes where this deduction is not possible.

2. Soy bean plants, which normally are sensitive to a certain herbicide, exhibit resistance against this herbicide. Different explanations are possible: the plants have, through genetic engineering, obtained a gene encoding a herbicide-degrading enzyme; alternately, the plants have undergone spontaneous mutations which either prevent uptake of the herbicide into the plant or alter the target of the herbicide within the plant cells. To exclude spontaneous mutations, different options for confirmation of the genetic modification are possible. The gene encoding the enzyme, the mRNA transcribed from the gene, the protein expressed, or the herbicide-degradation products specific for the enzyme used may be detected by suitable assays.
3. Cells of a bacterial strain, which normally are able to grow on lactose as sole carbon source, turn into a lactose-negative phenotype; that is, they are no longer capable of utilizing lactose. Different explanations are possible: the bacteria have acquired, through either genetic engineering or spontaneous mutation, defects in genes encoding lactose-transport proteins or in genes encoding en-

zymes necessary for lactose or sugar metabolism. Since neither specific degradation products nor proteins nor mRNAs are present for analysis, the only way to exclude spontaneous mutations is to analyze the gene loci possibly affected by the genetic modification. However, as will be discussed later as an example for "self-cloning," even determination of the DNA sequence of the gene affected may not be sufficient to distinguish between spontaneous mutation and genetic engineering.

When designing assays for the detection of modifications introduced by genetic engineering, the meaningfulness of the assays has to be considered. In example (1), the determination of the phenotype is already a convincing assay. However, in example (2) determination of the phenotype is of no value: a possible assay must be based on the determination of either specific degradation products, enzymes, mRNA, or DNA. In example (3) only assays based on the determination of specific DNA sequences will yield reliable results.

Another fact that must be considered is the degeneration suffered by the genetic information during conversion into structural and functional traits. The only biological process resulting in an exact 1:1 copy of the DNA is replication. Transcription usually yields 1:1 copies of the transcribed DNA regions (Fig. 7.1). However, nontranscribed DNA regions will never show up at the RNA level. Furthermore, especially in higher organisms, the primary transcript produced by the step of transcription may be altered by an editing process in which specific sequences – called introns – are deleted from the primary RNA to form the actual mRNA. During translation, further information gets lost or is obscured. This has several consequences.

1. As within the process of transcription, only part of the mRNA is translated into protein. The regions translated are called open reading frames.
2. A frame of three mRNA nucleotides (a codon) is required to encode one amino acid (Fig. 7.1). Three nucleotides out of four offer the possibility to form 64 different combinations. However, since only 20 amino acids are used for protein synthesis, several codons code for the same amino acid. Actually, each of the three amino acids serine, leucine and arginine is encoded by six different codons. Only methionine and tryptophan are each encoded by just one codon. Thus, the amino acid sequence of a protein is only partly suitable for deducing the nucleotide sequence of the mRNA.
3. Many proteins are subject to post-translational processing. One result of this processing may be the removal of part of the polypeptide chain. It is obvious that no information on the mRNA or DNA sequence of the removed parts can be deduced from the mature protein.

While sequence analyses of RNA and protein still allow to draw some conclusions on the DNA sequence, analyses of other constituents of the cell (lipids, carbohydrates, metabolites and solutes; Fig. 7.2) and of the phenotypes do not provide any clues on the DNA sequence. As shown above, such analyses may indicate the presence of a genetic modification. However, no information on the true nature of the modification is provided.

Thus, the ideal target molecule for detecting genetically engineered modifications is the DNA. This is even more true, since the most sensitive, fastest and powerful methods are available for DNA (see Chapter 8). Furthermore, the introduction of a foreign gene into the DNA of an organism can be unambiguously detected only at the level of DNA. The presence for example of a bacterial enzyme within an extract of a plant may be the result of a contamination. As long as the enzyme has not been altered by a post-translational process specific for bacteria, the protein itself will not reveal whether it was expressed in a plant or in a bacterium. However, the corresponding gene, cloned in a vector construct, transformed into the plant, and integrated into the plant DNA, can always be identified as a foreign gene, because it is flanked by DNA sequences which do not naturally flank this gene. An assay targeting the fusion sites of two DNA sequences of different origin, therefore, unambiguously identifies a product of a genetic engineering process: unique DNA sequences which are exclusively present in the specific recombinant DNA construct and nowhere else.

7.3
Food-grade Modifications

The considerations made in the previous section are especially important for the detection of very slight modifications in bacteria; these are the organisms for which the most advanced genetic engineering methods are available. In bacteria, foreign DNA is normally introduced via plasmids; subsequently, it may be inserted directly into the DNA of the organism at exactly predetermined positions with concomitant loss of all unnecessary vector sequences (see Fig. 12.1). However, the modification may not even involve the introduction of foreign DNA but simply rearrangement of the existing DNA, the latter of which is known as "self-cloning." These slight modifications, or the introduction of foreign DNA without any vector sequences (provided that the foreign DNA comes from a related, food-grade organism), are so called "food-grade" modifications. The concept of food-grade modifications demands that the exact DNA sequence of the entire modification is known, that no antibiotic resistance genes – which are used as selection markers for the introduction of foreign DNA into a host cell – remain in the genetically modified organism, and that in the case of introducing plasmids these have to have a narrow host range and are nonconjugative and nonmobilizable; that is, the chances of the plasmids being transferred to other organisms are extremely low. The idea for the food-grade concept stems from Council Directive 90/219/EEC on the contained use of genetically modified microorganisms. In Annex I of this directive, self-cloned nonpathogenic naturally occurring microorganisms fulfilling the criteria of Group I for recipient microorganisms are excluded from the directive. The criteria for classifying microorganisms in Group I are listed in Annex II of the directive and involve some of the above mentioned prerequisites for classifying organisms as "food-grade modified." The reason to exclude self-cloned microorganisms from directive 90/219/EEC was that such organisms are basically indistinguishable from and, thus, pose no larger risk than their nonmodified counterparts. However, it must be noted that the self-cloned microorganisms are not excluded from directive 90/220/EEC and, thus, not from regulation (258/97/EC) concerning novel foods and novel food ingredients.

Modifications in bacteria may, in the extreme, involve just the deliberate replacement of one nucleotide for another. Such minor modifications are indistinguishable from naturally occurring mutations. Nevertheless, since they have been constructed by genetic engineering, they will be subject to the procedure-related labeling (see Chapter 6, Section 6.2.3). Detection methods capable of detecting such minor alterations – DNA-sequencing or high-sensitivity hybridization (see Chapter 8) – will have to be developed. However, they will only reveal the presence of the specific modification, which could have also resulted from a natural mutation. Only with the knowledge that somebody constructed this modification will it and the corresponding food be classified as genetically engineered.

7.4
Detection of Unknown Modifications

It goes without saying that unknown modifications of the type presented at the end of the previous section can never be detected as modifications caused by genetic engineering. In principle this is true for all deliberate modifications produced by self-cloning techniques. Only if a considerable piece of foreign DNA is introduced can this piece be identified because of its unique sequence, which is not normally present in the organism.

Although self-cloning techniques are readily available for bacteria, bacteria are at present not in the focus for the development of detection methods, as no food containing living genetically engineered bacteria has yet been admitted.

For plants, the situation with respect to the detection of unknown modifications is somewhat more promising. For a foreign gene to be expressed in a plant, it must be fused to transcription signals at least, which are recognized and employed by the plant cell machinery. The number of suitable transcription elements is limited and can easily be used for the screening for unknown modifications (see Chapter 10). However, as the method relies on known elements, it is clearly not a 100% safe method. To cut a long story short, even considering all currently existing highly sophisticated DNA analysis methods, one must conclude that no reliable method is available that is capable of detecting any unknown modification in any given food.

8
DNA-based Methods for Detection of Genetic Modifications

Ralf Einspanier and Stephanie Rief

8.1
Introduction

The accelerated development of genetically modified organisms (GMOs) during the past ten years has raised a new set of questions concerning the release and traceability of such GMOs, as well as their possible adverse effects on the safety of both the environment and consumer.

When genetically modified crops were recently introduced into the market, it appeared as though the reliable recognition of such transgenes would inevitably become essential. Today, the main source of any novel genetic alteration that may be found in food is genetically modified plants and microorganisms. Although transgenic farm animals have been considered for specialized pharmaceutical production, they have not been introduced for large-scale agricultural production, except for aquaculture. It was considered that, in addition to food products, animal feed would be the major source of materials derived from transgenic plants.

The identification of GMO-fed animals and elucidation of the safety of so-called "secondary products" will surely lead to many very demanding technical problems in the future. If the specific analysis of animal-derived food products (meat, milk, eggs) produced with GMOs – as might be demanded by future governmental regulations – becomes routine with the introduction of highly sophisticated biochemical methods, then it is likely that close observation of any such end product will become mandatory. At present, as only limited scientific data are available relating to the biological effects of GMO food and feed, the need for continued environmental research is clear. If ultimately this becomes the cases, then any route by which genetic modification may be spread and appear in the environment will need to be identified and followed. Moreover, the specific monitoring of transgenic products will be obligatory in order to comply with national or international labeling regulations, as well as to identify any complex metabolic interactions.

DNA has been isolated from almost any given organic material on the basis of its remarkable chemical stability [1, 2]. The almost unlimited access to DNA as a result of the multiplication of nucleic acids using polymerase chain reaction (PCR) amplification technique [3] has made this DNA-based technology the most favorable technique for GMO analysis. When compared with established protein-based detection methods [4, 5], this nucleic acid technology has a major advantage in that it provides low detection limits (<1%), with great reliability. To date, in comparison with other methods (e.g., protein-ELISA), PCR is the technique of choice for monitoring GMOs in nucleic acid material isolated from foodstuff [6]. It is the exponential amplification of DNA that results in the exquisite sensitivity of the PCR method, and this is an especially important prerequisite if GMO analyses are to be seen as reliable, controlled, and available under certified working conditions in all laboratories where transgenic samples are examined. As a result of these advances, several programs have been established to ensure that the high proficiency demanded of analytical laboratories, is achieved. The programs include the food analysis performance assessment scheme (FAPAS), which operates under international harmonized protocols (ISO, IUPAC, AOAC), the intention being to develop DNA-based standard methods for the reliable monitoring of GMO materials in food and feed.

The detection methods used to identify genetic material derived from genetically modified organisms are reviewed in the following sections, with particular emphasis placed on the tracing of transgenic crops.

8.2
Recent DNA Methodology

The monitoring, detection and ultimate quantification of GMOs in food and feed samples by use of DNA techniques follows a more or less ubiquitous series of sampling, extraction, amplification, and scientific evaluation of results (Fig. 8.1). Clearly, each of these steps has its own limitations, but the extremely complex starting materials in particular may require treatment on an individual basis (for reviews, see Refs. [7, 8]).

8.2.1
Sampling Procedure

The most essential prerequisite for subsequent DNA analysis is to take a representative sample that will guarantee a statistically secure conclusion. Due to the great heterogeneity of highly diverse food material, it has been necessary to develop special sampling plans [9]. As a logical consequence, the outcome of subsequent analysis methods is only as good as the quality of sampling. Two possible approaches are often discussed, and either should lead to an adequate sampling plan:

- To take a large sample with further subsampling.
- To take multiple samples.

In addition, the particle size and distribution of the raw material must be taken into account, as well as the time at which samples are taken within the production

Figure 8.1. General GMO identification scheme providing the major steps from sample collection until the final judgment.

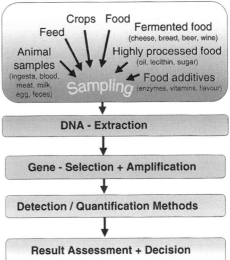

process. Careful attention to prevent contamination has led to essential physical precautions being taken when handling any transgenic material. From the sampling to the laboratory analysis, aerosols as well as polluted equipment and consumables must be excluded. From a practical standpoint, authentic reference materials are made available for selected transgenes (e. g., Bt 176 maize), although international standards and DNA procedures must be developed and provided for each GMO as well as for food and processed food individually.

8.2.2
Extraction and Purification of DNA

By taking advantage of the ubiquitous genetic material of DNA that is present in every organism, all transgenic material will be identified if residual, extractable DNA is present and the gene sequence sought is known. Therefore, the suitability of different methods for sample extraction and DNA purification for material from different sources must be considered. The isolation and purification of probable GMO-DNA from samples of different origin is the initial key step in the whole procedure. Hence, individual extraction procedures that lead to a wide variety of purification protocols form a vital part of the process.

Typically, the samples are mechanically disrupted and/or treated enzymatically to dissolve DNA with the help of organic compounds (phenol, chloroform), detergents, enzymes (RNase, protease), and chaotropic salts. Subsequent alcohol precipitation of the DNA results in its concentration and purification. In addition, high-throughput procedures are able to use ion-exchange or silica material, as well as paramagnetic particles which selectively bind nucleic acids under narrow-range pH conditions, thereby enabling the capture, washing and elution of DNA within the same cartridge. These extraction methods, which may be adapted to routine robotic applications, result in a highly standardized DNA purification process (with less cross-contamination and reduced individual sample variability, but similar DNA purity and homogeneity). However, the method should be selected to favor a comparative high quality of DNA, despite sometimes producing a lower yield. Indeed, a number of situations where the extraction of DNA has been limited (e. g., problematic matrices such as soybean oil [10]) have now been successfully overcome by the use of individually established protocols for difficult samples [11].

8.3
Specific Detection of Genetic Material

The unique biochemical feature of nucleic acids is the occurrence of a highly specific and predictable interaction between complementary single-stranded DNA fragments. The formation of hydrogen bonds between the corresponding heterocyclic bases (A-T, G-C) results in the formation of a double-stranded complex between probe and target DNAs. Most of the following applications which are used to detect DNA are founded on the basic molecular interaction known as hybridization. The main prerequisite of any successful transgene analysis is an exact knowledge of the genetic modification within each GMO. Therefore, free access to detailed sequence information, not only of the modified gene but also of the flanking regions, is absolutely necessary for subsequent GMO analysis to be conducted, as unknown or cross-bred constructs are difficult or even impossible to detect.

8.3.1
DNA Hybridization-based Detection Technique (Southern Blot)

One landmark event in the characterization of nucleic acids was the development of a DNA:DNA hybridization immobilized on a solid support [12]. This procedure, known as "Southern blot", includes the size separation of DNA by gel electrophoresis, followed by immobilization on a membrane with subsequent probe hybridization and detection through either radiolabeling or well-developed nonradioactive methods (e. g., chemiluminescence). Based on DNA:DNA hybridizations, the generation of a specific signal is highly dependent on variable parameters such as transfer efficiency, sequence homology, buffer condition, temperature, and incubation time. However, due to such limitations this technique alone does not provide the necessary performance to detect low abundant events as are normally found in GMO material. Only single copy gene detection methods have proved suitable, using massive concentrations of native plant DNA [13]. Although this technique is still useful for elucidation of the flanking genomic areas of an integrated transgene, low sensitivities (>0.5 pg DNA) have been the major restrictions. Southern blotting methods may support common DNA amplification methods (e. g., following PCR; see below) by verifying investigated DNA sequences through restriction enzyme digestion and subsequent hybridization to GMO-specific probes (restriction fragment-length polymorphism; RFLP) [14].

8.4
Nucleic Acid Amplification Methods using PCR

A frequent disadvantage of food inspection is the limited availability of target molecules to be analyzed in the sample material. When the requisite prior extraction procedures have been carried out, the low abundancy of DNA material can be further enriched by introducing selective amplification reactions that take advantage of naturally occurring nucleic acid polymerization.

8.4.1
The Common PCR

The polymerase chain reaction (PCR) is a relatively simple – but very effective – method of detecting minute quantities of DNA sequences. The technique permits exponential amplification of DNA *in vitro*, based on the unique properties of the heat-stable Taq-DNA-polymerase. The recognition of specifically short oligonucleotides (primers) hybridized on a single-stranded target DNA leads to subsequent DNA-polymerization and selective amplification of previously known DNA fragments (Fig. 8.2). In general, one single reaction cycle comprises heat-denatura-

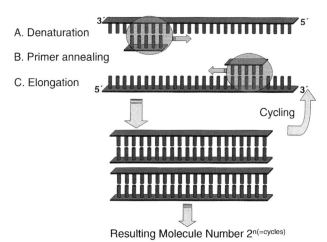

Figure 8.2. General principle of the polymerase chain reaction (PCR). The schematic cycle comprises three steps: (A) Denaturation of the double-stranded target DNA by heat separation of base-pair hydrogen bonds. (B) The forward and reverse primers anneal to their complementary single-stranded target. (C) Subsequent $5'\rightarrow 3'$ elongation of each primer by means of a heat-stable DNA polymerase generating a new double-stranded DNA. Each subsequent cycle will, in theory, double the initial number of DNA fragments and therefore enable an exponential increase of starting DNA molecules.

tion, primer annealing, and an enzymatic elongation step; the result is an approximate doubling of the number of DNA molecules. Detailed descriptions and special technical hints of this widely used PCR technology have been described in a flood of publications (for reviews, see Refs. [15, 16]). The conducting of multiple succeeding cycles (up to forty) leads to an exponential accumulation of the desired DNA fragment (e. g., CryIA, pat, 35S) up to a detectable amount of the desired amplicon. In this way, it may be possible to amplify every gene of interest (e. g., transgenes) from an appropriately purified residual DNA. Following each PCR, the amount and specificity of the resulting amplicons must be verified using three common methods:

1. Product length verification after electrophoretic separation in gels with nucleic acid-specific dyes intercalating in the double-stranded PCR product (e. g., ethidium-bromide).
2. Restriction enzyme digestion, followed by electrophoretic separation and Southern blotting by means of target-specific hybridization probes using either radioactive or non-radioactive labeling.
3. Exact sequence determination by DNA sequencing.

One of the earliest uses of standardized PCR was in medical research handling tissue biopsies, and this has led to the development of very reliable protocols under GLP (Good Laboratory Practice) conditions. In addition to the well-known common PCR assay, several specialized modifications have been introduced during the past decade, including asymmetric, allele-specific, nested, multiplex, differential, and competitive PCR. Each of these methods possesses unique features that have been reviewed elsewhere [17]. Hence, typical PCR methods may be adopted for GMO detection within food and feed, while methods suitable for the analysis of PCR products include high-pressure liquid chromatography (HPLC) [18] or capillary gel electrophoresis coupled with UV or laser-induced fluorescence detection [19]. Recently, feasible multiplex approaches capable of detecting up to seven genetically modified maize lines in one sample have been described as suitable for GMO monitoring [20, 21]. For example, one multiplex nested PCR assay has been commercialized which claims to prevent the occurrence of false-positive results (Biosmart Alin 1.0 GMO Screening System, Promega, WI, USA). In coping with internally standardized techniques, GMO methods based on triprimer and competitive PCRs were developed for detecting transgenic rice, soybean, and maize [22–24]. Individual protocols are now also available for rapid GMO screenings with the introduction of PCR-ELISA [25].

Such putative quantitative PCR assays are based on end-point measurements and have always to cope with the following restrictions:

- End-point measurement of the performed PCR is expected to represent the exponential range of amplification.
- Doubtful detection and adequate amplicon quantification methods.
- The laborious verification of each PCR product generated.

In order to overcome such limitations, new real-time PCR technologies have recently been developed which generate reliable PCR data within several hours.

8.4.2
Real-time PCR

A novel real-time PCR technique specifically to detect and quantify DNA based on conventional PCR principles (sequences of denaturation, annealing, elongation) has been successfully introduced. In overcoming the difficulties of the previously described PCR technique, this method permits direct online measurement of PCR products generated during the amplification procedure after each cycle. Therefore, increasing DNA concentrations are directly measured via excitation of selected fluorescence dyes by use of highly specialized instruments (e. g., TaqMan®, LightCycler®). (The interested reader may refer to further publications providing detailed information about real-time PCR systems and evaluating available instruments and important probe chemistries [26, 27]; distinct real-time approaches are also detailed in Chapters 10 and 11.)

Currently, three approaches to online detection are available. Each is based on fluorescence detection, and uses: (i) a DNA double-strand-specific fluorescence dye (SYBR-green); (ii) exonuclease probes (TaqMan Probes); and/or (iii) reversible hybridization probes (HybProbes, Beacons). As a specific fluorescence signal is generated and collected after each cycle, subsequent quantification can be made continually during the logarithmic phase, independently of any uncertain and probable arrested end-points of the reaction that characterize the former PCR amplification techniques.

The simplest approach is to measure synchronously the increase in generated PCR product using SYBR-green [28]. A disadvantage of this method is its lack of discrimination between amplicon-specific and unspecific double-stranded DNA in the reaction tube. Therefore, the sequence of the resulting product is verified via a reasonable melting curve analysis that separates potential unspecific side products from the desired known amplicon. The benefits of this method are its easy application together with relatively low cost.

Another method of measuring the increase in resultant PCR products is to use additional hybridizing oligonucleotides as well as the two PCR primers. The main advantage of internally hybridizing fluorescence-labeled probes is the synchronous detection of sequence specificity as well as amplification efficiency after each PCR. The first established real-time PCR detection system uses two dyes coupled to one probe [29]. This probe, which is internally hybridized to the generated PCR product, is labeled with both a reporter dye and a quencher dye; this results in suppressed light emission. By employing the exonuclease activity of the Taq-polymerase used, the reporter dye is liberated from the quencher during the elongation, thereby generating a fluorescence signal that is proportional in size to the number of templates present (Fig. 8.3).

A third technique has been introduced which successfully applies the fluorescence resonance energy-transfer (FRET) principle [28]. In detail, two closely adja-

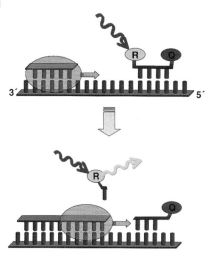

Figure 8.3. Principle of the 5' nuclease assay (TaqMan). An internal probe labeled with a reporter (R) at the 5' end and a quencher (Q) at the 3' end is split by the exonuclease-activity of the Taq polymerase during the elongation phase of each PCR cycle, leading to an increased fluorescence emission.

cent dye molecules are able to modulate the light emission if they are located within the range of one to five bases distance. In reality, two internal oligonucleotides lying adjacent were labeled with a sensor or an anchor dye, respectively. After homologous hybridization of both HybProbes, a specific light emission is detected (Fig. 8.4A). An alternative example represents the molecular beacon, a double-labeled oligonucleotide that fluoresces only during self-hybridization (Fig. 8.4B). Additional developments are in the pipeline that provide specialized real-time PCR applications (e.g., Scorpions). Several practical applications to detect GMOs in food have been reported based on this technique, and these might lead to a harmonized international modus operandi [30, 31]. Newly developed real-time PCR protocols developed to analyze food-borne pathogens and GMOs have been reviewed recently [32]. It is likely that absolute quantification of GMO gene fragments will be advisable when referring to obligatory threshold levels, and this will lead to specific methodological requirements with regard to accuracy and standardization [33; see also Chapters 10 and 11]. The extreme power of real-time PCR is well illustrated by introducing multiplex PCR approaches that enable synchronous detection of up to 27 genetic variations. In this respect, a new two-dimensional array composed of color and melting temperature (T_m) data was recently described [34].

In summary, recently developed real-time techniques are based on the intensity of the fluorescent dye-modulated light being directly proportional to the amplified DNA-product. The most recent real-time PCR approaches permit the generation of both qualitative and quantitative (absolute as well as relative) GMO observation within several hours, including enhanced sensitivity, high throughput, and the ability to perform simultaneous multiplex reactions as well as minimize unnecessary post-PCR manipulations. However, the absolute quantification of gene fragments using PCR will remain a challenge that depends mainly on the technique employed, on the instrumental hardware and software used, and finally on the availability of highly calibrated standards for each genetic modification.

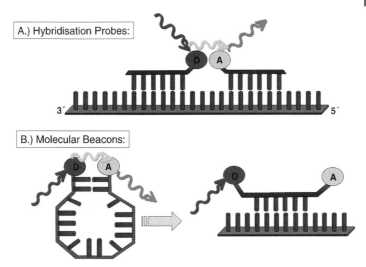

Figure 8.4. Application of further real-time quantification principle based on the fluorescence resonance energy transfer (FRET) method. (A) Hybridization probes: two hybridization probes only generate a specific fluorescence signal when hybridizing side by side on the target DNA. The first dye (donor = D) transfers its energy onto the second dye (acceptor = A). The resulting fluorescence is directly proportional to the number of target molecules. (B) Molecular beacons: a single end-labeled oligonucleotide (D, donor; A, acceptor) emits light only during self-hybridization via FRET. After specific target recognition, the dyes will depart and the specific fluorescence disappears.

8.4.3
Important Bioinformatic Considerations

Without sufficient bioinformatic support, a reliable conclusion and final decision concerning the presence of GMOs within food would be intolerable. Finally, the availability of valuable computer software, gene databases, statistical requirements and normalization procedures should be considered when generating and handling PCR results. Experience has proved that not all selected amplicons will produce good assay results, thereby stressing the importance of the initial search for the most suitable amplicon and probe combinations. Several software packages are available which support primer/probe selection and homology searches in gene databases. Success in both theoretical and practical specificity predictions enables highly reliable PCR-mediated GMO monitoring to be carried out. The desired relative and absolute quantification is not only generated by applying sophisticated mathematical models and normalization strategies, and by shifting genome equivalents (e.g., considering multiple chloroplast versus single nuclear gene copy numbers), but contaminating environmental DNA must also be taken carefully into account. The absolute number of GMO-specific DNA molecules should exceed the number of twenty copies in the isolate to provide for a good statistical probability [35]. As a consequence, standardization of GMO analysis using PCR has been extensively discussed, and several international trials have been reported [7, 36, 37] with acceptable limits of detection that serve the regulatory requirements of each authority [38, 39].

8.5
Alternative and Promising DNA Detection Techniques

In addition to the above-described PCR-based techniques, several alternative methods of DNA detection have now been developed which are based on DNA augmentation. Among these, a number of non-PCR methodologies have been described for the amplification of desired nucleic acids. These rely mainly on primer-based amplification, and variations involving the replication and detection of gene fragments are described in the following text. Although at present data relating to DNA-detection techniques are limited with regard to GMO analysis, future developments will undoubtedly extend the viability of these techniques.

8.5.1
Thermal Cycling Procedures

Ligase chain reactions (LCR) [40] have been introduced to employ either linear or exponential amplification by covalently bridging two adjacent primers through a heat-stable DNA ligase. A subsequent haptene-mediated detection of the ligated oligonucleotides enables the analysis of fewer than 100 target molecules [41]. Reliable monitoring approaches have been established using a combination of PCR, LCR and enzyme-immunoassay detection (PCR-LCR-EIA) when testing for dairy product adulteration [42]. This technique is most advantageous for discrimination between different genotypes using a single base exchange, and may be most valuable in screening transgene DNA. The possible robotic application of such combined assay systems should provide additional verification systems that utilize not only single point mutations but also genetically modified regions within GMOs.

8.5.2
Isothermic Amplification

The nucleic acid sequence-based amplification (NASBA) technique mimics the retroviral strategy of RNA replication to accumulate cDNA as well as RNA [43]. After introducing an artificial T7 promoter sequence through the first primer, NASBA operates continuously by taking advantage of the isothermal mode of action of a T7-DNA-dependent RNA-polymerase amplifying the gene of interest determined by the primers. This procedure will finally achieve rapid, exponential nucleic acid amplification (10^{10}-fold) within 1–2 h [44]. Comparable new techniques have been introduced as rolling cycle amplification (RCA) [45] or ramification amplification (RAM) [46] and strand displacement amplification (SDA), all of which possess their own advantages in the process of multiplying DNA or RNA using isothermal nucleic acid polymerization.

8.5.3
DNA-micro-arrays

The need for rapid high-throughput diagnostic systems has led to the development of miniaturized hybridization and detection technologies that have created a new industrial standard, the micro-array. Highly specific oligonucleotides, fixed *in situ* on carriers suitable for microscopic investigations, were introduced [47]. Further developments now enable a broad range of unique nucleic acid profiling. In order to screen for and to specify any genetically modified ingredient, newly developed micro-array platforms permit a profiling of different GMOs in one step [48]. A specific hybridization of the labeled sample DNA onto fixed capture nucleotides provides redundant information about the quality as well as the quantity of potential transgenes, most of which were analyzed using fluorescence tags (Fig. 8.5A).

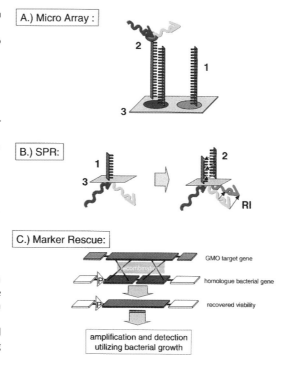

Figure 8.5. Alternative detection methods applicable to monitor genetic modifications. (A) Micro array system representing the selective hybridization of dye-labeled nucleic acids (2) onto gene-specific oligonucleotides (1) fixed in micro-spots on a glass-slide (3). A very high number of distinct genetic information will be provided after a single experiment, due to the possibility of placing numerous probes on one carrier. (B) Surface plasmon resonance (SPR). Surface-immobilized probes (1) interact with free unlabeled sample DNA (2), resulting in a detectable change of the refractive index (RI) of reflected light from a sensor chip surface (thin metal film, 3). (C) Marker rescue using homologue recombination techniques. Genetic recombination between the target gene and an inactive promoter possessing a bacterial homologue leads to the renovation of vital function. Specific growth of transformed bacteria indicates presence of GMO DNA. P = bacterial promoter.

However, several recently developed electrical, enzymatic, scanning force microscopy, capillary electrophoresis and optical detection methods have been considered and successfully applied [49] (for reviews, see Refs. [50, 51]). For example, several microfluidic "Lab-on-a-chip" technologies are currently available for the simultaneous verification and quantification of PCR products [52]. In particular, specially developed functional piezoelectric affinity sensors can detect GMO-DNA-hybridization directly by oligonucleotides which are immobilized on electrode surfaces generating piezoelectric signals [53], and thus specifically indicating the presence of transgenes. Preceding DNA target amplifications are still necessary however to provide sufficient assay sensitivity. Such nano-techniques may provide the future platform for the detection and quantification of GMO DNA in a rapid and (hopefully) cost-effective manner.

8.5.4
Mass Spectrometry (MS) of DNA

PCR products can be rapidly analyzed using mass spectrometry-based methods such as electrospray-MS or MALDI-TOF (matrix-assisted laser desorption/ionization-time-of-flight), both of which were originally developed to investigate other organic polymers such as proteins. Currently, DNA fragments are characterized on the basis of their mass and charge, which permits the sequence determination of short gene fragments of ~200 bp [54]. The application of this rapid and sensitive technique to short DNA sequences, as well as to genetic changes in plant and viral genomes [55], should produce a dramatic reduction in analysis time [56]. Comparison of transgene and conventional genomes by using species-specific genomic fragment libraries as reference should allow rapid discrimination to be made between GMO and non-GMO samples. Recently, short DNA repeats were successfully analyzed using MS techniques for forensic purposes [57]. Although MS analysis of DNA is currently limited by the molecular size of nucleic acids, these detection methods will become increasingly important as high throughput applications become common.

8.5.5
Supplementary Photon-driven Monitoring Methodologies

The method of choice for the analysis of biomolecular nanostructures is fluorescence correlation spectroscopy (FCS), which was developed in the mid-1970s. The principle of the method is the fluctuation analysis of fluorescence intensity, and the technique has been described as a reliable and rapid tool to detect PCR products [58]. In principle, FCS is based on resolution of the thermal fluctuation of single molecules (one partner must be dye-labeled) via fluorescence measurement that is auto-correlated with the particle concentration. FCS-based quantitative PCR has the advantages of being sensitive and both time- and labor-saving, and hence is becoming increasingly important for single molecule detection-based diagnosis (SMDD). Ultimately, the FCS method will most likely be used in a wide variety of biological applications (for a review, see Ref. [59]). More recently, FCS-based PCR has been used in screening experiments, where it utilizes the inter-molecular aggregations of pathogenic nucleic acids [60]. On the basis of recent SMDD data, this detection method should allow the analysis of genetic material, but without prior extensive PCR amplification [61, 62].

Another prominent yet label-free photon-technology suitable for monitoring inter-molecular relationships in very small sample volumes is that of surface plasmon resonance (SPR). This optical technique permits the detection and quantification of changes in refractive index (RI) in the vicinity of a sensor chip's surface to which ligands have been immobilized; this allows any interaction of biomolecules with the ligand to be detected [63]. In detail, one side of a glass prism is coated with a very thin layer of metal to which a DNA probe (single strand) is attached. When a light beam passes through the prism, the energy of the beam interacts with the cloud of electrons in the metal film, causing an oscillation of electrons (= plasmon) [64]. Reflection of the beam occurs at a certain resonance angle. When DNA binds to the probe at the metal film, the composition of the molecular complex on the metal surface changes, thus causing a shift in reflection angle that is characterized as the RI (Fig. 8.5B). By using SPR, the numbers of molecules bound in each complex can be determined. In addition to its use in detecting different DNA molecules after PCR amplification [65], this technology has been successfully applied to detect GMOs in food samples [66]. A major advantage of the method is that it can be combined directly with miniaturized biosensor technologies, thereby enabling DNA measurements to be made without previous probe labeling, and in reusable continuous flow systems.

8.5.6
Novel Biological Monitoring Approaches

A technique which takes advantage of naturally occurring homologue recombination between two distant, but homologous, DNA fragments was developed to detect transgenic targets by artificial microbial gene loci. Following the so-called "marker rescue" method, the finally surviving transformed microorganisms were grown and amplified in culture media, indicating a positive event (Fig. 8.5C) [67]. In principle, deficient mutant strains containing a homologue gene of interest, and a selection medium in which only the reconverted wild-type can survive, are provided. Subsequently, the GMO target can be detected by homologue recombination with the mutant possessing a functional promoter; this leads, after successful recombination, to bacterial growth in selective media. This technique was recently used to detect antibiotic resistance marker genes derived from environmental GMO plant material [68]. In the future, the technique – by using transformation events to detect GMO DNA from different sources – may lead to an interesting monitoring approach in which prior PCR amplification steps can be omitted.

Furthermore, the integration of synthetic genes may interact through positioning effects with the physiology of each GMO. Therefore, remarkable metabolic changes might occur that could be recorded via differential measurement of random expression parameters or distinct candidate transcripts. This would apply to the differential display reverse transcription (DDRT)-PCR technique used to detect changing cDNA-patterns within the GMO plant, or also to seek metabolic changes. New methods are currently under development to monitor such interactions in comparison with the conventional physiological reaction after the transgene event takes place. This will call for sensitive, comprehensive, high-throughput systems that are capable of detecting multiple cellular parameters, and should lead to a holistic insight that might enlighten the multiple biological interactions mediated by any desired genetic modification.

8.6
Conclusions and Future Prospects of GMO Detection Applying DNA-analysis

During the past decade, a range of universal and reliable detection methods based on residual DNA have been developed that are applicable both for the screening and quantification of GMOs in food or feedstuffs. Today, the extensive requirements to seek low yet abundant amounts of GMO DNA in diverse sample materials, as stipulated in the most recent legislative regulations, are carried out exclusively by the use of PCR methods.

It is clear that DNA is the ideal molecule to establish international GMO screening/quantifying systems that lead to a reliable certification, though the development of such ubiquitous methods is currently hindered by uncertainty and the increasing number of genetic modifications in released GMOs. Moreover, due to expected legislative regulations, products from fed animals should also be taken into account when attempting to prove the existence of residual recombinant DNA derived from transgenic feed [69, 70]. Ongoing progressive research should permit the development of techniques to detect traces of GMO material based on DNA methods.

In conclusion, the introduction of DNA-based techniques provides the key advantages of high sensitivity, and robust and rapid operation whilst providing the prerequisites of careful experimental design that avoids both false-negative and/or false-positive results.

Today, real-time PCR seems to be the best developed and most suitable GMO monitoring system, with highly sensitive, specific, reliable, quantitative and rapid detection available (see Chapters 10, 11 and 12). While related methods will clarify food production processes in relation to GMO components, natural DNA is present in virtually all daily food components, and any toxicological fears concerning such food-borne DNA can be excluded. On this basis, a number of safety considerations of transgenic DNA have been discussed [71] and are of further scientific interest. In addition, developed assay methods will not only be used to detect residual GMO DNA, but also to examine their biological potency with regard to possible environmental interactions. Prospectively, the rapidity and sensitivity of DNA-based detection methods used to identify the GMO content of food will, through flexibility and frugality, become the favored method of screening and quantification. In particular, care must be taken when calculating absolute amounts of GMO-specific DNA fragments with regard to sample variability, for example altering natural copy numbers of target genes in the nucleus versus mitochondria or chloroplasts. Currently, the routine quantification of GMO in food or feedstuffs is only reliably performed in specialized laboratories under highly standardized and controlled procedures.

These DNA-based assay systems are cost-intensive and require sophisticated instrumentation and laboratory equipment for their operation. Further perspectives may focus on simple as well as cost-reduced DNA screening methods adjusted to market requirements; these might include rapid test-strip kits which will ease the burden on somewhat limited governmental food-monitoring budgets. Finally, the most convincing advantage of DNA-based methods for detecting GMOs remains their robust nature when used in numerous sample degradations, together with the possibility of massive signal amplification within a short time period if residual DNA is shown to be present. In the future, the analysis of GMOs by use of DNA technology will continue to evolve, with mergers of the most innovative techniques such as micro- and nano-systems satisfying the great demand for food inspections.

References

1 Pääbo, S. (1993) Ancient DNA. *Sci. Am.* **269**, 86–92.
2 Hoss, M. (1995) Ancient DNA. *Horm. Res.* **43**, 118–120.
3 Saiki, R. K., Scharf, S., Faloona, F., Mullis, K. B., Horn, G. T., Erlich, H. A., Arnheim, N. (1985) Enzymatic amplification of beta-globin genomic sequences and restriction site analysis for diagnosis of sickle cell anemia. *Science* **230**, 1350–1354.
4 Lipp, M., Anklam, E., Stave, J. W. (2000) Validation of an immunoassay for detection and quantitation of a genetically modified soybean in food and food fractions using reference materials: interlaboratory study. *J. AOAC Int.* **83**, 919–927.
5 van Duijn, van G. J., van Biert, R., Bleeker-Marcelis, H., Van Boeijen, I., Ada, A. J., Jhakrie, S., Hessing, M. (2002) Detection of genetically modified organisms in foods by protein- and DNA-based techniques: bridging the methods. *J. Am. OAC Int.* **85**, 787–791.
6 Hemmer, W., Pauli, U. (1998) Labelling of food products derived from genetically engineered crops. A survey on detection methods. *Eur. Food Law Rev.* **9**, 27–38.
7 Anklam, E., Gadani, F., Heinze, P. (2002) Analytical methods for detection and determination of genetically modified organisms in agricultural crops and plant-derived food products. *Eur. Food Res. Technol.* **214**, 3–26.
8 Kuiper, H. A. (1999) Summary report of the ILSI Europe workshop on detection methods for novel food derived from genetically modified organisms. *Food Control* **10**, 339–349.
9 Gilbert, J. (1999) Sampling of raw materials and processed foods for the presence of GMOs. *Food Control* **10**, 363–365.
10 Pauli, U., Liniger, M., Zimmermann, A. (1988) Detection of DNA in soybean oil. *Food Res. Technol.* **207**, 264–267.
11 Gryson, N., Ronsse, F., Messens, K., Loose de, M., Verleven, T., Dewettinck, K. (2002) Detection of DNA during the refining of soybean oil. *J. Am. Oil Chem. Soc.* **79**, 171–174.
12 Southern, E. M. (1975) Detection of specific sequences among DNA fragments separated by gel electrophoresis. *J. Mol. Biol.* **98**, 503–517.
13 McCabe, M. S., Power, J. B., de Last, A. M., Davey, M. R. (1997) Detection of single-copy genes in DNA from transgenic plants by nonradioactive Southern blot analysis. *Mol. Biotechnol.* **7**, 79–84.
14 Sabelli, P. A. (1998) Southern blot analysis. In: *Molecular Biomethods Handbook* (Rapley, R., Walker, J. M.), Humana Press, Totowa, NJ, USA.
15 Innis, M. A., Gelfand, D. H., Sninsky, J. J. (1990) *PCR Protocols – A Guide to Methods and Applications.* Academic Press, San Diego.
16 Rapley, R. (1998) Polymerase chain reaction. In: *Molecular Biomethods Handbook* (Rapley, R., Walker, J. M.), Humana Press, Totowa, NJ, USA.

17 Rolfs, A., Schuller, I., Finckh, U., Weber-Rolfs, I. (**1992**) *PCR: Clinical Diagnostics and Research*. Springer, Berlin.

18 Röpke, R., Sauerwein, H., Stoffel, B., Hagen-Mann, K., Ahnen von, S., Meyer, H. H. D. (**1994**) Extended aspects of growth hormone (GH) action: GH receptor (GHR) mRNA expression in various bovine tissues and effects of a long-term treatment with GH on GHR mRNA and on GH binding in liver. *J. Reprod. Dev.* **40**, 227–234.

19 Garcia Canas, V., Gonzalez, R., Cifuenes, A. (**2002**) Detection of genetically modified maize by the polymerase chain reaction and capillary gel electrophoresis with UV detection and laser-induced fluorescence. *J. Agric. Food Chem.* **50**(5), 1016–1021.

20 Matsuoka, T., Kuribara, H., Akiyama, H., Miura, H., Goda, Y., Kusakabe, Y., Isshiki, K., Toyoda, M., Hino, A. (**2001**) A multiplex PCR method of detecting recombinant DNAs from five lines of genetically modified maize. *Shokuhin Eiseigaku Zasshi* **42** (1), 24–32.

21 Matsuoka, T., Kuribara, H., Takubo, K., Akiyama, H., Miura, H., Goda, Y., Kusakabe, Y., Isshiki, K., Toyoda, M., Hino, A. (**2002**) Detection of recombinant DNA segments introduced to genetically modified maize (*Zea mays*). *J. Agric. Food Chem.* **50**, 2100–2109.

22 Lee, D. K., Seok, S. J., Jang, I. C., Nahm, B. H., Kim, J. K. (**1998**) Triprimer-PCR method: rapid and reliable detection of transgenes in transgenic rice plants. *Mol. Cell* **8**, 101–106.

23 Wurz, A., Bluth, A., Zeltz, P., Pfeifer, C., Willmund, R. (**1999**) Quantitative analysis of genetically modified organisms (GMO) in processed food by PCR-based methods. *Food Control* **10**, 385–389.

24 Hupfer, C., Hotzel, H., Sachse, K., Moreano, F., Engel, K. H. (**2000**) PCR-based quantification of genetically modified Bt maize: single-competitive versus dual competitive approach. *Eur. Food Res. Technol.* **212**, 95–99.

25 Brunnert, H. J., Spener, F., Boerchers, T. (**2001**) PCR-ELISA for the CaMV-35S promoter as a screening method for genetically modified Roundup Ready soybeans. *Eur. Food Res. Technol.* **213**, 366–371.

26 Schmittgen, T. D. (**2001**) Real-Time Quantitative PCR. *Methods* **25**, 383–385.

27 Giulietti, A., Overbergh, L., Valckx, D., Decallonne, B., Bouillon, R., Mathieu, C. (**2001**) An overview of real-time quantitative PCR: applications to quantify cytokine gene expression. *Methods* **25**, 386–401.

28 Wittwer, C. T., Herrmann, M. G., Moss, A. A., Rasmussen, R. P. (**1997**) Continuous fluorescence monitoring of rapid cycle DNA amplification. *BioTechniques* **22**, 130–138.

29 Holland, P. M., Abramson, R. D., Watson, R., Gelfand, D. H. (**1991**) Detection of specific polymerase chain reaction product by utilizing the 5-3 exonuclease activity of *Thermus aquaticus* DNA polymerase. *Proc. Natl. Acad. Sci. USA* **88**, 7276–7280.

30 Pietsch, P., Waiblinger, H. U. (**2001**) Quantification of genetically modified soybeans in food with the LightCycler system. In: *Rapid cycle real-time PCR* (Meuer, S., Wittwer, C., Nakagawara, K. eds), Springer, Berlin.

31 Dahinden, I., Zimmermann, A., Liniger, M., Pauli, U. (**2001**) Variation analysis of seven LightCycler-based real-time PCR systems to detect genetically modified products (RRS, Bt176, Bt11, Mon810, T25, Lectin, Invertase). In: *Rapid Cycle Real-time PCR, methods and Applications – Microbiology and Food Analysis* (Reischl, U., Wittwer, C., Cackerill, F., eds), Springer Berlin.

32 Reischl, U., Wittwer, C., Cackerill, F. (**2002**) *Rapid Cycle Real-time PCR methods and Application – Microbiology and food analysis*. Springer, Berlin.

33 Niesters, H. G. (**2001**) Quantification of viral load using real-time amplification techniques. *Methods* **25**, 419–429.

34 Wittwer, C. T., Herrmann, M. G., Gundry, C. N., Elenitoba-Johnson, K. S. J. (**2001**) Real-time Multiplex PCR assays. *Methods* **25**, 430–442.

35 Kay, S., Van de Eede, G. (**2001**) The limits of GMO detection. *Nature Biotechnol.* **19**, 405.

36 Roseboro, K. (**2000**) Standardizing GMO testing. *Food Proc. USA* **61**, 61–63.

37 Pöpping, B. (**2001**) Methods for the detection of genetically modified organisms: precision, pitfalls, and proficiency. *Int. Lab.* **31**, 23–29.

38 Huebner, P., Waiblinger, H. U., Pietsch, K., Brodmann, P. (**2001**) Validation of PCR methods for quantitation of genetically modified plants in food. *J. Am. OAC Int.* **84**, 1855–1864.

39 Einspanier, R. (**2001**) Quantifying genetically modified material in food: searching for a reliable certification. *Eur. Food Res. Technol.* **213**, 415–416.

40 Barany, F. (**1991**) Genetic disease detection and DNA amplification using cloned thermostable ligase. *Proc. Natl. Acad. Sci. USA* **88**, 189–193.

41 Kratochvil, J., Laffler, T. G. (**1994**) Nonradioactive oligonucleotide probes for detecting products of the ligase chain reaction. In: *Methods in Molecular Biology 28* (Isaac, P. G., ed.), Humana Press, Totowa, NJ, USA.

42 Klotz, A., Einspanier, R. (**2001**) Development of a DNA-based screening method to detect cows milk in ewe, goat and buffalo milk and dairy products by PCR-LCR-EIA-technique. *Milk Science International – Milchwissenschaft* **56** (2), 67–70.

43 Guatelli, J. C., Whitfield, K. M., Kwoh, D. Y., Barringer, K. J., Richman, D. D., Gingeras, T. R. (**1990**) Isothermal, in vitro amplification of nucleic acids by a multienzyme reaction modeled after retroviral replication. *Proc. Natl. Acad. Sci. USA* **87**, 1874–1878.

44 Malek, L., Sooknanan, R., Compton, J. (**1994**) Nucleic acid sequence-based amplification (NASBA). In: *Methods in Molecular Biology 28* (Isaac, P. G., ed.), Humana Press, Totowa, NJ, USA.

45 Lizardi, P. M., Huang, X., Zhu, Z., Bray-Ward, P., Thomas, D. C., Ward, D. C. (**1998**) Mutation detection and single-molecule counting using isothermal rolling-circle amplification. *Nature Genet.* **19**, 225–232.

46 Zhang, D. Y., Brandwein, M., Hsuih, T. C. H., Li, H. B. (**1998**) Amplification of target-specific, ligation-dependent circular probe. *Gene* **211**, 277–285.

47 Southern, E., Maskos, U., Elder, R. (**1992**) Analyzing and comparing nucleic acid sequences by hybridization to arrays of oligonucleotides: evaluation using experimental models. *Genomics* **13**, 1008–1017.

48 Aarts H. J., van Rie J. P., Kok E. J. (**2002**) Traceability of genetically modified organisms. *Expert Rev. Mol. Diagn.* **2**(1) 69–76.

49 Hintsche, R., Paeschke, M., Uhlig, A., Seitz, R. (**1997**) Microbiosensors using electrodes made in Si-technology. *EXS* **80**, 267–283.

50 Möller, R., Csaki, A., Köhler, M., Fritsche, W. (**2000**) DNA probes on chip surfaces studied by scanning force microscopy using specific binding of colloidal gold. *Nucleic Acids Res.* **28**, e91.

51 Müller, O., Hahnenberger, K., Dittmann, H., Yee, H., Nagle, R., Iisley, D. (**2000**) A microfluidic system for high speed reproducible DNA sizing and quantification. *Electrophoresis* **21**, 128–134.

52 Birch, L., Archard, C. L., Parkes, H. C., McDowell, D. G. (**2001**) Evaluation of LabCGip technology for GMO analysis in food. *Food Control* **12**, 535–540.

53 Minunni, M., Tobelli, S., Pratesi, S., Mascini, M., Piatti, P., Bogani, P., Buiatti, M., Mascini, M. (**2001**) A piezoelectric affinity biosensor for genetically modified organisms (GMOs) detection. *Anal. Lett.* **34**, 825–840.

54 Chapman, J. R. (**1998**) Mass spectrometry. In: *Molecular Biomethods Handbook* (Rapley, R., Walker, J. M., eds), Humana Press, Totowa, NJ, pp. 669–696.

55 Amexis, G., Oeth, P., Abel, K., Ivshina, A., Pelloquin, F., Cantor, C. R., Brau, A., Chumakov, K. (**2001**) Quantitative mutant analysis of viral quasispecies by chip-based MALDI-TOF mass spectrometry. *Proc. Natl. Acad. Sci. USA* **98**, 12097–12102.

56 Larsen, L. A., Christiansen, M., Vuust, J., Andersen, P. S. (**2001**) Recent developments in high-throughput muta-

57. Carey, L., Mitnik, L. (2002). Trends in DNA forensic analysis. *Electrophoresis* 23, 1386–1397.
58. Björling, S., Kinjo, M., Földes-Papp, Z., Hagman, E., Thyberg, P., Rigler, R. (1998) Fluorescence correlation spectroscopy of enzymatic DNA polymerization. *Biochemistry* 37, 12971–12978.
59. Thompson, N. L. (1991) Fluorescene Correlation Spectroscopy In: *Topics in Fluorescence Spectroscopy* (Lakowicz, J. R., ed.) Plenum Press, NY, pp. 337–378.
60. Walter, N. G., Schwille, P., Eigen, M. (1996) Fluorescence correlation analysis of probe diffusion simplifies quantitative pathogen detection by PCR. *Biochemistry* 93, 12805–12810.
61. Kinjo, M., Rigler, R. (1995) Ultrasensitive hybridization analysis using fluorescence correlation spectroscopy. *Nucleic Acids Res.* 23, 1795–1799.
62. Kask, P., Palo, K., Ullmann, D., Gall, K. (1999) Fluorescence-intensity distribution analysis and its application in biomolecular detection technology. *Proc. Natl. Acad. Sci. USA* 96, 13756–13761.
63. McDonnell, J. M. (2001) Surface plasmon resonance: towards an understanding of the mechanisms of biological molecular recognition. *Curr. Opin. Chem. Biol.* 5, 572–577.
64. Alberts, B., Johnson, A., Lewis, J., Raff, M., Roberts, K., Walter, P. (2002) *Molecular Biology of the Cell*. Garland Science, Taylor & Francis Group, NY.
65. Kai, E., Sawata, S., Ikebukuro, K., Iida, T., Honda, T., Karube, I. (1999) Detection screening. *Pharmacogenomics* 2, 387–399.
   tion of PCR products in solution using surface plasmon resonance. *Anal. Chem.* 15, 796–800.
66. Feriotto, G., Borgatti, M., Mischiati, C., Bianchi, N., Gambari, R. (2002) Biosensor technology and surface plasmon resonance for real-time detection of genetically modified roundup ready soybean gene sequences. *J. Agric. Food Chem.* 27, 955–962.
67. Contente, S., Dubnau, D. (1979) Marker rescue transformation by linear plasmid DNA in *Bacillus subtilis*. *Plasmid* 2, 555–571.
68. de Vries, de J., Wackernagel, W. (1998) Detection of nptII (kanamycin resistance) genes in genomes of transgenic plants by marker-rescue transformation. *Mol. Gen. Genet.* 257, 606–613.
69. Einspanier, R., Klotz, A., Kraft, J., Aulrich, K., Poser, R., Schwaegele, F., Jahreis, G., Flachowsky, G. (2001) The fate of forage plant DNA in farm animals: a collaborative case-study investigating cattle and chicken fed recombinant plant material. *Eur. Food Res. Technol.* 211, 129–134.
70. Klotz, A., Mayer, J., Einspanier, R. (2002) Carry over of feed-DNA into farm animals: first investigations with pigs, chicken embryos and commercial available chicken samples. *Eur. Food Res. Technol.* 214, 271–275.
71. Jonas, D. A., Elmadfa, I., Engel, K. H., Heller, K. J., Kozianowski, G., König, A., Müller, D., Nnarbonne, J. F., Wackernagel, W., Kleiner, J. (2001) Safety considerations of DNA in food. *Nutr. Metab.* 45, 235–254.

9
Genetic Engineering of Fishes and Methods for Detection

Hartmut Rehbein

9.1
Introduction

After the end of the second world war, the fisheries industries experienced a steady, strong and longlasting increase in its activities, but this was brought to an end during the 1990s. In fact, in 2000, according to FAO statistics, the total catch of fish, shrimps and mollusks amounted to only 130×10^6 tons, of which 27% was delivered by aquaculture [1]. Fish consumption varies considerably between countries, but from a global viewpoint fishery products are an important source of protein (~16%) for human nutrition [2]. The growing demand for high-quality seafood protein can be met only by the future rise of aquaculture, as the resources of capture fisheries are becoming maximally exploited [3].

The genetic engineering of aquatic organisms may have great potential for the expansion of the farming of fish and shellfish to be used for human consumption [4], and also offers interesting possibilities for the use of transgenic fish as biofactories in the production of pharmaceuticals [5].

During the past decade, extensive progress has been made in the development and farming of transgenic fish, including Atlantic salmon (*Salmo salar*). Indeed, Atlantic salmon carrying an "all-fish-gene cassette" consisting of an ocean pout (*Macrozoarces americanus*) antifreeze protein promoter coupled to the growth-hormone gene from the Pacific Chinook salmon (*Oncorhynchus tshawytscha*) [6, 7] may be the first marketable transgenic animal for human consumption. The US/Canadian company Aqua Bounty Farms has applied to the US Food and Drug Administration for permission to market this salmon [8], and several other transgenic fish species have also now been patented [9, 10].

Research into transgenic shellfish and mollusks is still in its early stages, as the genetic and metabolic background of these animal groups is much less well understood than for fish [11].

Nonetheless, excellent progress has been made recently in the transfer of genes into oyster embryos [12], and into the gonads of crustaceans [13] and other species [11].

This chapter will focus on the production, detection, and food safety of genetically modified food fish. Topical information on ecological problems caused by the farming of transgenic fish can be found elsewhere [14].

9.2
Development and Production of Transgenic Fish

Genetic engineering of fish has substantial advantages compared with traditional breeding technology. The ability to introduce specific genetic elements offers the possibility of achieving a desired goal in a relatively short time (i.e., within a few years), with fast-growing transgenic salmon and tilapia being prominent examples. Using this approach, stable integration inheritance of the construct can be achieved within three to four generations.

In addition to growth enhancement, the use of transgenic fish provides several other advantages (Table 9.1), including better resistance against diseases and superior utilization of feed — both factors of major importance in the improvement of farmed fish production.

During the past decade, more than thirty fish species have been genetically modified in different countries such as the USA, Canada, UK, Cuba, Israel, China, and Finland. Some examples of transgenic fish are listed in Table 9.2.

Table 9.1. Benefits expected from transgenic fish [5, 17].

- Acceleration of fish growth
- Increase in overall size
- Higher yield of fillet
- Increase of food conversion efficiency
- Superior utilization of carbohydrates as low-cost diet
- Cold tolerance
- Freeze resistance
- Control of sex and reproductio
- Stress resistance
- Improved sensory properties (color, odor, flavor, taste, texture)
- Improvement of nutritional properties

Table 9.2. Examples of transgenic fish.

Fish species	Characteristic	Reference(s)
Carp, *Cyprinus carpio*	Enhanced growth	18–20
Channel catfish, *Ictalurus punctatus*	Enhanced growth	21
Mud loach, *Misgurnus mizolepis*	Enhanced growth	22
Tilapia, *Oreochromis hornorum* hybrid	Enhanced growth	23
Tilapia, *Oreochromis niloticus*	Enhanced growth	24
Rainbow trout, *Oncorhynchus mykiss*	Enhanced growth	25
Coho salmon, *Oncorhynchus kisutch*	Enhanced growth	26, 27
Cutthroat trout, *Oncorhynchus clarki*	Enhanced growth	27
Chinook salmon, *Oncorhynchus tshawytscha*	Enhanced growth	27
Atlantic salmon, *Salmo salar*	Enhanced growth	7, 28
Arctic charr, *Salvelinus alpinus*	Enhanced growth	29
Channel catfish	Disease resistance	9
Atlantic salmon	Antifreeze protein	30
Rainbow trout	Biosynthesis of ascorbic acid	31
Rainbow trout	Better carbohydrate utilization	31
Arctic charr	Better carbohydrate utilization	32

9.2.1
Structure of Gene Cassettes

A typical "gene cassette" or "construct" consists of the following elements:

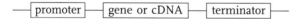

Two types of gene cassettes have been used for the transformation of fish. During the initial phases of these experiments, the elements of viral, bacterial, and cold- or warm-blooded animals were combined (for a review, see Ref. [15]). Although originally the human growth hormone (GH) gene was introduced into more than ten fish species, more recently it was found preferable to fuse elements taken from other fish, or even from the same species, for two reasons: (i) more efficient expression of the transgene; and (ii) better consumer acceptance of transgenic fish. Some examples of constructs are listed in Table 9.3. Promoters of the following fish genes were found eventually to be highly effective in most cases:

- the metallothionein promoter of sockeye salmon (*Oncorhynchus nerka*), which could drive growth hormone transgene expression in liver and many other cell types [16];
- the antifreeze protein promoter of ocean pout [6]; and
- the sockeye salmon histone H3 promoter [34].

Growth enhancement was the first motive for the construction of genetically modified fish, and as a consequence growth hormone genes (respectively their cDNA) formed the central part of most of the gene cassettes. Terminators, which are necessary for correct termination of transcription, either came from the genes introduced, or had to be inserted in case of cDNA [6].

9.2.2
Methods of Gene Transfer

Microinjection into fertilized eggs was chosen by most groups to introduce constructs. The pros and cons of other techniques, such as electroporation, sperm-mediated transfer and high-velocity microprojectile bombardment, have also been discussed [35]. Uptake of microinjected DNA was found to be considerably enhanced by using noncovalent DNA-nuclear transport peptide complexes [36].

9.2.3
Evidence for Gene Transfer and Expression

The methods of gene transfer currently applied do not allow site-specific introduction of a single copy of a construct into the genome of the fish to be modified. Thus, an unpredictable number of constructs may be integrated into the genome of the host. Another complicating effect in the production of transgenic fish is mosaicism in the first generation [5]. As the constructs normally do not integrate into the host genome prior to the first division of the egg, a substantial number of cells of the embryo normally lack the transgene [5].

Hybridization techniques and polymerase chain reaction (PCR)-based methods have been used to follow the fate of introduced gene cassettes. For example, evidence was obtained by means of Southern blotting for antifreeze protein (AFP) gene transfer to Atlantic salmon. DNA from two out of a group of thirty-two fingerlings gave positive hybridization signals when probed with a labeled AFP gene sequence [37]. The restriction pattern shown by the DNA was consistent with that of the injected DNA, indicating that nondegraded construct was present in the salmon.

In another case, Southern blotting was performed to identify uptake and integration of the OPAFPcsGH construct by tilapia, and to estimate the number of copies present in specimens of transgenic fish [24].

PCR offers many possibilities to control gene transfer by placing primers into different elements of the construct. The event of integration of the construct into the genome of the host can be detected by locating one primer into the

Table 9.3. Gene cassettes used for construction of transgenic food fish.

Gene cassette	Fish species	Reference
P: Rous sarcoma virus long terminal repeat G: Rainbow trout growth hormone 1 cDNA	Carp, *Cyprinus carpio*	18
P: Mouse metallothionein G: Human growth hormone 1 gene	Carp, *Cyprinus carpio*	19
P: Carp β-actin G: Chinook salmon growth hormone cDNA	Carp, *Cyprinus carpio*	20
P: Giant silk moth Cecropin B G: Giant silk moth Cecropin B gene	Channel catfish, *Ictalurus punctatus*	9
P: Rous sarcoma virus long terminal repeat G: Rainbow trout growth hormone cDNA G: Coho salmon growth hormone cDNA	Channel catfish, *Ictalurus punctatus*	21
P: Cytomegalovirus G: Tilapia growth hormone cDNA T: Simian virus 40 poly A site	Tilapia, *Oreochromis hornorum* hybrid	23
P: Ocean pout antifreeze protein G: Chinook salmon growth hormone cDNA T: Ocean pout antifreeze protein 3' region	Tilapia, *Oreochromis niloticus*	24
P: Mud loach β-actin G: Mud loach growth hormone gene	Mud loach, *Misgurnus mizolepis*	22
P: Cytomegalovirus G: Sockeye salmon growth hormone 1	Arctic charr, *Salvelinus alpinus*	29
P: Cytomegalovirus G: Human glucose transporter type 1 cDNA T: Bovine growth hormone polyA site	Arctic charr, *Salvelinus alpinus*	32
P: Sockeye salmon metallothionein G: Rat hexokinase type II cDNA T: Sockeye salmon growth hormone 1	Arctic charr, *Salvelinus alpinus*	32
P: Sockeye salmon histone 3 G: Rat hexokinase type II cDNA T: Sockeye salmon growth hormone 1	Arctic charr, *Salvelinus alpinus*	32
P: Sockeye salmon metallothionein G: Human glucose transporter type 1 cDNA T: Sockeye salmon growth hormone 1	Arctic charr, *Salvelinus alpinus*	32
P: Sockeye salmon histone 3 G: Human glucose transporter type 1 cDNA T: Sockeye salmon growth hormone 1	Arctic charr, *Salvelinus alpinus*	32
P: Cytomegalovirus G: Human glucose transporter type 1 cDNA T: Bovine growth hormone polyA site	Rainbow trout, *Oncorhynchus mykiss*	32

Table 9.3. (continued)

Gene cassette	Fish species	Reference
P: Sockeye salmon metallothionein G: Rat hexokinase type II cDNA T: Sockeye salmon growth hormone 1	Rainbow trout, *Oncorhynchus mykiss*	32
P: Sockeye salmon histone 3 G: Rat hexokinase type II cDNA T: Sockeye salmon growth hormone 1	Rainbow trout, *Oncorhynchus mykiss*	32
P: Sockeye salmon metallothionein G: Human glucose transporter type 1 cDNA T: Sockeye salmon growth hormone 1	Rainbow trout, *Oncorhynchus mykiss*	32
P: Sockeye salmon histone 3 G: Human glucose transporter type 1 cDNA T: Sockeye salmon growth hormone 1	Rainbow trout, *Oncorhynchus mykiss*	32
P: Cytomegalovirus G: Rat gulonolactone oxidase cDNA	Rainbow trout, *Oncorhynchus mykiss*	31
P: Sockeye salmon metallothionein G: Rat gulonolactone oxidase cDNA	Rainbow trout, *Oncorhynchus mykiss*	31
P: Cytomegalovirus G: Rat hexokinase type II cDNA	Rainbow trout, *Oncorhynchus mykiss*	33
P: Cytomegalovirus G: Human glucose transporter type 1 cDNA	Rainbow trout, *Oncorhynchus mykiss*	33
P: Sockeye salmon metallothionein-B G: Sockeye salmon growth hormone 1 gene	Rainbow trout, *Oncorhynchus mykiss*	25
P: Ocean pout antifreeze protein G: Chinook salmon growth hormone cDNA	Rainbow trout, *Oncorhynchus mykiss*	27
P: Ocean pout antifreeze protein G: Chinook salmon growth hormone cDNA	Cutthroat trout, *Oncorhynchus clarki*	27
P: Ocean pout antifreeze protein G: Chinook salmon growth hormone cDNA	Chinook salmon, *Oncorhynchus tshawytscha*	27
P: Sockeye salmon metallothionein B G: Sockeye salmon growth hormone 1 gene	Coho salmon, *Oncorhynchus kisutch*	26
P: Ocean pout antifreeze protein G: Chinook salmon growth hormone cDNA	Coho salmon, *Oncorhynchus kisutch*	27
P: Ocean pout antifreeze protein G: Chinook salmon growth hormone cDNA	Atlantic salmon, *Salmo salar*	7, 28
G: Winter flounder antifreeze protein gene	Atlantic salmon, *Salmo salar*	30

P = Promoter; G = gene; T = terminator.

genome of the host, and the second one into an element of the construct. For example, several PCR systems were used to screen Atlantic salmon for growth hormone transgenesis. The construct injected into the salmon consisted of an ocean pout antifreeze protein gene promoter linked to Chinook salmon growth hormone cDNA, and as the terminator of the ocean pout antifreeze protein gene 3' region (OPAFPcsGHc) [7]. By using different pairs of primers located in the promoter, gene, and terminator, amplicons ranging in size from 199 to 855 base pairs (bp) could be produced. Furthermore, PCR results were confirmed by hybridization with a GHcDNA probe. These PCR systems proved to be very convenient for identifying positive transgenic fish which possessed the elements of the construct in correct sequence [37].

In several experiments the transgenic fish did not have the characteristics expected from gene transfer. Typical examples included Atlantic salmon possessing AFP genes [37], or Arctic charr and rainbow trout, which were supplied with two genes for more efficient utilization of carbohydrates [38]. Expression of the transferred growth hormone gene or antifreeze protein gene was analyzed using Northern blotting and reverse-transcribed (RT)-PCR of mRNA. Using the latter technique, GHmRNA was detected in liver, gill and spleen, as well as in the pituitary of transgenic Atlantic salmon [37].

The final product of the transgene – a peptide or protein – can be detected and characterized by commonly used methods of protein or enzymatic analysis. In order to test the efficiency of injection techniques or promoters, reporter genes coding for fluorescent protein [39], chloramphenicol acetyl transferase or β-galactosidase [40] have been used, as these gene products are easily detectable.

Antifreeze protein was measured in transgenic Atlantic salmon [37], and the level of growth hormone was determined in genetically modified tilapia [41].

9.3
Examples of Successful Production of Transgenic Fish

The details and characteristics of five types of transgenic fish, all of which have been reared over several generations, are provided in the following sections.

9.3.1
Atlantic Salmon

The transgenic Atlantic salmon is a prominent representative of the "Blue Revolution", which refers to the enhanced productivity of aquaculture by biotechnology for sustainable food supply to feed the world's growing population. Doubts have been expressed as to salmon being the correct species for this "revolution", or whether other species such as carp or tilapia, utilizing low-cost plant feed, are better candidates for transgenic farmed fish in developing countries [17].

About ten years ago it was first reported that transgenic Atlantic salmon carrying an "all-fish-gene cassette" showed a dramatic increase in growth enhancement, on average three- to five-fold [7]. This fish reaches commercial size (3–4 kg) in approximately half the time needed by standard nontransgenic salmon. The genetic modification, OPAFPcsGHc has been published in detail, as mentioned in the previous section. In the meantime, several generations of homozygous transgenic fish have been produced. Recently, in a series of three reports, the growth rate, body composition and feed digestibility/conversion and other physiological properties of the F_2 generation of growth-enhanced transgenic salmon have been reported [43, 44].

Over a presmolt growth interval of 8 to 55 g, the transgenic fish exhibited an almost three-fold greater rate of growth relative to nontransgenic salmon, accompanied by a more than two-fold increase in daily feed consumption. Coefficients of protein and energy digestibility were in the same range for transgenics and controls, but transgenic salmon relative to control fish exhibited a 10% improvement in gross feed conversion efficiency. Body protein, dry matter, ash, lipid and energy were significantly lower in the transgenic salmon compared with controls, while moisture content was significantly higher.

9.3.2
Pacific Salmon

Coho salmon (*O. kisutch*) is another example of extraordinary growth enhancement being achieved by the introduction of a growth hormone gene, in this case an "all-salmon gene cassette" [26]. The construct (pOnMTGH1) consisted of the metallothionein-B promoter fused to the full-length type 1 growth hormone gene, both from sockeye salmon, *O. nerka*. Different aspects of physiology and metabolism of transgenic coho have been studied during recent years to obtain more insight into the effects of elevated growth hormone levels. Extreme growth enhancement was accompanied by phenotypic effects. Transgenesis led to changes in body and head shape, with the dorsal caudal peduncle and abdominal regions being distinctly enlarged compared with controls [45]. It was also observed that transgenic coho salmon possessed an enlarged surface area of the intestine compared with nontransgenics of the same size [46].

A study of muscle biochemistry and physiology in growth-enhanced transgenic coho salmon relative to older controls of the same size revealed a number of differences between the two groups [47]. Compared with the nontransgenic fish, the transgenics had more red muscle, higher numbers of small-diameter muscle fibers in both the dorsal and lateral regions of the somitic muscle, and higher activities of phosphofructokinase and cytochrome oxidase in white muscle. Subtractive hybridization of muscle RNA of transgenic fish from control fish provided a library of cDNAs, the expression of which was up-regulated in the transgenic fish. By sequencing of some of the fragments, cDNAs similar to mRNAs of myosin light chain 2, β-parvalbumin, skeletal alpha-actin and myosin heavy chain were identified. These differences in gene expression, together with the above-mentioned changes in muscle architecture and enzyme activities may have a significant influence on meat quality.

9.3.3
Tilapia (*O. hornorum* hybrid)

Tilapia, which belong to the family *Cichlidae*, are warm-water fish that feed mainly on plants, but also on small invertebrates. At least ten of more than sixty tilapia species are used for human consumption. Intensive research on transgenic tilapia has been carried out during the past decade in Cuba, and this has resulted in transgenic lines with moderate growth acceleration, without causing any detrimental effects to the fish [48]. The tilapia were genetically modified using a construct containing the human cytomegalovirus 5′ regulatory sequences linked to the tilapia growth hormone cDNA and the polyadenylation site from simian virus 40: CMV-tiGh-CAT-SV40. The transgene was stably transmitted to F_4 generations in a Mendelian fashion, indicating that the gene cassette was integrated into the host genome.

Under optimal rearing conditions, transgenic tilapia grew 60–80 % faster than nontransgenic siblings, but under less favorable experimental conditions either small or no differences in growth rate were observed, as was the case in the study discussed here [49]. Transgenic tilapia exhibited about 3.6-fold less relative food consumption than nontransgenic controls, but had a significantly higher (290 %) food conversion efficiency. The efficiency of growth, protein synthesis retention, anabolic stimulation and average protein synthesis were also enhanced in transgenic fish. In juvenile tilapia, but not in adult fish, glutamate oxaloacetate transaminase (GOT) and glutamate pyruvate transaminase (GPT) activities in transgenic fish were enhanced in muscle and reduced in liver, compared with controls. It was concluded that the GH-transgenic fish utilized the energy released by oxidation of amino acids more efficiently [49].

9.3.4
Tilapia (*O. niloticum*)

Another tilapia species, *O. niloticum*, was genetically modified by injecting the construct OPAFPcsGHc, which had also been used to produce transgenic Atlantic salmon (see section 9.3.1) [24]. Expression of Chinook salmon growth hormone resulted in considerable growth enhancement. The average weight of transgenic fish was three- to four-fold greater than that of their nontransgenic siblings. Two growth and nutritional trials were performed to obtain a deeper insight into the physiological effects of growth hormone gene transfer [50]. At 7 months, the mean mass of transgenic tilapia was 2.5-fold that of control fish. Significant increases in head:total length ratio, viscera-somatic index and hepato-somatic index were observed in transgenic fish. Female and male gonado-somatic indices were also found to differ between transgenics and controls.

In a shorter-term growth performance trial, transgenic tilapia were shown to be more efficient utilizers of protein, dry matter and energy.

9.3.5
Carp

Similar observations were made for transgenic carp, C. carpio, produced by microinjection of a construct consisting of the mouse metallothionein promoter fused to the human growth hormone 1 gene into fertilized eggs [19]. The F_4 generation of transgenic carp fed diets of 20, 30 or 40% crude protein showed 2- to 2.5-fold higher growth rates than the nontransgenic controls. Amounts of recovered protein were significantly higher in transgenics than in controls fed all diets, whereas recovered energy was significantly higher in transgenic fish which were fed the 40% protein diet. For fish fed each diet, the transgenics had higher body contents of dry matter and protein, but lower content of lipids than the controls.

9.4
Methods to Detect Processed Transgenic Fish

In case of the five examples of transgenic fish described in the previous section, PCR-based methods have been published for the detection of transgenics. These methods have been used to differentiate those fish carrying the construct from nontransgenics, and can also be used for the identification of transgenic processed fish. However, as DNA may be degraded in the fish products, the fragments to be amplified should not be too long, i.e., less than 500 bp [51].

For example, a recently reported PCR-based method to identify genetically modified coho salmon has been developed and evaluated in a collaborative study [52]. The coho salmon contained an "all-salmon" gene construct (OnMTGH1) (see section 9.3.2). Two PCRs were performed; the first served as a control for amplifiable PCR (463 bp amplicon) present in the sample, and at the same time as a tool for species identification using the PCR-restriction fragment length polymorphism (RFLP) test. The second PCR, working with primers located in construct (one of the primers located in the promoter, the other within the growth hormone gene) detected the genetic alteration. The amplicon of the second PCR was 427 bp in length, and this was confirmed by RFLP or sequencing.

In a collaborative study conducted in fifteen laboratories, in each case the species and the genetic modification had been assigned correctly.

9.5
Food Safety of Transgenic Fish

For the evaluation of food safety of growth hormone-transgenic fish, the principle of "substantial equivalence" may be applied. The OECD's group of National Experts of Safety in Biotechnology came to the conclusion "that no issue could be identified which reduced or invalidated the application of the principle of substantial equivalence to food or food components derived from modern aquatic biotechnology" [53]. This means that wild or farmed fish of the same species should be used for comparison. Fishery products made from transgenic fish should be evaluated on the basis of products made from nontransgenics. It was concluded that the application of bioengineering does not result in itself in a special risk to the consumer.

This position was recently confirmed by the results of a study of safety considerations of DNA in food [54]. It is clear that transgenic fish containing "all-fish gene cassettes" will not pose any risk to the consumer arising from uptake of the construct. However, in cases where constructs consist partly of nonfish elements, special risks (e. g., by integration of DNA into human or gut microflora genome) could not be identified. Uptake, degradation, and metabolism of considerable amounts (100–1000 mg per person per day) of foreign DNA is a normal process in human life.

The food safety of transgenic fish has been addressed in two publications [55, 56]. The consequences of eating foreign DNA has been considered above, but other aspects which should be considered in respect of health risks for consumers are: (i) the gene product; and (ii) potential pleiotropic effects.

9.5.1
The Gene Product

Until now, in most cases of transgenic food fish the gene product has been a fish growth hormone, either from the same species (e. g., in the case of tilapia, tiGH) or from another fish species. The evaluation of safety of consuming tiGH-transgenic tilapia included a study of the effects of injection of recombinant tiGH into macaques, and a test of *in vitro* activity of tiGH on rabbit xiphoid cartilage bioassay [57]. From the results of both experiments, the conclusion was drawn that tiGH had no effect of metabolism of higher vertebrates, and especially no biological activity in nonhuman primates. Furthermore, no adverse effects were detected in human healthy volunteers after short-term consumption of transgenic tilapia.

While the concentration of growth hormone is low in organisms, other proteins such as lysozyme or antifreeze proteins need a much higher concentration for proper fulfillment of their function [37]. Transgenic fish possessing enhanced amounts of these proteins should be analyzed in respect of the allergic properties of these proteins [58]. These proteins are naturally occurring in a number of fish species, but may not be expected in others. If there is any suggestion that they have allergenic potential, then correct labeling of transgenic fish will be mandatory.

9.5.2
Pleiotropic Effects

These are defined as indirect changes of the phenotype (metabolism, composition, appearance) in consequence of altering a gene. Pleiotropic effects have been found in studies concerning the physiology of transgenic coho salmon (see section 9.3.2). The observation of enhanced gene expression of parvalbumin, a well-known major human allergen in food fish, was identified as a potential risk to consumer health [47].

Pleiotropic effects should be considered very seriously if transgenic fish are to be produced from species which possess (naturally) any undesirable or even toxic compounds, or if such compounds are produced during processing and storage. For example, gadoid fish (e. g., cod, *Gadus morhua*) possess trimethylamine oxide demethylase activity, and this may lead to the formation of formaldehyde and dimethylamine during the frozen storage of products [59]. Some other fish species (e. g., scombroids) contain high amounts of precursors (imidazole compounds) of biogenic amines in the fillet.

Pleiotropic effects may be less serious than anticipated with regard to the presence of toxins [56]. Some toxic fish species (e. g., pufferfish) do not produce toxins endogenously, although this has not been proven for all species used for human consumption. For example, redfish (*Sebastes* spp.) or eel (*Anguilla* spp.) possess toxic proteins in the skin or mucus, the origin and properties of which are largely unknown [60].

References

1. FAO *Fishstat Plus*. Rome, **2002**.
2. FAO *Aquaculture Production Statistics* 1984–1993. Rome, **1995**.
3. G. Hubold. *Infn. Fischwirtsch. Fischereiforsch.* **2000**, *47*, 163–179.
4. J. de la Fuente, F. O. Castro (eds), *Gene Transfer in Aquatic Organisms*. Springer, Berlin, **1998**.
5. N. Maclean, R. J. Laight. *Fish and Fisheries* **2000**, *1*, 146–172.
6. S. J. Du, Z. Gong, C. H. Tan, G. L. Fletcher, C. L. Hew. *Mol. Mar. Biol. Biotech.* **1992**, *1*, 290–300.
7. S. J. Du, Z. Gong, G. L. Fletcher, M.A, Shears, M. J. King, D. R. Idler, C. L. Hew. *Bio/Technology* **10**, 176–181.
8. T. Reichhardt. *Nature* **2000**, *406*, 10–12.
9. R. K. Cooper, F. M. Enright, US Patent No. 5,998,698.
10. N. G. McKeown Sherwood, D. B. Parker, J. E. McRory, D. W. Lescheid, US Patent No. 5,695,954.
11. M. Gomez-Chiarri, G. J. Smith, J. de la Fuente, D. A. Powers, in: *Gene Transfer in Aquatic Organisms* (J. de la Fuente, F. O. Castro, eds), Springer, Berlin **1998**, 107–125.
12. J. T. Buchanan, A. D. Nickens, R. K. Cooper, T. R. Tiersch. *Mar. Biotechnol.* **2001**, 322–335.
13. A. Sarmasik, C. Z. Chun, I. K. Jang, J. K. Lu, T. T. Chen. *Mar. Biotechnol.* **2001**, *3*, S177–S184.
14. E. Hallerman, D. King, A. Kapuscinski. *Aquaculture* **1999**, *173*, 309–318.
15. T. J. Pandian, L. A. Marian. *Curr. Sci.* **1994**, *66*, 635–649.
16. T. Mori, R. H. Devlin. *Mol. Cell. Endocrinol.* **1999**, *149*, 129–139.
17. R. A. Dunham. *J. World Aquacult. Soc.* **1999**, *30*, 1–11.
18. N. Chatakondi, R. T. Lovell, P. L. Duncan, M. Hayat, T. T. Chen, D. A. Powers, J. D. Weete, K. Cummins, R. A. Dunham. *Aquaculture* **1995**, *138*, 99–109.
19. C. Fu, Y. Cui, S. S. O. Hung, Z. Zhu. *J. Fish Biol.* **1998**, *53*, 115–129.
20. Y. Hinits, B. Moav. *Aquaculture* **1999**, *173*, 285–296.
21. R. A. Dunham, A. C. Ramboux, P. L. Duncan, M. Hayat, T. T. Chen, C. M. Lin, K. Kight, I. Gonzalez-Villasenor, D. A. Powers. *Molec. Mar. Biol. Biotechnol.* **1992**, *1*, 380–389.
22. Y. K. Nam, J. K. Noh, Y. S. Cho, H. J. Cho, K.-N. Cho, C. G. Kim, D. S. Kim. *Transgenic Res.* **2001**, *10*, 353–362.
23. J. de la Fuente, I. Guillen, R. Martinez, M. P. Estrada. *Genetic Analysis: Biomolecular Engineering* **1999**, *15*, 85–90.
24. M. A. Rahman, N. Maclean. *Aquaculture* **1999**, *173*, 333–346.
25. R. H. Devlin, C. A. Biagi, T. Y. Yesaki, D. E. Smailus, J. C. Byatt. *Nature* **2001**, *409*, 781–782.
26. R. H. Devlin, T. Y. Yesaki, C. A. Biagi, E. M. Donaldson, P. Swanson, W.-K. Chan. *Nature* **1994**, *371*, 209–210.
27. R. H. Devlin, T. Y. Yesaki, E. M. Donaldson, A. J. Du, C.-L. Hew. *Can. J. Fish. Aquat. Sci.* **1995**, *52*, 1376–1384.
28. C. L. Hew, G. L. Fletcher. **1996**, US Patent No. 5,545,808.

29 A. Krasnov, J. J. Agren, T. I. Pitkänen, H. Mölsä. *Genetic Analysis: Biomolecular Engineering* **1999**, *15*, 99–105.
30 C. L. Hew, P. L. Davies, G. Fletcher. *Mol. Mar. Bio. Biotechnol.* **1992**, *1*, 309–317.
31 A. Krasnov, T. I. Pitkänen, H. Mölsä. *Genetic Analysis: Biomolecular Engineering* **1999**, *15*, 315–319.
32 T. I. Pitkänen, A. Krasnov, M. Reinisalo, H. Mölsä. *Aquaculture* **1999**, *173*, 319–332.
33 A. Krasnov, T. I. Pitkänen, M. Reinisalo, H. Mölsä. *Marine Biotechol.* **1999**, *1*, 25–32.
34 W.-K. Chan, R. H. Devlin. *Mol. Mar. Biol. Biotechnol.* **1993**, *2*, 308–318.
35 T. J. Pandian, L. A. Marian. *Curr. Sci.* **1994**, *66*, 635–649.
36 M.-R. Liang, P. Aleström, P. Collas. *Mol. Reprod. Dev.* **2000**, *55*, 8–13.
37 C. L. Hew, G. L. Fletcher, P. L. Davies. *J. Fish Biol.* **1995**, *47* (Supplement A), 1–19.
38 A. Krasnov, T. I. Pitkänen, H. Mölsä. *Genetic Analysis: Biomolecular Engineering* **1999**, *15*, 115–119.
39 K. Hamada, K. Tamaki, T. Sasado, Y. Watai, S. Kani, Y. Wakamatsu, K. Ozato, M. Kinoshita, R. Kohno, S. Takagi, M. Kimura. *Mol. Mar. Biol. Biotechnol.* **1998**, *7*, 173–180.
40 I. Guillen, R. Lleonart, D. Garcia del Barco, R. Martinez, F. Herrera, A. Morales, M. T. Herrera, R. Morales, J. de la Fuente. *Biotecnologica Aplicada* **1996**, *13*, 279–283.
41 R. Martinez, M. P. Estrada, J. Berlanga, I. Guillen, O. Hernandez, E. Cabrera, R. Pimentel, R. Morales, F. Herrera, A, Morales, J. C. Pina, Z. Abad, V. Sanchez, P. Lelamed, R. Lleonart, J. de la Fuente. *Mol. Mar. Biol. Biotechnol.* **1996**, *5*, 62–70.
42 J. T. Cook, M. A. McNiven, G. F. Richardson, A. M. Sutterlin. *Aquaculture* **2000**, *188*, 15–32.
43 J. T. Cook, M. A. McNiven, G. F. Richardson, A. M. Sutterlin. *Aquaculture* **2000**, *188*, 33–45.
44 J. T. Cook, M. A. McNiven, G. F. Richardson, A. M. Sutterlin. *Aquaculture* **2000**, *188*, 47–63.
45 T. O. Ostenfeld, E. McLean R. H. Devlin. *J. Fish Biol.* **1998**, *52*, 850–854.
46 E. D. Stevens, R. H. Devlin. *J. Fish Biol.* **2000**, *56*, 191–195.
47 J. A. Hill, A. Kiessling, R. H. Devlin. *Can. J. Fish. Aquat. Sci.* **2000**, *57*, 939–950.
48 J. de la Fuente, R. Martinez, I. Guillen, M. P. Estrada, R. Lleonart, in: *Gene Transfer in Aquatic Organisms* (J. de la Fuente, F. O. Castro, eds), Springer, Berlin **1998**, 83–105.
49 R. Martinez, J. Juncal, C. Zaldivar; A. Arenal, I. Guillen, V. Morera, O. Carrillo, M. Estrada, A. Morales, M. P. Estrada. *Biochem. Biophys. Res. Commun.* **2000**, *267*, 466–472.
50 M. A. Rahman, A. Ronyai, B. Z. Engidaw, K. Jauncey, G.-L. Hwang, A. Smith, E. Roderick, D. Penman, L. Varadi, N. Maclean. *J. Fish Biol.* **2001**, *59*, 62–78.
51 P. Bossier. *J. Food Sci.* **1999**, *64*, 189–193.
52 H. Rehbein, R. H. Devlin, H. Rüggeberg. *Eur. Food Res. Technol.* **2002**, *214*, 352–355.
53 Anonymous. *Aquatic Biotechnology and Food Safety,* **1994**, OECD, Paris.
54 D. A. Jonas, I. Elmadfa, K.-H. Engel, K. J. Heller, G. Kozianowski, A. König, D. Müller, J. F. Narbonne, W. Wackernagel, J. Kleiner. *Ann. Nutr. Metab.* **2001**, *45*, 235–254.
55 D. B. Berkowitz, I. Kryspin-Sorensen. *Bio/Technology* **1994**, *12*, 247–252.
56 I. Kryspin-Sorensen, D. Berkowitz. *Intern. J. Food Sci. Nutr.* **1993**, *44* (Suppl. 1), S17–S21.
57 I. Guillen, J. Berlanga, C. M. Valenzuela, A. Morales, J. Toledo, M. P. Estrada, P. Puentes, O. Hayes, J. de la Fuente. *Mar. Biotechnol.* **1999**, *1*, 2–14.
58 S. L. Taylor, J. A. Nordlee. *Food Technol.* **1996**, *50*, 231–234, 238.
59 H. Rehbein, W. Schreiber. *Comp. Biochem. Physiol.* **1984**, *79B*, 447–452.
60 G. Marcuse, F. Marcuse. *Giftige und gefährliche Tiere der Meere*, Landbuch-Verlag, 1989.

10
Detection Methods for Genetically Modified Crops

Rolf Meyer

10.1
Introduction

With the introduction of foods derived from genetically modified organisms (GMO) – the first products entered the European market in 1996 – there has been an increasing need for appropriate analytical methods to comply with the strict regulations in the European Union (EU) and other regions [1]. Today, more than 50 different GM plants have been generated and produced, mostly in the USA, with the trend increasing. For most purposes, a screening or identification step might be sufficient to determine whether a product contains a GMO but the so-called "threshold regulation" specifies that foodstuffs must be subjected to labeling where material derived from these GMOs is present in food ingredients in a proportion above 1% of the food ingredients individually considered. This requires a semi-quantitation of GMOs in each ingredients to ascertain whether they are present at >1%, which would require labeling, or <1%, which would not require labeling. Therefore, different procedures must be considered to distinguish "identity preserved" (IP) nonGM soya, for example, from soya containing proportions of GM soya in processed food products. An accurate determination of the proportion of GM soya is required and a quantitative analysis would need to be carried out on the ingredients themselves before incorporation in the processed food.

In general, three distinct steps are needed: detection; identification; and quantitation. Screening methods that provide a positive/negative statement can be used to determine if a product contains a GMO. The assays on raw material are generally performed with the polymerase chain reaction (PCR) or with immunological assays such as the enzyme-linked immunoassay (ELISA). The immunological assays are based on the specific binding between the expressed transgenic protein, for example neomycin phosphotransferase II (*Npt*II), the enzyme 5-pyruvylshikimate-3-phosphate synthase (EPSPS) or numerous Cry proteins (Bt, *Bacillus thuringiensis*, insecticidal endotoxin) and an antibody. If no protein is expressed from introduced DNA, this technique cannot be applied. Proteins are denatured during food processing (i.e., thermal treatments) and any conformational change in the epitope structure of the protein renders the test ineffective. Immunological methods are therefore only suitable for raw material analysis. DNA is much more thermally stable than protein and can survive food processing; hence, processed foods are generally analyzed with PCR methods. A food product might even have undergone such extensive processing that ingredients of potential GMO origin are no longer detectable. The detection of DNA and/or protein become difficult if highly refined ingredients, such as starch, lecithin, sugar or vegetable oils are considered. Prior to GMO identification or screening the presence of amplifiable DNA in a food matrix needs to be determined using universal or plant species-specific PCR assays (i.e., soya, maize, potato, tomato, rapeseed). If plant-DNA is present and the screening result is positive, further analysis is required to determine whether the GMO is authorized within the EU or regions with different regulations (e.g., Switzerland, Japan) and, at what concentration the GMO is present in the food ingredients (determination of the level of GMO is required) [2, 3].

In the area of identification and quantification of GM crops there are a number of difficulties. In particular, the status of ploidy degree of the chromosomes in cells and the number of integrated copies of the genetic modification are often unknown. The same transgenic construct ("cassette") could be used in different plant species such as maize and soybean, and the use of different "cassettes" in the same GM crop to combine GM-traits (called "stacked" genes; e.g., maize Mon810 × T25 from Pioneer) could lead to ambiguous results. The development of methods is hampered by the availability of certified GM crop reference material and the access of GM free material (e.g., from the same parent line).

In this chapter, an overview is presented on DNA analytical methods that may be needed to determine the presence of GM crops, or for analysis of GM crop derived-produce. The principle of detection of recombinant DNA in food by PCR is discussed following the three main steps: DNA-extraction; amplification by PCR; and verification of PCR products [4].

10.2
Isolation of Plant DNA

10.2.1
Sampling

One of the major problems in analytical testing is the sampling procedure. A sample has to be representative of the batch/lot of the product from which it was taken. The sampling plan and sample size have to meet statistical requirement with respect to homogeneity and threshold limit up to which the result should be reliable. Sampling plans for cereals and pulses are described in ISO 13690 [5]. Raw materials (soya beans, maize kernels) are often not systematically mixed during harvest, storage, transportation, etc., whereas processed ingredients (flour, protein isolate, lecithin, and starch) already present a restricted degree of variance. However, batch-to-batch variations have also to be considered. The degree of heterogeneity of a given sample and the actual threshold, which is set for the acceptance of the presence of GM material, will define both the number of samples taken and the appropriate sample size. The higher the degree of heterogeneity, the more critical will be the choice of the appropriate sampling plan. For the detection of low levels of GM material the required sample size will increase accordingly in order to be representative. The United States Department of Agriculture (USDA), Grain Inspection, Packers and Stockyards Administration (GIPSA, Washington D.C.) has established sampling guidelines for diagnostic testing for GM grains [6]. A harmonization of sampling procedures for GMOs and derived products is expected in future (CEN working group 11) [7].

10.2.2
Sample Preparation

Practical approaches for homogenization of samples are described in the Swiss Food Manual [8] using soya and maize products as example. A sample (700 g) of soy beans (whole beans or dehulled) or about 1 kg of maize kernels should be incubated with 1500, or 2100 mL, respectively, of sterile water for up to 20 h and then homogenized in a blender (Table 10.1). For dry samples (soya flakes, soya flour), 30 g should be incubated with 60 mL sterile water and homogenized in a Stomacher or blender. For wet samples (tofu, soya sausage), 30 g should be homogenized directly, whereas liquid samples (e.g., soya drink) should be shaken thoroughly before weighing. Homogenous reference material, for example, the Certified Reference Material (CRM) in the range from 0.1–5% GM soya or GM maize flour produced by the Institute for Reference Materials and Measurements (IRMM), Belgium, does not require further homogenization.

In order to avoid cross-contamination between samples, disposable material and decontamination solution (e.g., hypochlorite solution and HCl or commercial products that destroy DNA) are recommended. Cross-contamination by dust should be avoided by physical separation of the sample preparation areas.

Table 10.1. Recommended size of laboratory sample in case of homogenous distribution of GM particles in the investigated lot (3500 particles) at 1% threshold for GMOs according to Ref. [8].

Crop	Kernel (particles) weight [mg]	Sample size [g]
Soy beans	200	700
Maize kernels	300	1050
Soya or maize grits	50	126
Soya or maize flour	<1	<2.6

10.2.3
DNA Extraction and Analysis

Assuming that the laboratory sample is representative for the field sample and that it has been adequately homogenized, even small aliquots of vegetal material or products are sufficient for DNA extraction, usually between 100 and 350 mg. The efficiency of the PCR depends on DNA quality and purity. DNA quality is determined by its fragment length and its degree of damage due to the exposure to heat, low pH and/or nucleases that cause hydrolysis, depurination and/or enzymatic degradation. Therefore, DNA quality varies according to the material under examination, the degree of processing of the food sample and the DNA extraction method applied. DNA isolated from processed ingredients and foods is of low quality, with available targets sequences being rather short, in the range of 100–400 base pairs (bp) for soya protein preparations and processed tomato products. Thus, an appropriate choice of primers in order to obtain short amplicons should be made. Various components of the food matrix affect the purity of the DNA, for example polysaccharides, lipids and polyphenols or chemicals used during the DNA extraction procedure. The key enzyme of the PCR, *Taq* polymerase, is inhibited by polysaccharides, EDTA, phenol, sodium dodecyl sulfate (SDS) and many other compounds.

A vast range of methods is available for DNA isolation, and many of them have been evaluated for their applicability to GMO detection in plant material and plant-derived food products [3]. Currently, three different approaches to DNA isolation from plant material and plant-derived products are favored for GMO detection: the cetyl-trimethylammonium bromide (CTAB) method [9], DNA-binding silica columns (various commercial available kits), and a combination of these two (for example, see Fig. 10.1). The use of CTAB as detergent in the DNA extraction buffer is efficient for a wide range of plant materials and plant-derived foods, especially due to the good separation of polysaccharides from DNA. Adapted protocols for starch (enzymatic treatment with α-amylase) and lecithin (using a hexane extraction step to remove lipids) have been developed to improve the yield and quality of DNA from these matrices [10, 11].

The quantity and purity of the DNA is usually estimated by measuring the absorbance at 260 nm and at 280 nm in a spectrophotometer. An absorbance of 1 unit at 260 nm corresponds to a concentration of 50 μg double-stranded DNA per mL or 37 μg single-stranded DNA per mL (when denatured with 0.2 M NaOH) [8]. The ratio between the absorbance values at 260 nm and 280 nm gives an estimate of DNA purity: generally a pure DNA has an $A_{260/280}$ ratio ranging from 1.7 to 1.9. A higher $A_{260/280}$ ratio is indicative of an RNA contamination, whereas a lower $A_{260/280}$ ratio is encountered when a contamination with proteins occurred. Alternatively, the DNA concentration is measured by fluorescence induced by reagents such as Pico Green (Molecular Probes, USA). The approximate amount of DNA and the average fragment size of DNA isolated from processed food samples is estimated on a agarose gel stained with ethidium bromide or fluorescent nucleic acid gel stains. DNA might be contaminated with RNA, leading in

Sample preparation

1. Weigh 30 g of dry samples (soya flakes, soya flour) and incubate with 60 mL sterile water for 20 min.
2. Homogenize in an blender or stomacher for 5 min.

DNA isolation and purification

1. Weigh 100 mg sample and add 1.5 mL CTAB buffer.
2. Vortex and incubate at 65 °C for at least 60 min.
3. Centrifuge at 15,000 × g for 10 min.
4. Add 700 µL supernatant to 700 µL chloroform and vortex.
5. Centrifuge at 15,000 × g for 10 min.
6. Pipette 400 µL supernatant to 2 mL binding buffer.
7. Mix and load on silica-membrane column.
8. Wash twice with 750 µL wash buffer.
9. Dry columns by centrifugation at 12,000 × g for 5 min.
10. Elute DNA with 50 µL elution buffer by centrifugation at 12,000 × g for 5 min after 5 min incubation.

PCR analysis and verification

Figure 10.1. Typical example of a sample preparation and DNA isolation protocol. The analytical flow sheet describes the way to extract DNA from plant raw materials and processed food samples. Based on a modification of the CTAB-DNA extraction method according to [9], it combines the convenience of spin-column technology with the selective binding properties of silica gel-membranes. The DNA adsorbs to the silica-membrane in the presence of high salt concentration while potentially interfering substances pass through the column. Impurities are washed away and the DNA is eluted.

some cases to overestimation of the DNA content if no RNase treatment is performed. Usually, 5 to 100 ng DNA is used in PCR analysis. The DNA content in low DNA containing matrices such as lecithin and starch cannot be measured directly.

10.3
Detection Strategies

PCR methods in its different formats (single or nested-PCR; PCR-ELISA, QC-PCR, real-time PCR) are available today, and have found broad application in GMO detection (as reviewed by Anklam et al. [3]).

10.3.1
Screening

Introduced DNA elements common to several GM crops are, for example, the cauliflower mosaic virus (CaMV) promoter (P-35S), the 5-enol-pyruvylshikimat-3-phosphate synthase (*EPSPS*) gene, the gene coding for *CryIA(b)* toxin from *Bacillus thuringiensis*, phosphinothricin acetyltransferase (*pat/bar*) genes from *Streptomyces* spp., the *Agrobacterium tumefaciens* nopaline synthase terminator (*nos3'*), and the neomycin phosphotransferase gene (*nptII*), a selection marker for transformation of plant cells that confers resistance to kanamycin. For routine screening purposes, control laboratories are focusing on target sequences that are present in many (but not in all) GM crops currently on the market such as the CaMV 35S promoter, the *nos3'* terminator, and *nptII*. Table 10.2 provides an overview

Table 10.2. Applicability of potential screening methods for the detection of genetically modified crops (selection) based on Agbios database (www.agbios.com) [12] and additional data.

Crop	Trade name Event or line	Company	Genetic modification	Presence of 35S-promoter	Presence of nos 3'-terminator	Presence of nptII gene
Maize[1]	Maximizer Bt-176	Syngenta Seeds	IR	Yes	No	No
Maize[1]	Bt-11	Syngenta Seeds	IR	Yes	Yes	No
Maize	Bt Xtra DBT 418	Dekalb Genetics	IR, HT	Yes	No	No
Maize[1]	LibertyLink T25	Aventis CropScience (formerly AgrEvo)	HT	Yes	No	No
Maize[1]	StarLink CBH-351	Aventis	IR	Yes	Yes	No
Maize	Roundup Ready GA21	Monsanto	HT	No	Yes	No
Maize	Roundup Ready NK603	Monsanto	HT	Yes	Yes	No
Maize[1]	Yieldgard Mon 810	Monsanto	IR	Yes	No	Yes[2]
Soybean[1]	Roundup Ready GTS 40-3-2	Monsanto	HR	Yes	Yes	No

10.3 Detection Strategies

Table 10.2. (continued)

Crop	Trade name Event or line	Company	Genetic modi-fication	Presence of 35S-promoter	Presence of nos 3'-terminator	Presence of nptII gene
Cotton	BXN cotton	Calgene	HT	No	No	Yes
Cotton		Calgene	HT, IR	Yes	No	Yes
Cotton	19-51A	Du Pont	HT	No	No	No
Cotton	Bollgard, Ingard	Monsanto	IR	Yes	Yes	Yes
Cotton	Roundup Ready	Monsanto	HT	Yes	Yes	Yes
Rapeseed	LibertyLink Innovator, HCN92	Aventis	HT	Yes	No	Yes
Rapeseed	MS8 × RF3	Aventis (formerly Plant Genetic Systems)	HT, MS	No	Yes	Yes
Rapeseed	23-18-17, 23-198	Calgene	PQ_1	No	No	Yes
Rapeseed	Roundup Ready	Monsanto	HR	No	No	No
Potato[1]	NewLeaf	Monsanto	IR	Yes	Yes	Yes
Potato[1]	NewLeaf Y	Monsanto	IR, VR	No	Yes	Yes
Potato[1]	Russet Burbank, NewLeaf Plus	Monsanto	IR, VR	No	Yes	(Yes)
Tomato[1,3]	Flavr Savr (MacGregor)	Calgene	PQ_2	Yes	No	Yes
Tomato[1,3]		Zeneca Seeds	PQ_2	Yes	Yes	Yes
Tomato[3]	1345-4	DNA plant Technology	PQ_2	Yes	Yes	Yes
Squash	Freedom ZW20	Upjohn/ Seminis Vegetable	VR	Yes	No	No
Squash	Freedom CZW-3	Asgrow/ Seminis Vegetable	VR	Yes	No	Yes

Table 10.2. (continued)

Crop	Trade name Event or line	Company	Genetic modification	Presence of 35S-promoter	Presence of nos 3'-terminator	Presence of nptII gene
Papaya		Cornell University	VR	Yes	Yes	Yes
Chicory	Radicchio Rosso	Bejo Zaden BV	MS, HT	No	Yes	Yes

1   Based on experimental data obtained from reference material.
2   Size of amplicon is about 325 to 350 bp.
3   Tomato concentrate or tomato ketchup are highly acidic which may lead to negative results due to DNA degradation.
( ) Not present in all lines.
IR Insect-resistant.
HT Herbicide-tolerant.
MS Male-sterile.
VR Virus-resistant.
PQ_1 New product quality: altered fatty acid profile.
PQ_2 New product quality: delayed fruit ripening.

on the applicability of potential screening methods for the detection of a selection of GM crops [12]. Screening for the CaMV 35S promoter alone is not sufficient, and additional target sequences, a combination of gene-specific genes (nptII, nos, EPSPS, pat/bar, CryIA(b), are needed to screen for a wider range of GM crops. Primers for screening methods should cover possible variants of target sequences used in GM crops.

10.3.2
Specific Detection

Unequivocal identification of GM crops has to be based on target sequences that are characteristic for each individual transgenic organism. This can be reached by selection of primers to target introduced DNA that span over the boundary of two adjacent genetic elements (e.g., promoters, structural genes, terminators) or cross-border (junction regions) between insert and plant DNA (event-specific). However, detailed knowledge of the gene organization and DNA sequences of the inserted transgenic constructs is required to design appropriate primers for a PCR-based detection strategy. Three different detection strategies: screening; construct-; and event-specific detection were used to detect and identify Roundup Ready soybeans. A schematic presentation of these strategies is given in Fig. 10.2. Numerous detection methods have been introduced in a database on commercialized genetically modified foods so far that provides detailed information about analytical parameters and the references to the original publication and validation data [13].

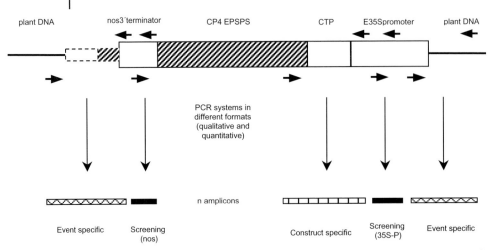

Figure 10.2. Detection strategies for transgenic soya (Roundup Ready soybean). Position of primers for screening (35S-CaMV promoter, *nos* 3'-terminator), construct- and event-specific detection methods.

Table 10.3. PCR-based detection methods used as control systems for amplifiable DNA from food matrices.

Target organisms, specificity	Target sequence (gene)	Reference
Eukaryotes	18S rRNA	21
Vertebrates	mtDNA (*cytb* gene)	22
Plants	cpDNA (tRNA gene)	23
Soya	Lectin gene, *le*1	14
Soya	Lectin gene, *le*1	11
Soya	Soy heat-shock protein gene, HSP	24
Maize	Zein (methionine-rich storage protein)	15
Maize	Invertase, *ivr*1	16
Potato	Patatin	17
Wheat	25S-18S rRNA	21
Rapeseed	Acetyl-CoA carboxylase (BnACCg8)	18
Rapeseed	Cruciferin and Napin	19
CaMV	ORF V and ORF VI	20

CaMV = cauliflower mosaic virus; cpDNA = chloroplast DNA; mtDNA = mitochondrial DNA; ORF = open reading frame.

Prior to any screening or GM crop identification it is recommended to demonstrate the presence of amplifiable DNA in a food matrix using universal or plant species-specific PCR assays [14–24], as shown in Table 10.3. The presence or absence of DNA extracted from highly processed food originated from soya can be checked by a soya-specific PCR systems such as lectin *Le*1 (single copy gene) [14]. Other control PCR systems have been described for maize DNA (zein gene or maize invertase) [15, 16], potato (patatin gene) [17], rapeseed (acetyl-CoA carboxylase, cruciferin and napin) [18, 19] and also for the natural presence of CaMV in infected vegetables [20], for example broccoli (see Table 10.3). Plant-specific genes are used as a reference for the qualitative and quantitative PCR detection of GM crops in mixed food samples.

10.3.2.1 Example for Qualitative Detection

An example of a qualitative approach for the detection of transgenic soya is shown in Fig. 10.3. A soybean-specific test (nested-PCR for soya lectin gene *Le*1) is followed by a Roundup Ready soya (RRS) -specific test (nested-PCR, cross-border sequence: CaMV 35S promoter/CTP sequence/EPSPS gene) [25]. A band of 118 bp in the nested soya *Le*1-gene PCR indicates the presence of soya DNA. Since no amplifiable DNA was detectable in the chocolate sample, the purified lecithin and the soy sauce (lanes 6, 7, and 10), it can also be expected that no DNA from GM-soya will be present, which is indicated by the amplification of a 169 bp product in the RRS-specific test. All RRS reference samples (lanes 3, 4, 5, and 11) must give this signal. Regarding the two raw lecithin samples (lanes 8 and 9) that were positive for soya-DNA, one of them (lane 9) was also found positive for RRS. The test is only qualitative and not quantitative. The detection limit is usually less than 0.1% (w/w) RRS, but this depends on the food matrix, the average size of soya DNA, and the presence of PCR inhibitors.

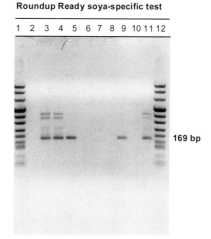

Figure 10.3. Agarose gel electrophoresis of nested-PCR products obtained from soya-derived raw materials and products for analysis of soya (left) and Roundup Ready soya (right); lanes 1 and 12, molecular weight marker; lane 2, negative control: without DNA; lane 3, positive control: 100% Roundup Ready soybeans (RRS); lane 4, soybeans ground and de-fatted 100% RRS; lane 5, soya protein isolate 100% RRS; lane 6, chocolate; lane 7, purified lecithin; lane 8, raw lecithin, non GMO; lane 9, raw lecithin, containing RRS; lane 10, soy sauce; lane 11, raw oil from 100% RRS; bp = base pair.

10.3 Detection Strategies

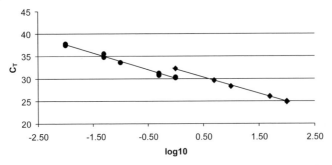

Figure 10.4. Example of quantification by real-time polymerase chain reaction. Standard curves from 1% Bt11 maize, threshold cycle (C_T) values versus starting quantity. One curve (circles) represents the C_Ts (35S-CaMV promoter) plotted against the known quantity of 35S-CaMV promoter/maize, the other curve (squares) represents the C_Ts (zein) plotted versus the known quantity of Bt11 maize (log format). According to the C_Ts of unknown samples, the amount of 35S-CaMV promoter/zein can be determined by comparison of the values with those from the standard curves [26].

10.3.3
Quantification

Real-time PCR methods for the quantification of GM soya or GM maize normally rely on the amplification of transgenic specific sequences and their quantification relative to an endogenous (plant-specific) reference gene that gives an estimation of the total amount of target DNA in the sample. The quantification is based on standard curves prepared from reference material. An example of standard curves for real-time multiplex-PCR for the quantification of CaMV 35S-promoter in GM-maize is given in Fig. 10.4. A multiplex-PCR method has been applied based on the quantification of the 35S-CaMV promoter and of zein as reference gene for maize on an ABI Prism 7700 Sequence Detection System (Applied Biosystems, USA). The standard curve was prepared from 1.0% Bt11-maize (CRM standard, Fluka) and approximately 100 ng DNA was used for the first point of the curve (threshold cycle C_T of 23 for the zein/VIC labeled probe and a C_T of 29 for the 35S-CaMV/FAM labeled probe) [26].

10.3.4
Verification

Several methods are used to verify PCR results and they vary in reliability, precision, and cost. Specific cleavage of the amplification products by restriction endonuclease is the simplest method to identify the PCR products. The presence of the 35S-CaMV promoter, for example, is confirmed if the 195 bp fragment is cleaved by the restriction endonuclease *XmnI* to yield two fragments of 115 and 80 bp [8]. The more time-consuming, but also more specific, transfer of separated PCR-products onto membranes (Southern blot) followed by hybridization with a DNA probe specific for the target sequence is used in official methods listed in the German Food Act [9]. Alternatively, PCR products may be verified by direct sequencing. This is the most accurate proof of amplified DNA, but this option is not available in all laboratories and is not the method of choice for routine analysis. Nested-PCR assays combine high specificity and sensitivity. The use of two pairs of primers spanning the boundary of two or three genetic elements is regarded as sufficient specific for a GMO-product. In general, the increased sensitivity of nested-PCR systems allows low levels of GMOs to be detected in raw materials and finished products. Real-time PCR is the current method of choice for the determination of GMOs in food. The accumulation of PCR products can be followed using fluorogenic probes; these are oligonucleotides which have a reporter and quencher dye attached to either end. In the intact probe the proximity of the quencher greatly reduces the fluorescence emitted from the reporter dye (Förster resonance energy transfer). During PCR, the probe anneals to the target sequence – which makes the test very specific. As the primer extends along the target sequence it cleaves and fragments the probe, and the reporter dye emits fluorescence as it is no longer proximity to the quencher.

10.3.5
Validation

The objective of the validation of an analytical PCR method is to demonstrate that the successive procedures of sample extraction, preparation and analysis will yield acceptable accurate, precise and reproducible results for a given analyte in a specified matrix. The process of validation allows the independent use of methods and results, which are comparable among each other. For the validation of a qualitative analytical test system specificity/selectivity, sensitivity [matrix effects/inhibition, limit of detection (LOD), accuracy/precision (repeatability RSD_r), intermediate precision, reproducibility (RSD_R) and robustness] have to be established. In addition to these parameters, limit of quantification (LOQ), accuracy/trueness and linearity/working range have to be evaluated for a quantitative analytical test system [3]. An increasing number of food control laboratories are adopting PCR as the technology of choice for GMO detection and have validated the methods in-house or in ring-trials. A limited number of detection methods for GM crops in raw materials or plant-derived foods have been validated in inter-laboratory studies so far [27–35]. A non exhaustive list of methods tested in ring-trials is given in Table 10.4. Any laboratory using such methods is expected to carefully validate every single step of the analytical procedure and participate in performance schemes (e.g., FAPAS), especially for quantitative methods [36].

Table 10.4. List of methods for GM crop detection validated in interlaboratory studies [27].

Food source/target	Primers, antibodies (Ab)	Size of amplicon [bp]	Reference(s)
CaMV 35S promoter			
35S-1/35-S-2	f: 5′-GCTCCTACAAATGCCATCA-3′ r: 5′-GATAGTGGGATTGTGCGTCA-3′	195	EC-JRC, [28]
35S-cf3/35S-cr4	f: 5′-CCACGTCTTCAAAGCAAGTGG-3′ r: 5′-TCCTCTCCAAATGAAATGAACTTCC-3′	123	EC-JRC, [29]
Maize Bt11			
IVS2-2/PAT-B	f: 5′-CTGGGAGGCCAAGGTATCTAAT-3′ r: 5′-GCTGCTGTAGCTGGCCTAATCT-3′	189	BgVV, Germany
Maize event 176 (Maximizer)			
Cry03/Cry04	f: 5′-CTCTCGCCGTTCATGTCCGT-3′ r: 5′-GGTCAGGCTCAGGCTGATGT-3′	211	BgVV, Germany, [30]
Cry05/Cry06	f: 5′-CCGCAGCCGATCCAACAATG-3′ r: 5′-GCTGATGTCGATGGGGGTGTAG-3′	134	BgVV, Germany, [30]
Maize Roundup Ready (GA21)			
GA21 3-5′/GA21 3-3′	f: 5′-GAAGCCTCGGCAACGTCA-3′ r: 5′-ATCCGGTTGGAAAGCGACTT-3′	133	NFRI, Japan, [31]
Maize T25 (Liberty Link)			
Q T25 1-5′/T25-1-3′/ T25Ta q	f: 5′-GCCAGTTAGGCCAGTTACCCA-3′ r: 5′-TGAGCGAAACCCTATAAGAACCCT-3′	149	NFRI, Japan, [31]
Maize Mon810 (Yield Gard)			
VW01/VW03	f: 5′-TCGAAGGACGAAGGACTCTAACG-3′ r: 5′-TCCATCTTTGGGACCACTGTCG-3′	170	BgVV, Germany
Maize StarLink (CBH 351)			
CaM03-5′/CBH02-3′	f: 5′-CCTTCGCAAGACCCTTCCTCTATA-3′ r: 5′′-GTAGCTGCTGGTGTAGTCCTCGT-3′	170	NIHS, Japan, [32]
Cry9C-5′/35Ster-3′	f: 5′-CCTATAGCTTCCCTTCTTCC-3′ r: 5′-TGCTGTAATAGGGCTGATGA-3′	171	NIHS, Japan, [32]
Protein Cry9C	MAb and polyclonal Ab against Bt Cry9C		(AACC), USDA, USA
NOS-Terminator			
HA-NOS118-f/ HA-NOS118r	f: 5′-GCATGACGTTATTTATGAGATGGG-3′ r: 5′-GACACCGCGCGCGATAATTTATCC-3′	118	EC-JRC, [29]
NOS-1/NOS-3	f: 5′-GAATCCTGTTGCCGGTCTTG-3′ r: 5′-TTATCCTAGTTTGCGCGCTA-3′	180	EC-JRC, [28]
Papaya 55-1, 66-1			
CaM3-5′/GUSn-3′	f: 5′-CCTTCGCAAGACCCTTCCTCTATA-3′ r: 5′-TCGTTAAAACTGCCTGGCAC-3′	250	NIHS, Japan, [33]
NosC-5′/CaMVN-3′	f: 5′-TTACGGCGAGTTCTGTTAGG-3′ r: 5′-CATGTGCCTGAGAAATAGGC-3′	207	NIHS, Japan, [33]

Table 10.4. (continued)

Food source/target	Primers, antibodies (Ab)	Size of amplicon [bp]	Reference(s)
Potato NewLeaf Plus			
p-FMV02-5'/ PLRV01-3'	f: 5'-AAATAACGTGGAAAA-GAGCTGTCCTGA-3' r: 5'-AAAAGAGCGGCATATGCGG-TAAATCTG-3'	234	NIHS, Japan, [34]
PLRV-rep1-5'/ PLRV-rep1-3'	f: 5'-CTTCTTTCACGGAGTTCCAG-3' r: 5'-TCGTCATTAAACTTGACGAC-3'	172	NIHS, Japan, [34]
Potato New Leaf Y			
p-FMV05-5'/ PVY02-3'	f: 5'-AAAAGAGCTGTCCTGACAGC-3' r: 5'-TCCTCCTGCATCAATTGTGT-3'	225	NIHS, Japan
PVY01-5'/PVY01-3'	f: 5'-GAATCAAGGCTATCACGTCC-3' r: 5'-CATCCGCACTGCCTCATACC-3'	161	NIHS, Japan
Soya Roundup Ready			
35S-af2/Petu-r1	f: 5'-TGATGTGATATCTCCACTGACG-3' r: 5'-TGTATCCCTTGAGCCATGTTGT-3'	171	BgVV, Germany
Q-RRS1-F/RR1-R/ Probe	f: 5'-CATTTGGAGAGGACACGCTGA-3' r: 5'-GACCCATGTTGTTAATTTGTGCC-3'	74	BgVV, Germany, GenScan Europe AG
Protein EPSPS	mAb and polyclonal Ab against CP4 EPSPS		EC-JRC, [35]

f: forward primer, r: reverse primer; mAb: monoclonal antibody.
EC-JRC: European Commission, Joint Research Center.
BgVV: Federal Institute for Health Protection of Consumers and Veterinary Medicine, Germany.
NIHS: National Institute of Health Sciences, Japan.
NFRI: National Food Research Institute, Japan.
USDA: United States Department of Agriculture, USA.
AACC: American Association of Cereal Chemists.

10.4
Outlook and Conclusions

Although one could question the scientific basis for parts of the current and pending EU GMO regulations, and whether they will achieve the desired objective of restoring consumer confidence, they are reality in the European Union. Therefore, industry must cope with their requirements and find solutions that are acceptable to consumers and economically feasible. Thresholds and IP play a very important role in finding these solutions. The current threshold has initiated considerable activity along the chain, and there are chances that newly proposed regulations will not reduce these efforts. In an average industrial setting, incoming raw materials will be tested for the presence of (known) GM varieties. Immunoassays can be used for primary materials, whereas for each subsequent point of analysis (e.g., after transport, processing steps, etc.) PCR methods will be used. ELISA-based techniques were initially used as the method of analysis for the determination of GM crops in food ingredients (e.g., soya flour in flour), but these were found to be less appropriate for the determination of GMOs in compound foods. If raw materials are analyzed, then protein-based methods could be an alternative to DNA-based PCR methods. The latter will be applied both for qualitative as well as quantitative purposes. Although the use of Real-time PCR for the quantification of DNA is the method of choice, the method leads to difficulties in measuring low levels of GM material in processed food.

Although several question marks remain about the reliability of the various methods, as well as about the availability of probes for those GMOs that can reasonably be expected to be present in a given lot, the overall experience is that for practical purposes they are workable. New GM traits of various crops are under development and will enter the market sooner or later, whereas existing commercialized GM products will disappear. Therefore, continuous monitoring of the level of GM materials and the identification of variety genotypes will be pre-requisite for the verification of nonGMO status in the supply chain, and so validated methods are required.

The question that remains is whether the efforts spent on this issue are justified in relation to other issues that the agro-food chain is facing at present.

References

1 E. Anklam and D. A. Neumann, *J. AOAC Int.*, **2002**, 85, 754–756.
2 G. Van den Eede, S. Kay, E. Anklam, H. Schimmel, *J. AOAC Int.*, **2002**, 85, 757–761.
3 E. Anklam, F. Gadani, P. Heinze, H. Pijnenburg, and G. Van den Eede, *Eur. Food Res. Technol.* **2002**, 214, 3–26.
4 R. Meyer, *Food Control*, **1999**, 10, 391–399.
5 ISO 13890: Cereals, pulses and milled products – Sampling of static batches.
6 http://www.usda.gov/gipsa/biotech/sampling_grains_for_biotechnolog.htm.
7 Detection of genetically modified organisms and derived products – sampling. CEN/TC 275/WG11 Doc N 0009, Version May 2002.
8 Swiss Food Manual (2002), chapter 52B, (CD-ROM available from BBL-EDMZ, CH-3030 Bern, Ed. Bundesamt für Gesundheit, Facheinheit Lebensmittel und Gebrauchsgegenstände, Bern. E-mail: edmz@bbl.admin.ch.
9 Official Collection of Test Methods (1998) Detection of a genetic modification of soybeans by amplification of the modified DNA sequence by means of the polymerase chain reaction (PCR) and hybridisation of the PCR product with a DNA probe. German Federal Foodstuffs Act – Food Analysis, article 35, L 23.01.22–1. Beuth, Berlin, Köln.
10 K. Meyer, C. Rosa, C. Hischenhuber, R. Meyer, *J. AOAC Int.*, **2001**, 84, 89–99.
11 A. Wurz, H. Rüggeberg, P. Brodmann, H. U. Waiblinger, K. Pietsch, *Dt. Lebensm. Rundsch.*, **1998**, 94, 159–161.
12 AGBIOS, Agriculture & Biotechnology Strategies Inc., Canada. (http://www.agbios.com).
13 DMIF-GEN (1999) Development of methods to identify foods produced by means of genetic engineering. EU-Project SMT4-CT96-2072. DMIF-GEN Final Report, 15 December 1999. DMIF-GEN database (http://biotech.jrc.it).
14 R. Meyer, F. Chardonnens, P. Hübner, J. Lüthy, *Z. Lebensm. Unters. Forsch.* **1996**, 203, 339–344.
15 E. Studer, C. Rhyner, J. Lüthy, P. Hübner, *Mitt. Gebiete Lebensm. Hyg.* **1997**, 88, 515–524.
16 B. Ehlers, E. Strauch, M. Goltz, D. Kubsch, H. Wagner, H. Maidhof, B. Bendiek, B. Appel, H.-J. Buhk, *Bundesgesundheitsbl.*, **1997**, 4, 118–121.
17 C. Wolf, P. Hübner, J. Lüthy, *Mitt. Gebiete Lebensm. Hyg.* **2001**, 92, 159–167.
18 M. Hernández, A. Río, T. Esteve, S. Prat, M. Pla, *J. Agric. Food Chem.*, **2001**, 49, 3622–3627.
19 C. Wolf, J. Lüthy, *Mitt. Lebensm. Hyg.* **2000**, 91, 352–355.
20 C. Wolf, M. Scherzinger, A. Wurz, U. Pauli, P. Hübner, J. Lüthy, *Eur. Food Res. Technol.*, **2000**, 201, 367–372.
21 M. Allmann, U. Candrian, J. Lüthy, *Z. Lebensm. Unters. Forsch.* 1993, 196, 248–251
22 R. Meyer, Ch. Höfelein, J. Lüthy, U. Candrian, *J. AOAC Int.*, **1995**, 78, 1542–1551.
23 P. Taberlet, L. Gielly, G. Pautou, J. Bouvet, *Plant Mol. Biol.* **1991**, 17, 1105–1109.

24 G. van Duijn, *Chemisch Magazine*, **1997**, 411–413.
25 R. Meyer, E. Jaccaud (1997) Detection of genetically modified soya in processed food products; development and validation of a PCR assay for the specific detection of Glyphosate-Tolerant Soybeans. In: R. Amado, R. Battaglia (eds) Proceedings 9th European Conference on Food Chemistry. Authenticity and adulteration of food – The analytical approach, vol. 1. Interlaken, Switzerland 24–26 September 1997, pp. 23–28.
26 M. Höhne, C. Rosa Santisi, R. Meyer, *Eur Food Res. Technol.*, **2002**, 215, 59–64.
27 Codex Alimentarius Commission. Codex Committee on Methods of Analysis and Sampling, 24th Session, Budapest, Hungary, 18–22 November 2002. (http://www.codexalimentarius.net/reports.asp)
28 M. Lipp, E. Anklam, P. Brodmann, K. Pietsch, J. Pauwels, *Food Control* 10, **1999**, 379–383.
29 M. Lipp, A. Bluth, F. Eyquem, L. Kruse, H. Schimmel, G. Van den Eede, E. Anklam, **2001**, *Eur Food Res. Technol*, 212, 497–504.
30 C. Hupfer, H. Hotzel, K. Sachse, K.-H. Engel, **1998**, *Z. Lebensm. Unters. Forsch.* 206, 203–207.
31 Y. Shindo, H. Kuribara, T. Matsuoka, S. Futo, C. Sawada, J. Shono, H. Akiyama, Y. Goda, M. Toyoda, A. Hino, 2002, *J. AOAC Int.*, 85, 1119–1126.
32 T. Matsuoka, H. Kuribara, S. Suefuji, H. Miura, Y. Kusakabe, H. Akiyama, Y. Goda, K. Isshiki, M. Toyoda, A. Hino, **2001**, *J. Food Hyg. Doc. Japan*, 42, 197–201.
33 Y. Goda, T. Asano, M. Shibuya, A. Hino, M. Toyoda, 2001, *J. Food Hyg. Japan* 42, 231–236.
34 H. Akiyama, K. Sugimoto, M. Matsumoto, K. Isuzugawa, M. Shibuya, Y. Goda, M. Toyado, *J. Food Hyg. Japan*, **2002**, 43, 24–29.
35 M. Lipp, E. Anklam, *J. AOAC Int.*, **2000**, 83, 919–927.
36 P. Hübner, H.-U. Waiblinger, K. Pietsch, P. Brodmann, *J. AOAC Int.*, **2001**, 84, 1855–1864.

11
Methods to Detect the Application of Genetic Engineering in Composed and Processed Foods

Karl-Heinz Engel and Francisco Moreano

11.1
Introduction

The advantages and potential risks arising from the application of genetic engineering in the production of foods and feeds have been the subjects of many controversial discussions [1]. Public skepticism has resulted in the implementation of regulatory frameworks that require the monitoring and labeling of genetically modified organisms (GMO) and their derivatives in the food chain, and this has initiated tremendous research activities in order to develop the necessary detection techniques [2–4].

In the European Union, the so-called "Novel Foods Regulation" (Regulation (EC) No. 258/97) [5] provides the legal basis for labeling of foods and food ingredients containing or consisting of GMO. Labeling is required if any characteristic or food property (such as composition, nutritional value or nutritional effect, and intended use of the food) renders the novel food or food ingredient no longer equivalent to an existing food or food ingredient. According to Council Regulation (EC) 1139/98 [6], mandatory labeling is triggered by the detection of recombinant DNA or protein expressed due to the genetic engineering. Although this regulation only applies to Roundup Ready® soybeans and Bt-176 maize, it served as the general model for labeling of GMO-derived foods and food ingredients. As a result, proteins and DNA have become the major targets for detection strategies.

The labeling provisions have been further specified by extending them to foods and food ingredients containing additives and flavorings derived from GMO (Regulation (EC) 50/2000 [7]) and by introducing a *de minimis* threshold (Regulation (EC) 49/2000) [8]. It was acknowledged that adventitious contamination of foods with DNA or protein resulting from genetic modification cannot be excluded. Therefore, labeling is not required if appropriate steps to avoid such contaminations have been taken and if material derived from GMO is present in food in proportions no higher than 1 % related to specific food ingredients. In consequence, appropriate techniques for the quantitation of GMO in raw and processed foods had to be developed.

The European Commission has proposed a new legislative package concerning traceability and labeling of GMO and traceability of food and feed products produced from GMO [9]. It foresees labeling requirements independent from the detection of DNA or proteins, but intends to keep a threshold strategy for consideration of adventitious contaminations with GMO material.

This chapter reviews current strategies for the detection of GMO in foods, focusing especially on issues related to composed and processed foods. The suitability of protein- and DNA-based methods for the detection and quantitation of GMO is discussed, taking into account that commonly used food technological processes may result in significant changes of these target compounds.

11.2
Challenges Specific to the Detection of GMO in Composed and Processed Foods

Mechanical stress, heat treatment, pH variations, enzymatic activities and fermentations are common parameters in food processing, which may result in significant degradation of proteins and nucleic acids. In food ingredients and in foods made thereof, the presence of analytes suitable for GMO detection will clearly depend on the degree of treatment to which the raw materials have been exposed. In addition to degradation, intentional or unintentional removal of DNA or protein in the course of food processing is also important.

Aspects of food manufacturing which might influence critically the potential to detect the application of genetic engineering in composed and processed foods are outlined in Fig. 11.1.

Raw materials (e. g., crops) may be subject to a spectrum of processes which differ in the degree of treatment, ranging from simple mechanical procedures such as milling to complex sequences of chemical reactions, for example the refinement of edible plant oils. The products obtained may be used as foods or serve as ingredients for the production of composed foods. The formulation of several food ingredients to create composed foods will result in additional challenges to GMO analysis, as it is highly likely that the target molecules will not show the same degree of degradation in each ingredient. Considering that each ingredient possesses a characteristic matrix and composition, it is highly probable that analytes incorporated into the different matrices will not be equally accessible to analysis. Subsequent processing of composed foods may lead to chemical and structural changes and interactions between ingredients used for food preparation, again resulting in increased analyte degradation. The processes involved in the production of such complex foods can be limited to rather simple steps such as mixing, but they may also involve severe thermal treatments which induce chemical changes, such as baking. It is unclear whether degradation will progress equally in every ingredient, or if ingredient matrices may have an effect on degradation rates.

Approaches for the analysis of GMO in foods must cope with these challenges that are set by the complexity of analyte degradation and the additional problems related to composed foods. All steps of GMO analysis, including the detection,

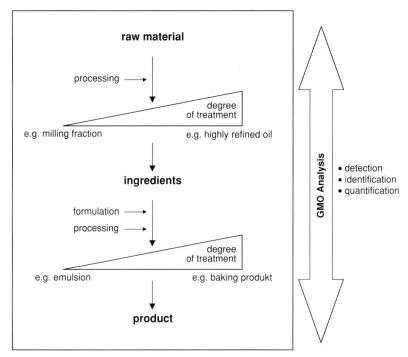

Figure 11.1. Challenges specific to GMO analysis in the course of food processing.

identification and quantification [4], are influenced by these technological parameters.

11.3
Degradation of Proteins and DNA

Both proteins and DNA are subject to degradation as a result of the physical, enzymatic and chemical treatments that may occur during food processing.

11.3.1
Proteins

Denaturation of proteins is initiated by reversible or irreversible changes of the native conformation (secondary, tertiary, or quaternary structure). These changes may result from the breakage of stabilizing disulfide bonds or hydrogen bonds, the neutralization of ionic groups, or hydrophilic/hydrophobic interactions. Denaturation may result from the application of shear forces, or it can be initiated by changes in temperature or pH, by enlargement of interface areas or by addition of organic solvents, salts or detergents. Reversible denaturation generally takes place when the unfolded molecule is stabilized by interactions with the denaturing agent, for example an organic solvent. Removal of the denaturing agent consequently allows the recovery of native protein configuration. Irreversible denaturation occurs when unfolded proteins are stabilized by formation of intra- or intermolecular covalent bonds, usually disulfide bonds between free thiol groups.

Food manufacturing, maturation or spoilage processes involve a number of enzymatic reactions, and this leads to major changes in the structural properties of proteins. In addition to functional group transfer reactions and redox reactions, the protease-catalyzed cleavage of peptide bonds is seen as the most important reaction contributing to protein degradation [10].

Hydrolytic fragmentation of proteins occurs in acidic milieus, especially when combined with heat treatment. Food processing may induce further chemical modifications of native proteins depending on the presence of available functional groups (intra- and intermolecular cross-linking via disulfide, isopeptide, and ester bonds), on the food composition (presence of reducing sugars or free oxygen) and on the applied process parameters (e.g., temperature, pH, or high pressure).

Proteins possess linear and conformational epitopes that may be recognized by specific antibodies. Immunoassays have been used to follow structural changes of proteins during food processing as result of mechanical and thermal treatments in the form of extrusion, cooking, roasting, or autoclaving [11–15].

11.3.2
DNA

Although DNA, as a macromolecule, exhibits relatively high chemical stability there is a broad spectrum of chemical and enzymatic reactions that result in DNA modification and/or degradation [16]. Food manufacturing which involves mechanical processes, fermentation steps or heat treatment may contribute to such reactions.

DNA fragmentation may be initiated by the application of shear forces [17]. The disruption of plant material results in the destruction of cell compartments, which in turn brings formerly separated enzymes and substrates into contact. As a result, DNA extracted from plant tissues is susceptible to digestion by endogenous nucleases [18]. DNA preparation involves the use of proteases or strong protein-denaturing agents to eliminate these activities of nucleases and allow the isolation of high molecular-weight DNA. If no denaturants are used, DNA may be degraded to fragments of less than 500 base pairs (bp) within an hour [19].

Klein et al. [20] followed the fate of DNA during the industrial extraction of sugar from sugar beet. When purified nucleic acid was added to raw juice (one of the intermediate products) at 70 °C, a rapid degradation of DNA was observed which indicated the presence of nucleases in the plant material.

Nucleic acids undergo spontaneous non-enzymatic hydrolysis in solution, with RNA being more vulnerable than DNA. At low pH conditions, depurination of the N-glycosidic link between purine bases and deoxyribose in the nucleic acid backbone is the first step in the degradation of DNA, and this is followed by

Figure 11.2. DNA fragmentation within a heating process.

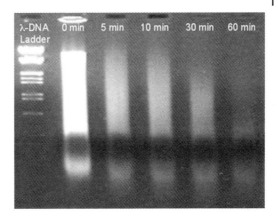

hydrolysis of adjacent 3′,5-phosphodiester linkages at the depurinated site. This acid-catalyzed reaction results in measurable shortening of DNA strands and is accelerated by simultaneous heat treatment, resulting in a random cleavage of DNA molecules [21].

Several studies have described the effects of food processing on the fragmentation of DNA. For instance, the mean fragment length of DNA extracted from heat-treated pork was reduced from 1.1 to 0.3 Kb [22]; similar effects were observed in DNA from processed tomato products [23]. Kingombe et al. [24] demonstrated that proteins as well as DNA are severely degraded during treatment of meat meal at 133 °C for 20 min at 3 bar.

The shift of the mean fragment length of DNA extracted from maize flour subjected to heat treatment at 95 °C is shown in Fig. 11.2 [25]. Quantitative data describing the recovery of DNA sequences in material treated under such conditions are presented in section 11.4.2.2.

Ensiling is another example for a process which creates a harsh environment for plant DNA via a combination of reactions. Chopping of the plant tissue results in disruption of cell walls and membranes, release of DNA and eventually in the degradation by endogenous nucleases of the plant and/or exogenous nucleases of the microflora. In addition, lowering of the pH as a result of lactic acid fermentation accelerates DNA degradation. These reactions were reflected in different contents of DNA in extracts obtained from non-ensiled and ensiled Bt-maize [26].

11.4
Analytical Approaches

11.4.1
Protein-based Methods

For the detection of GMO in raw materials, various methods based on the recognition of recombinant proteins have been developed. The most common test formats are enzyme-linked immunosorbent assays (ELISA) and immunochromatographic (lateral flow) strip tests [27]. These methods are available for the proteins expressed in the most important GMO, including insect-resistant maize, potatoes, cotton or herbicide-resistant maize, potatoes, cotton, soybean and canola [28–30].

The stability of recombinant CryIA(b) protein during ensiling of Bt maize has been studied [31]. The disruption of the maize tissue, the lowering of pH due to lactic acid fermentation and the action of plant and microbial proteases result in degradation of the recombinant protein during the ensiling process. After four months of ensilage, no CryIA(b) protein could be detected by ELISA using immunoaffinity-purified polyclonal rabbit and protein G-purified goat antibodies specific for the protein.

Using herbicide-tolerant soybeans as an example, the importance of selecting appropriate antibody reagents for the form of protein to be detected in the sample has been demonstrated [27]. One ELISA was able to recognize the CP4 EPSPS protein in soybeans and defatted soybean flakes, but not in soybean toasted meal. Hence, a second test was necessary which involved antibodies that reacted specifically with the denatured protein as it is present in the heat-treated material.

The limitation of an immunoassay to detect either native or denatured proteins represents a major drawback to the application of protein-based methods for the analysis of processed foods. As protein denaturation advances continuously during the course of processing, the quantitative character of immunoassays would be lost unless innumerable quantitation standards were developed to describe correctly the amounts of targeted proteins at each processing stage.

With regard to composed foods, it must be considered that immunoassays are not capable of discriminating between similar recombinant proteins expressed by different transgenic crops. Quantitative analysis of food samples, containing different ingredients from recombinant crops, would require the development of event-specific assays allowing a differentiation of all recombinant proteins.

11.4.2
DNA-based Methods

Methods for the detection of specific DNA sequences are based on polymerase chain reaction (PCR) techniques. Principles of this methodology have been described in detail [32]. Aspects essential for qualitative and quantitative PCR in composed and processed foods are outlined in the following paragraphs.

11.4.2.1 Qualitative PCR

Extraction of DNA and inhibition of PCR DNA extraction protocols and commercially available kits currently used for routine analysis have been designed to guarantee high DNA quality and an effective removal of inhibiting compounds in processed foods or tissue samples [33–35]. The performance of several widely applied DNA extraction methods as regards DNA quality and yield has been compared [36].

Extraction methods have been developed and/or modified in order to recover amplifiable DNA from highly processed products. Strategies to facilitate the extraction of DNA from products such as lecithin have been reported [37]. Commercially available kits based on the use of DNA-binding resins deliver low DNA yields but high DNA quality, thereby allowing the detection of DNA sequences in even highly processed food additives such as lecithin [36].

Foods are complex systems which contain a broad range of compounds other than DNA, which may inhibit PCR reactions and thus lead to false-negative results. Inhibiting compounds may be inherent in tissues of analytical samples (e. g., polysaccharides, lipids, and polyphenols), or they may be chemicals applied during DNA extraction [38–43]. The routine use of homologous or heterologous internal positive controls in PCR reactions represents a simple but powerful tool to overcome incertitudes as regards false-negative results [44].

Sensitivity of PCR The amplification efficiency in PCR is diminished by factors such as inhibiting agents, reagent limitation and increasing viscosity of the mixture. In theory, qualitative PCR analysis may detect even a single DNA sequence in a reaction vessel. Routine testing demonstrated that detection limits ranged between one and ten copies of the targeted DNA fragment [45]. The sensitivity of PCR systems depends upon a number of factors, including the number of run cycles, the position of the reaction vessel in the thermocycler, the affinity of the primers to the targeted sequence, amplicon length, the presence of inhibiting compounds, the composition of reaction mixes, and DNA quantity and quality. Therefore, it is difficult to make a general statement on the sensitivity of PCR reactions, and these must be validated on a case-by-case basis [46].

Specificity of PCR Potentially, DNA-analysis methods offer a variety of targets for the detection of GMO; these include the transgene itself and regulatory and marker gene sequences, as well as overlapping and/or border region sequences. DNA target sequences must be chosen carefully in order to guarantee high specificity in PCR systems. Positive detection of genetic elements usually used for the transformation of crops, such as the Cauliflower Mosaic Virus (CaMV) 35S promoter or the NOS terminator originally from *Agrobacterium tumefaciens*, strongly indicate the presence of GMO-derived DNA in the sample. However, positive signals may also be generated as result of contamination of analytical samples with bacteria or viruses if the primer sequences chosen are not sufficiently specific [47, 48].

In contrast, unambiguous results are obtained if the chosen primers cover overlapping areas that comprise regulatory sequences and the transgene; that is, sequences that do not occur naturally [49]. The successful use of this strategy has been demonstrated for the Flavr Savr™ tomato, glyphosate-resistant crops (Roundup Ready cotton and soybeans) and several transgene maize lines [50–53].

Additional attention must also be paid to the fact that similar genetic constructs are being used in different crops. An overview of maize events containing the commonly used insect resistance (*cryIA(b)* gene) and/or herbicide tolerance (*EPSPS* gene) traits is provided in Table 11.1.

Even more specific detection systems are achieved when targets consist of sequences covering the integration site between the plant genome and the transgene; this results in event-specific methods suitable for the detection of single transgenic lines. At this level of specificity it is possible to discriminate between authorized and non-authorized GMO containing similar transgenic constructs. Event-specific methods have been described for the detection of Bt11, Mon810 maize and Roundup Ready soya [54–58].

DNA degradation The effects of using highly fragmented DNA as a template for PCR have been elucidated [59]. DNA fragmentation severely reduces the efficiency of PCR. In accordance with these findings, the influence of the size of the targeted sequence on the detection of insect-resistant Bt 176 maize in heat-treated products has been described [53]. The probability of detecting the GMO decreased rapidly during the course of heat treatment when targeting the entire 1914 bp sequence of the synthetic *cryIA(b)* gene. On the other hand, a shorter target sequence (211 bp), covering part of the CDPK promotor and the *cryIA(b)* gene, was detectable even after heating for 105 min.

Other investigations addressed the effects of fermentation and/or thermal processes, such as ensiling of transgenic maize or the distillation of ethanol from fermented transgenic invertase potato (B33-INV) mashes, on the degradation of DNA and consequently on the traceability of target sequences [26, 60]. The detectability of Bt-specific genetic modification in ensiled maize material was found to depend on the length of the genomic target region to be amplified. By amplifying a Bt maize-specific DNA sequence of 211 bp, the genetic modification was detected up to seven months after ensilage. On the other hand, detection of the transgene via a 1914 bp amplicon was only possible for up to five days of ensilage.

A 190 bp sequence of the *patatin* gene and 839 bp sequence of the *hygromicin phosphotransferase* gene were used as targets to follow the detection of DNA during the course of the distillation process converting potatoes into ethanol [60]. Using the 190 bp amplicon, potato DNA could be detected after all steps; positive results were obtained even in the splent. Detection of the 839 bp sequence was limited by the distillation step, however.

Klein et al. [20] followed the elimination of nucleic acids during the sugar manufacturing process. Intermediates and end products were analyzed for the presence of DNA via PCR, using the ADP-glucose pyrophosphorylase (*AGP*ase) gene as a target for sugar beet DNA and the genes for the beet necrotic yellow vein virus

Table 11.1. Maize events containing the commonly used transgenic constructs *cryIA(b)*-gene and/or *EPSPS*-gene.

Event	Company	Description
176	Syngenta Seeds, Inc.	Insect-resistant maize produced by inserting the *cryIA(b)* gene from *Bacillus thuringiensis* subsp. *kurstaki*. The genetic modification affords resistance to attack by the European corn borer (ECB).
BT11	Syngenta Seeds, Inc.	Insect-resistant and herbicide-tolerant maize produced by inserting the *cryIA(b)* gene from *Bacillus thuringiensis* subsp. *kurstaki*, and the phosphinothricin N-acetyltransferase (PAT) encoding gene from *S. viridochromogenes*.
MON80100	Monsanto Company	Insect-resistant maize produced by inserting the *cryIA(b)* gene from *Bacillus thuringiensis* subsp. *kurstaki*. The genetic modification affords resistance to attack by the European corn borer (ECB).
MON802	Monsanto Company	Insect-resistant and glyphosate herbicide-tolerant maize produced by inserting the genes encoding the CryIA(b) protein from *Bacillus thuringiensis* and the 5-enolpyruvylshikimate-3-phosphate synthase (EPSPS) from *A. tumefaciens* strain CP4.
MON809	Pioneer Hi-Bred Int. Inc.	Resistance to European corn borer (*Ostrinia nubilalis*) by introduction of a synthetic *cryIA(b)* gene. Glyphosate resistance via introduction of the bacterial version of a plant enzyme, 5-enolpyruvyl shikimate-3-phosphate synthase (EPSPS).
MON810	Monsanto Company	Insect-resistant maize produced by inserting a truncated form of the *cryIA(b)* gene from *Bacillus thuringiensis* subsp. *kurstaki* HD-1. The genetic modification affords resistance to attack by the European corn borer (ECB).
GA21	Monsanto Company	Introduction, by particle bombardment, of a modified 5-enolpyruvyl shikimate-3-phosphate synthase (EPSPS), an enzyme involved in the shikimate biochemical pathway for the production of aromatic amino acids.
MON832	Monsanto Company	Introduction, by particle bombardment, of glyphosate oxidase (GOX) and a modified 5-enolpyruvyl shikimate-3-phosphate synthase (EPSPS), an enzyme involved in the shikimate biochemical pathway for the production of aromatic amino acids.
NK603	Monsanto Company	Introduction, by particle bombardment, of a modified 5-enolpyruvyl shikimate-3-phosphate synthase (EPSPS), an enzyme involved in the shikimate biochemical pathway for the production of aromatic amino acids.

coat protein (*cp21*) and neomycin phosphotransferase (*aphA*) as specific targets for the virus-resistant transgenic beet DNA. Southern blot hybridization of the targeted sequences delivered positive signals in PCR samples from raw juice only, but not in those from carbonatation sludge I, carbonatation sludge II, thin juice, thick juice, or white sugar from transgenic beets. These results pointed to a severe degradation of nucleic acids already in the first steps of processing. This was verified by adding pUC18 DNA to fresh raw juice samples and incubating the mixtures for different periods of time. The resulting DNA degradation was ascribed to the enzymatic activity of sugar beet endogenous nucleases. Further disappearance of DNA was explained by its irreversible adsorption onto the sludge, precipitation, hydrolysis due to the high temperatures in the carbonatation and evaporation steps, and as result of the exclusion of DNA in the crystallization step.

Hellebrand et al. [61] investigated the presence of rapeseed DNA in cold-pressed and refined oils by using a nested PCR system. The starting point was the assumption that the thermostability of DNA under alkaline conditions should be sufficient to partially preserve DNA throughout the processing, and that the filtration steps (as used in the oil industry) are not capable of retaining DNA molecules. Amplifiable DNA could be isolated from cold-pressed oil samples. However, PCR analysis of extracts from refined oil samples delivered non-specific signals, which could not be unequivocally identified. Again, the importance of choosing short target fragments for a successful detection was emphasized.

The limits of the PCR-based detection of genetically modified soya and in the course of bread production were studied in different approaches showing a strong dependence of the analytical success on the individual processing parameters. Straub et al. [62] employed an official method according to § 35 of the German Food Law, which had been previously validated in an interlaboratory study. It was shown that high molecular-weight DNA was only present in baking aids and flour samples, whereas DNA isolated from dough and bread samples had been partially degraded and exhibited average fragment sizes <500 bp and <300 bp, respectively. Although the content of genetically modified soya in the baking aid was diluted to a proportion of 0.4% with reference to the dry matter of the end product, positive detection of the targeted sequences was achieved at every stage of processing.

Moser et al. [63] assessed how baking parameters of different bread and pastry formulas affected DNA degradation and, in consequence, the traceability of target sequences. Genetically modified maize (used in whole grain bread and pastry formulas) and soybean (used in the toast bread formula) were added at concentrations of 0.5% and 0.3%, respectively. The maize-specific sequence of the *invertase* gene was successfully detected even in the end products, whereas no positive results were obtained for the transgenic target sequence at the final stage of processing of the whole grain bread and pastry formulas. This indicated that the combination of either acidic milieus or mechanical stress in the extrusion stages with the exposure to thermal treatment, completely prevented detection of transgenic DNA at the given concentration. In contrast, transgenic and isogenic soya DNA was detected at all manufacturing levels. Apparently, the diminished overall stress impact during the production of toast bread did not result in complete degradation of the targeted sequences.

Straub et al. [64] investigated the effect of storage on the fate of DNA from recombinant starter cultures in fermented, heat-treated sausages. Free recombinant DNA of the starter culture obtained from the meat matrix was shown to represent only a minute part of the total recombinant DNA content recovered after lytic treatment of the cell walls. Worst-case studies showed that minute amounts of free recombinant DNA were protected by the meat matrix against the activity of DNase and could be detected even after storage for over nine weeks. The other portion of recombinant DNA remained entrapped in the killed cells and so was even better protected against enzyme activity. Again, DNA deterioration was most clearly observed when long (1322 bp) rather than short (166 bp) target sequences were chosen.

DNA degradation due to stress factors in the course of food processing has been the subject of several studies examining the influence of fermentation, heating and other forms of processing on the DNA-based detection of GMO. These approaches have shown the suitability of PCR technology for detecting target DNA sequences from GMO or pathogenic microorganisms even at advanced stages of processing [26, 53, 60, 65–67]. High sensitivity can be achieved if adequate DNA extraction methods capable of eliminating potential inhibitors are applied. Internal positive controls are required to rule out false-negative results. The specificity of the GMO detection can be improved by the application of appropriate primer pairs. Event-specific discrimination between GMO with similar transgenic constructs is achievable. Degradation of DNA in the course of food processing has an adverse effect on the detection efficiency, especially when long target sequences are to be detected. Target sequences ≤ 100 bp have been shown to withstand harsh manufacturing processes.

11.4.2.2 Quantitative PCR

Standard PCR end-point analysis, as used for the qualitative detection of DNA sequences, cannot be applied for quantitations due to the unsteadiness of the amplification efficiency between different PCR reactions, as well as within the process of a PCR reaction itself. Currently available approaches for the quantitation of DNA sequences are based either on competitive PCR or on real-time PCR technologies.

Regardless of the technique used, determinations of ingredient-related GMO contents in composed and processed foods require the simultaneous assessment of DNA sequences specific for the genetic modification and for an endogenous reference gene. This strategy is based on the assumption that the breakdown of both target sequences occurs simultaneously. Therefore, this type of relative quantitation is considered to reflect the proportion of GMO during all stages of processing [68].

Sensitivity of quantitative PCR Detection limits of quantitative PCR, given as a percentage of GM material, depend on the amount of amplifiable genome copies from the plant species available in the sample. For instance, if 10^4 copies of plant DNA were extracted from the food sample, the detection limit would range between 0.01 % and 0.1 %. These values reflect the theoretical best case (when one single target sequence is detected) and the worst case (when ten copies of the target sequence are needed to yield a positive signal). The quantitation limit using TaqMan technology has been empirically set at ten copies of the targeted DNA sequence, considering that it is practically impossible to obtain reproducible data in triplicate or quadruplicate at lower copy levels. This means that if, again, 10^4 copies of the plant genome were extracted from the food sample, the quantitation limit would not allow a differentiation between GMO contents lower than 0.1 % (10 molecules of GM DNA within 10^4 molecules of species DNA) [45].

Specificity of quantitative PCR When developing methods for the quantitative detection of GMO in composed foods, special attention must be paid to the fact that similar genetic constructs are being used in different crops. Therefore, suitable target genes must be chosen to enable analysts to assign the detected DNA target sequence to the respective genetically modified crop. This objective could be achieved through the implementation of event-specific detection systems [69].

11.4.2.3 Competitive PCR

Quantitative competitive PCR is based on the co-amplification of the target DNA sequence and an exogenous standard (competitor) which is spiked into each sample at known concentration [70–74]. Both, target DNA and competitor, possess identical primer binding sites and similar lengths as well as internal sequences.

Throughout the PCR reaction, amplification of the target and competitor sequences underlie identical reaction conditions and compete for the same limiting reagents such as primers. Provided that equivalent amplification kinetics exist for both the target and competitor sequences, the ratio between molar amounts of both PCR products equals the ratio between the amounts of target DNA and competitor in the sample prior to amplification. Quantitation is most accurate at the point of equivalence; that is, where the molar ratio of target and internal standard sequences in the reaction probe equals one.

Sample analysis by means of competitive PCR requires the preparation of several samples, each containing a constant template DNA concentration but increasing competitor concentrations to allow accurate calculation of the competitor amount at the point of equivalence. Ratios between signal intensities of competitor and target amplicons, measured after gel electrophoresis, are plotted logarithmically against competitor amounts initially used in the sample. The point of equivalence is determined by the intersection of the linear regression curve with the abscissa [75].

On this theoretical basis, several methods have been developed for the quantitation of DNA sequences occurring in transgenic crops commonly used for the production of foods and food ingredients [58, 76–80].

Studer et al. [80] described the construction of internal standards for the quantitation of Roundup Ready soybean and for Bt 176 maize in food samples. The systems were calibrated with recombinant DNA mixtures and tested with certified, commercially available standards, yielding reproducible results and good linearity within the range of operation. A similar approach was used for the construction of internal standards allowing the quantitation of the 35S promoter and NOS terminator sequences, commonly used for regulating the gene expression in transgenic crops [79].

Despite the proven suitability of these methods for their application on certified reference materials, analysis of processed foods revealed that DNA degradation affects the amount of amplifiable target sequences, and this leads to diminished recoveries of DNA. Therefore, double competitive PCR systems were needed to additionally assess the amounts of plant-specific target sequences as an endogenous reference, allowing a normalization of the results and the analysis of processed and/or composed foods [76]. Several plant-specific genes were characterized and competitive PCR systems for the quantitation of the soy-specific lectin (*le1*) gene and the maize-specific invertase (*ivr1*) gene have been developed [77, 81]. Additional genes generally targeted as endogenous reference are the maize-specific zein (*ze1*), the high mobility group (*HMG*) protein genes and the rape-specific acetyl-CoA carboxylase (*BnAccg8*) gene [81–85].

Detailed studies of the effect of DNA degradation in the course of heat treatment on the quantitation of transgenic maize via competitive PCR and the normalization of results using an endogenous reference gene have been performed [25, 81]. In unprocessed maize flour, the quantitation of a single target sequence could be successfully used for the determination of transgenic maize proportions based on the parallel analysis of certified standard flour mixtures. However, the determination of DNA concentrations (via UV-absorption) in extracts of processed foods does not allow differentiation between amplifiable and degraded DNA, and this led to severe discrepancies in the results yielded by the single competitive quantitation. This drawback was overcome by the development of a complementary competitive PCR system designed for the quantitation of an endogenous reference gene.

Heat treatment continuously degrades DNA, and this results in a strongly reduced average fragment length (see Fig. 11.2). In order to demonstrate that this process affects the target sequence of Bt-maize to the same degree as any other region of the maize genome of comparable length, aliquots of a DNA mixture (10% Bt) were heat-treated at 95 °C. A transgenic specific as well as a reference target sequence were quantified by means of competitive PCR (Fig. 11.3).

The almost parallel decrease of the recovery of both target sequences forms the basis for dual competitive PCR. Thus, GMO proportions may be calculated based on the ratio between the determined amounts of DNA sequences targeting the transgene and the reference gene, delivering near-constant values for the Bt-proportions in the starting material and in the thermally treated samples. Quantitation of GMO proportions became consequently independent from: a) the presence of DNA other than from maize; and b) the degradation of target sequences throughout processing, since degradation of transgene and reference gene were shown to occur in a parallel manner.

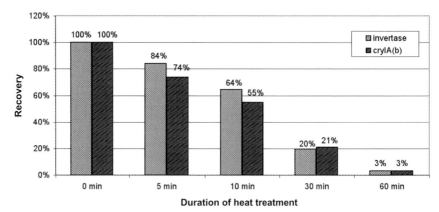

Figure 11.3. Quantification of a heat-treated DNA mixture from conventional and Bt-maize (10%). Recovery of a transgene specific target sequence (*crylA(b)*-gene, 212 bp) and an endogenous reference target sequence (*ivr1*-gene, 226 bp).

The development of a competitive PCR system for the event-specific quantitation of Bt11 corn was an important milestone to improve the specificity of quantitation systems. This approach decreases the ambiguity of GMO detection resulting from the presence of similar expression cassettes in different transgenic crops or within one crop [57]. The 5′ site of the integrated transgenic sequence in the Bt11 corn was characterized using inverse PCR technology and further used for the development of a competitive PCR system covering the integration border.

11.4.2.4 Real-time PCR

Real-time PCR technology is based on the use of a thermal cycler with an integrated optical unit, which allows the measurement of PCR product amounts at every stage of the reaction. This is achieved by monitoring the increase in fluorescence caused by intercalation of DNA-binding dyes in the resulting PCR products, or by the hydrolysis of hybridization probes labeled with a reporter and a quencher dye. The main disadvantage of using intercalating dyes is the unavoidable detection of non-specific PCR products such as primer dimers. Hybridization probes (e.g., TaqMan Probes) are synthetic oligonucleotides complementary to the target DNA. The principle of their detection relies upon the release of the reporter dye as hybridized probes are hydrolyzed by the 5′ → 3′ exonuclease activity of Taq DNA polymerase. The increase of fluorescence emitted by the reporter dye is proportional to the exponential amplification of target sequences. Other formats of hybridization probes (e.g., LightCycler Probes or molecular beacons) do not require hydrolysis to generate a signal and have also been used successfully in the quantitation of DNA sequences [39, 86].

Since, during the first stages of amplification, PCR kinetics can be described as a steady exponential process, the starting copy number of target sequences can be extrapolated on the basis of a standard curve which describes the correlation between starting copy numbers and threshold cycles (Ct-values) [87]. The construction of standard curves requires the analysis of external reference dilutions (absolute standards) with defined concentrations of target sequences. Several methods for the production of absolute standards have been described, though target sequences are usually cloned into a plasmid vector and used as template to generate a standard curve [88–90]. An accurate quantitation is only guaranteed when the amplification efficiency of the plasmid standards is equal to that of the target sequence within the genome of the host organism. This requirement is fulfilled when standard curves generated from plasmid standards and dilutions of genomic DNA render similar slopes. Efficiencies can be considered as equal if the difference between the slopes is < 0.1.

To date, various methods for the quantitation of GMO proportions in raw food materials have been presented and validated in international interlaboratory trials [56, 57, 77, 91–96]. Kits for the quantitation of transgenic soy (Roundup Ready™) and maize (Maximizer™ Bt176, Bt11, Liberty Link™ T25, Yield Guard™ MON810, Roundup Ready™ and StarLink™) are commercially available.

The first real-time PCR method to be applied for the quantitative detection of a genetically modified organism in foods was described by Wurz et al. [77]. The method was adapted to detect a Roundup Ready soya-specific sequence and a plant-specific sequence within the lectin (*le1*) gene. Special attention was paid to avoid significant differences in amplicon lengths and to keep them as short as possible, meeting basic prerequisites for its application on processed foods. The approach was subsequently tested with certified reference materials containing 0.1%, 0.5% and 2% transgenic soya, yielding results that were in good agreement with the expected output.

Approaches for the quantitative detection of Bt 176 "Maximizer" maize and Roundup Ready soybean, targeting the *cryIA(b)* and the CP4 *EPSPS* transgenes, respectively, have also been introduced [96]. Endogenous reference targets used were the maize-specific *zein* (*ze1*) and the soya-specific *lectin* (*le1*) genes. For the first time, PCR conditions were optimized to allow the quantitation of transgenic and isogenic targets in one tube, thus eliminating variations other than those inherent in the Ct measurements.

Pietsch and Waiblinger [93] described a real-time PCR method for the quantitation of genetically modified soybean using LightCycler technology. Primers developed specifically anneal to the overlapping sequence between the CMV 35S promoter and the *Petunia hybrida* 5-enolpyruvylshikimate-3-phosphate synthase signal peptide. The soybean-specific lectin gene was used as reference. The linearity of the system was shown by using DNA dilutions from certified standard materials. The analysis of a soybean diet sample was taken as an example for the application of this approach on manufactured foods. This method was subsequently validated [94], yielding diminishing relative confidence intervals at $P = 95\%$ from 37% to 9.3% with increasing transgenic contents in the reference materials (0.1% RRS and 5% RRS, respectively).

The German Federal Institute for Health Protection of Consumers and Veterinary Medicine (BgVV) recently presented results regarding the interlaboratory testing of commercially available kits for the quantitation of Roundup Ready® soybean and Bt 176 maize [95]. These kits allow the determination of GMO proportions in food samples, in the case of Roundup Ready® soya relying on the amplification of two 74 bp-long fragments from the functional insert and from the soybean-specific lectin (*le1*) gene. In the case of Bt 176, a 129 bp-long section of the Bt toxin gene and a 79 bp-long sequence of the maize high mobility group protein gene *(HMG)* are amplified [99].

Although these approaches are evidentially suitable for their application on reference materials, several questions remain open. In respect to the analysis of certified reference materials using the Bt 176 method, both real-time PCR systems used showed a slight trend to overestimate the GMO content in all samples. Nevertheless, the most questionable effect results from the analysis of heat-sterilized kernels, which resulted in a strong understatement of the set GMO proportion.

The influence of some technological parameters on the quantitation of GMO via real-time PCR was studied by Moreano et al. [97, 98]. In a first approach, the influence of the particle size composition of maize milling products on the determination of GMO contents was studied. Different milling fractions of maize – from coarse grits down to fine flours – are usually applied for the industrial production of foods and foodstuffs [97]. Similar fractions were prepared at laboratory scale by grinding whole maize kernels and subsequently sieving the material through standard sieves. The particle size compositions of the obtained fractions were characterized using a laser diffraction system. The application of two established DNA extraction methods for purposes of GMO analysis in foods [33] revealed a strong correlation between the particle size composition of the milling fractions and the DNA yields in the extracts: the DNA yield increased with decreasing median particle sizes (Table 11.2).

These data indicated that in composed products, which contain conventional and transgenic material in form of different ingredients, the ratios of DNA concentrations determined in the extracts may not reflect the proportions of GM material in the food sample. In order to confirm this assumption, four mixtures with a GMO content of 1 % were prepared by different combinations of coarse

Table 11.2. DNA extraction from maize milling products with different particle size distributions.

Milling fraction	Median particle size X_{50} [μm] ± $CI_{(95\%)}$	DNA concentration** [ng/μl]	
		CTAB-extraction*	Wizard-extraction*
Coarse grits	1049 ± 16	196	200
Regular grits	697 ± 41	173	236
Meal	287 ± 53	320	347
Flour	19 ± 2	527	359

* Starting from 300 mg per sample
** means of duplicate measurements

Table 11.3. Composition of sample mixtures with a Bt-maize content of 1%.

Sample	Bt 176 maize		Conventional maize	
	Coarse grits [mg]	Flour [mg]	Coarse grits [mg]	Flour [mg]
Mix 1	3	–	297	–
Mix 2	–	3	–	297
Mix 3	–	3	297	–
Mix 4	3	–	–	297

grits and flour made from conventional and transgenic maize Bt 176, respectively (Table 11.3).

The quantitation of DNA was performed using a commercially available real-time PCR quantitation kit, which had been previously tested within an international ring trial (GeneScan, Bt 176 Maximizer™ Corn DNA Quantitation System [95]). Table 11.4 shows how quantitation results significantly varied depending on the composition of the mixtures. An accurate quantitation of the adjusted GMO content was solely possible in DNA extracts from mixtures, in which conventional and transgenic material were present in form of analogous milling fractions. DNA extracted from mixtures containing fractions with different particle size distributions delivered a strong over- and underestimation of the adjusted GMO proportion. Distortions of quantitation results were lower but still substantial if DNA was extracted using the Wizard® extraction method, rather than applying the CTAB/chloroform method.

Table 11.4. Quantification of Bt-maize contents in mixtures of milling products with different particle size distributions.

Mixture	Bt-maize contents* ± $CI_{(95\%)}$ [%]	
	CTAB-Extraction**	Wizard-Extraction**
1	0.8 ± 0.5	0.9 ± 0.4
2	1.4 ± 0.4	1.0 ± 0.1
3	4.3 ± 1.1	2.6 ± 0.8
4	0.2 ± 0.1	0.3 ± 0.2

* Kit: GeneScan, Bt 176 Maximizer™ Corn DNA Quantification System
   Amplicon lenghts: reference gene 79 bp, transgene 104 bp [100]
** Starting from 300 mg sampling material

The second approach studied the influence of heat-induced DNA degradation on GMO quantitation results [98]. The distortion of the quantitation results observed in the interlaboratory testing of the commercially available kit for the quantitation of Bt 176 maize [95] permitted the assumption that the length differences between both target sequences might have been too high. This difference resulted in a faster degradation rate of the GMO specific target sequence during heat-treatment, thus yielding this method unsuitable for its application on processed foods. In contrast to the Bt 176 maize method, accurate results were delivered by the assessment of GMO contents in a textured vegetable protein product using the Roundup Ready soya method, which uses target sequences with exactly the same length.

In order to confirm this assumption, two different milling fractions obtained from Bt 176 maize (coarse corn grits and corn flour) were heated in boiling water for 60 and 120 min. The DNA extracts (Wizard® method) were quantified using the above-mentioned kit.

Recoveries of GMO contents calculated on the basis of the obtained C_t-values are listed in Table 11.5. The application of this kit for the determination of GMO contents via real-time PCR in heat-treated products resulted again in a drastic underestimation of the true GMO content. This distortion of the recovery of GMO contents in processed samples confirmed the observations made during the ring trial, and could be ascribed to higher degradation rates for the longer transgenic target sequence (104 bp) compared to the reference target sequence (79 bp). This effect was more pronounced for heat-treated maize grits than for heat-treated flour.

In order to support this finding which suggested that results from the relative quantitation of GMO in processed products via real-time PCR could be affected by length differences between reference and transgenic target sequences, a second real-time PCR quantitation system [91] was applied. In contrast to the previous method, the difference between the lengths of both target sequences was much smaller (16 bp) and the reference target sequence was longer than the transgene target sequence. Application of this method resulted in a slight overestimation of GMO contents (see Table 11.5). This is in agreement with the postulated higher degradation rates for the longer target sequence. Due to the smaller length differences between target sequences, the degree of distortion is in this case lower than as observed by the previous method. Distortion of quantitation results was again stronger in DNA from coarse maize grits than in DNA from maize flour.

In conclusion, particle size distributions of food components were found to have a significant effect on the efficacy of DNA extraction. Over- and underestimation of

Table 11.5. Distortion of the recovery of GMO contents in coarse grits and flour from Bt-maize in the course of heat-treatment.

Duration of heat-treatment [min]	Recovery of GMO contents[*] ± SD [%]		Recovery of GMO contents[**] ± SD [%]	
	Coarse grits	Flour	Coarse grits	Flour
0	86 ± 21	71 ± 11	112 ± 4	114 ± 8
60	56 ± 11	56 ± 13	120 ± 9	117 ± 4
120	26 ± 12	46 ± 7	137 ± 13	126 ± 1

[*] Bt 176 Maximizer™ Corn DNA Quantification System [100] Amplicon lengths: reference gene 79 bp, transgene 104 bp.

[**] Quantification method as described by Höhne et al. [91] Amplicon lengths: reference gene 84 bp, transgene 68 bp.

GMO contents may occur in composed foods, especially if conventional and transgenic components are added as fractions with different particle size distributions. DNA in milling fractions with different median particle sizes was found to undergo different degradation kinetics, delivering further problems for the analysis of composed and processed foods. In relation to the design of detection methods, length differences between reference and transgenic target sequences result in higher degradation rates for the longer sequence and consequently in a distortion of relative quantitation results.

11.5
Conclusions

Immunoassay methods represent rapid and cost-effective techniques for the detection and quantitation of proteins expressed in genetically engineered crops. They are being applied successfully to test raw agricultural materials, though their applicability in processed foods is strongly limited by the sensitivity of proteins to technological treatments. It is rather unlikely that protein-based assays will become available for the complex spectrum of composed and processed foods. However, these methods will remain important for traceability and identity preservation purposes and for the characterization of raw material placed into the food chain.

PCR-based methods for the detection of specific DNA sequences offer a significant increase in analytical sensitivity. The flexibility of PCR has been exploited for the development of highly specific methods that allow not only the detection but also the identification of DNA; that is, they offer the potential to discriminate authorized from non-authorized transgenic lines containing similar transgenic constructs.

DNA quantitation techniques allow the determination of ingredient-related GMO contents in composed and processed foods, on the basis of the simultaneous assessment of DNA sequences specific for the genetic modification and for an endogenous reference gene. The limitations imposed by the time-consuming double competitive techniques have been overcome by the development of real-time PCR methodologies. However, it has been shown that commonly used technological processes can strongly affect the accuracy of DNA quantification results, leading to distortions in the determination of GMO contents. As the content of amplifiable DNA decreases during the course of food processing, it is imperative to ensure that the breakdown of both reference and transgenic target sequences will occur in parallel. This requires the targeting of short amplicons with similar lengths (~100 bp) which may be detected even at stages of advanced processing.

As yet, quantitation methods have been tested almost exclusively on the basis of unprocessed raw materials. However, in the future, validations have to include trials which prove the accurate quantitation of GMO contents at several stages of food manufacturing. Not only the decrease in analytical sensitivity due to the degradation of DNA during processing but also the distortion of results due to technological parameters must be taken into account.

Another undeniable fact is that the copy number of a gene used as reference may vary, depending upon the species. The effects of further breeding of GMO – and especially the crossing of GMO with different traits leading to progeny containing two or more genetic constructs – represent major problems for quantitative analysis. "Gene stacking" will raise further questions as regards the use of correct reference materials and quantitation principles. The increasing complexity of GMO-derived material in composed and processed foods will make the quantitation of a specific GMO at the ingredient level and the test for compliance with a *de minimis* threshold increasingly difficult.

References

1 Robinson C. (**2001**) *Genetic modification technology and food; consumer health and safety.* ILSI Europe Concise Monograph Series.

2 Anklam E., Neumann D. (**2002**) Method development in relation to regulatory requirements for detection of GMOs in the food chain. *J. AOAC Int.* 85, 754–756.

3 ILSI Europe Novel Food Task Force in collaboration with the European's Joint Research Center (JRC) and ILSI International Food Biotechnology Committee. (**2001**) Method development in relation to regulatory requirements for detection of GMOs in the food chain. In: ILSI Europe report series, Summary report of a joint workshop held in December 2000; pp. 7–8.

4 Van den Eede G., Kay S., Anklam E. (**2002**) Analytical challenges: Bridging the gap from regulation to enforcement. J. AOAC Int. 85, 757–761.

5 Regulation (EC) No 258/97 of the European Parliament and of the Council of 27 January 1997 concerning novel foods and novel food ingredients. Official Journal of the European Communities – 14.02.**1997** – No L 043, P. 0001–0007.

6 Council Regulation (EC) No 1139/98 of 26 May 1998 concerning the compulsory indication of the labelling of certain foodstuffs produced from genetically modified organisms of particulars other than those provided for in Directive 79/112/EEC. Official Journal of the European Communities – 03.06.**1998** – No L 159, P. 0004–0007.

7 Commission Regulation (EC) No 50/2000 of 10 January 2000 on the labelling of foodstuffs and food ingredients containing additives and flavourings that have been genetically modified or have been produced from genetically modified organisms. Official Journal of the European Communities – 11.1.**2000** – No L 006, P. 0015–0017.

8 Commission Regulation (EC) No 49/2000 of 10 January 2000 amending Council Regulation (EC) No 1139/98 concerning the compulsory indication on the labelling of certain foodstuffs produced from genetically modified organisms of particulars other than those provided for in Directive 79/112/EEC. Official Journal of the European Communities – 11.1.**2000** – No L 6, P. 0013–0014.

9 Directive 2001/18/EC of the European Parliament and of the Council of 12 March 2001 on the deliberate release into the environment of genetically modified organisms and repealing Council Directive 90/220/EEC – Commission Declaration. Official Journal L 106, 17/04/**2001**, P. 0001–0039.

10 Belitz H.-D., Grosch W., Schieberle P. (**2001**) In: *Lehrbuch der Lebensmittelchemie.* Springer Publishing.

11 Quillien L., Gaborit T., Gueguen J., Melcion J. P., Kozlowski A. (**1990**) ELISA as a tool for evaluating the denaturating effect of extrusion cooking on pea legumin (*Pisum sativum L.*). *Sciences des Aliments* 10, 429–442.

12 Varshney G. C., Mahana W., Filloux A. M., Venien A., Paraf A. (**1991**) Structure of native and heat-denatured ovalbumin as revealed by monoclonal antibodies: Epitopic changes during heat treatment. *J. Food Sci.* 56, 224–233.

13 Marn M. L., Casas C., Cambero M. I., Sanz B. (**1992**) Study of the effect of heat (treatments) on meat protein denaturation as determined by ELISA. *Food Chem.* 43, 147–150.

14 Unglaub W., Müller R., Jemmi T., Stuker G. (**1998**) Erhitzungsnachweis an Fleisch-Knochenmehl aus Nebenprodukten der Lebensmittelgewinnung mittels Enzymimmunologie. *Fleischwirtschaft* 78, 1072–1078.

15 Roux K. H., Teuber S. S., Robotham J. M., Sathe S. K. (**2001**) Detection and stability of the major almond allergen in foods. *J. Agric. Food Chem.* 49, 2131–2136.

16 Lindahl T., Nyberg B. (**1972**) Rate of depurination of native deoxyribonucleic acid. *Biochemistry* 11, 3610–3618.

17 Adam R. E., Zimm B. H. (**1977**) Shear degradation of DNA. *Nucleic Acids Res.* 4, 1513–1537.

18 Ausubel F. M., Brent R., Kingston R. E., Moore D. D., Seidman J. G., Smith J. A., Struh K. (**1992**) *Current Protocols in Molecular Biology.* Greene Publishing Associates and John Wiley & Sons, New York.

19 Thomson J. (**2000**) Joint FAO/WHO Expert Consultation on Foods Derived from Biotechnology, Topic 11: Gene Transfer: Mechanisms and Food Safety Risks. Biotech. 00/13.

20 Klein J., Altenbuchner J., Mattes R. (**1998**) Nucleic acid and protein elimination during the sugar beet manufacturing process of conventional and transgenic sugar beets. *J. Biotechnol.* 60, 145–153.

21 Maniatis T., Fritsch E. F., Sambrook J. (**1982**) *Molecular Cloning.* CSH Press.

22 Ebbehy K. F., Thomsen P. D. (**1991**) Species Differentiation of Heated Meat Products by DNA Hybridization. *Meat Sci.* 30, 221–234.

23 Ford M., Du Prat E., Barallon R. V., Rogers H. J., Parkers H. C. (**1996**) Detection of Genetic Modifications in Food. In: *Authenticity ,96*, IFR, Norwich, September 1–3, 1996. Poster/Conference Abstract.

24 Kingombe C. I. B., Lüthy E., Schlosser H., Howald D., Kuhn M., Jemmi T. (**2001**) A PCR-based test for species-specific determination of heat treatment conditions of animal meals as an effective prophylactic method for bovine spongiform encephalopathy. *Meat Sci.* 57, 35–41.

25 Moreano F., Hotzel H., Sachse K., Engel K.-H. (**2000**) Quantification of DNA in thermally treated products. In: *Book of Abstracts of the joint workshop on method development in relation to regulatory requirements for the detection of GMOs in the food chain.* ILSI Europe Novel Food Task Force in collaboration with the European Commission's Joint Research Centre (JRC) and ILSI International Food Biotechnology Committee. December 2000, Brussels, Belgium.

26 Hupfer Ch., Mayer J., Hotzel H., Sachse K., Engel K.-H. (**1999**) The effect of ensiling on PCR-based detection of genetically modified Bt maize. *Eur. Food Res. Technol.* 209, 301–304.

27 Stave J. W. (**2002**) Protein immunoassay methods or detection of biotech crops: Applications, Limitations, and practical considerations. *J. AOAC Int.* 85, 780–786.

28 Lipp M., Anklam E., Stave J. W. (**2000**) Validation of an immunoassay for detection and quantitation of a genetically modified soybean in food and food fractions using reference materials: interlaboratory study. *J. AOAC Int.* 83, 919–927.

29 Adamczyk, J. J., Adams, L. C., Hardee, D. D. (**2000**) Quantification of cryIA(c)-endotoxin in transgenic BT cotton: Correlating insect survival to different protein levels among plant parts and varieties. *Pro.-Beltwide Cotton Conf.* 2, 929–932.

30 Bookout, J. T., Joaquim, T. R., Magin, K. M., Rogan, G. J., Lirette, R. P. (**2000**) Development of a Dual-Label Time-Resolved Fluorometric Immunoassay for the Simultaneous Detection of Two

Recombinant Proteins in Potato. *J. Agric. Food Chem.* 48, 5868–5873.

31 Fearing P. L., Brown D., Vlachos D., Meghji M., Privalle L. (**1997**) Quantitative analysis of CryIA(b) expression in Bt maize plants, tissues, and silage and stability of expression over successive generations. *Molecular Breeding* 3, 169–176.

32 Baumforth K. R., Nelson P. N., Digby J. E., O'Neil J. D., Murray P. G. (**1999**) Demystified… the polymerase chain reaction. *Mol. Pathol.* 52, 1–10.

33 Anklam E., Gadani F., Heinze P., Pijnenburg H., Van den Eede G. (**2002**) Analytical methods for detection and determination of genetically modified organisms in agricultural crops and plant-derived food products. *Eur. Food Res. Technol.* 214, 3–26.

34 Gadani F., Bindler G., Pijnenburg H., Rossi L., Zuber J. (**2000**) Current PCR methods for the detection, identification and quantification of genetically modified organisms (GMOs) A brief review. *Beitr. Tabakforsch. Int.* 19, 85–96.

35 Barallon R. (**1999**) Review of DNA extraction methods. Proceedings of the European Research Project SMT4-CT96-2072: Development of methods to identify foods produced by means of genetic engineering. BGVV-Hefte, 05/1999.

36 Zimmermann A., Lüthy J., Pauli U. (**1998**) Quantitative and qualitative evaluation of nine different extraction methods for nucleic acids on soya bean food samples. *Z. Lebensm. Forsch.* 207, 81–90.

37 Wurz A., Rüggeberg H., Brodmann P., Waiblinger H. U., Pietsch K. (**1998**) DNA-Extraktionsmethode für den Nachweis gentechnisch veränderter Soja in Sojalecithin. *Z. Lebensm. Rundsch.* 94, 159–161.

38 Scheu P. M., Berghof K., Stahl U. (**1998**) Detection of pathogenic and spoilage micro-organisms in food with the polymerase chain reaction. *Food Microbiol.* 15, 13–31.

39 Lantz P.-G., Abu Al-Soud W., Knutsson R., Hahn-Hägerdal B., Radström P. (**2000**) Biotechnical use of polymerase chain reaction for microbiological analysis of biological samples. *Biotechnology Annual Review*, Volume 5, 87–130.

40 Wilson I. (**1997**) Inhibition and facilitation of nucleic acid amplification. *Appl. Environ. Microbiol.* 63, 3741–3751.

41 Mizushina Y., Yoshida S., Matsukage A., Sakaguchi K. (**1997**) The inhibitory action of fatty acids on DNA polymerase. *Biochim. Biophys. Acta* 1336, 509–521.

42 De Boer S. H., Ward L. J., Li X., Chittaranjan S. (**1995**) Attenuation of PCR inhibition in the presence of plant compounds by addition of BLOTTO. *Nucleic Acids Res.* 23, 2567–2566.

43 Rossen L., Norskov P., Holmstrom K., Rasmussen O. F. (**1992**) Inhibition of PCR by components of food samples, microbial diagnostic assays and DNA-extraction solutions. *Int. J. Food Microbiol.* 17, 37–45.

44 Englund S., Ballagi-Pordany A., Bölske G., Johansonn K. E. (**1999**) Improved reliability of diagnostic PCR with an internal positive control molecule. *Proc. 6th Intl. Coll. Paratuberculosis*: Manning E. J. B., Collins M. T. (eds). Section 4 Diagnostic Applications and Approaches.

45 Spiegelhalter F., Lauter F. R., Russell J. M. (**2001**) Detection of genetically modified food products in a commercial laboratory. *J. Food Sci.* 66, 634–640.

46 Lipp M., Bluth A., Eyquem F., Kruse L., Schimmel H., Van den Eede G., Anklam E. (**2001**) Validation of a method based on polymerase chain reaction for the detection of genetically modified organisms in various processed foodstuffs. *Eur. Food Res. Technol.* 212, 497–504.

47 Wolf C., Scherzingern M., Wurz A., Pauli U., Hübner P., Lüthy J. (**2000**) Detection of cauliflower mosaic virus by the polymerase chain reaction: testing of food components for false-positive 35S-promoter screening results. *Eur. Food Res. Technol.* 210, 367–372.

48 Stadler M., Hübner P. (**1998**) Probleme bei der Anwendung der 35S- und NOS-PCR für die GVO-Analytik. *Mitt. Gebiete Lebensm. Hyg.* 89, 308–317.

49 Pöpping B. (**2001**) Methods for the detection of genetically modified organisms: Precision, pitfalls and proficiency. *International Laboratory* 31, 23–29.

50 Matsuoka T., Kuribara H., Takubo K., Akiyama H., Miura H., Goda Y., Kusakabe Y., Toyoda M., Hino A. (**2002**) Detection of recombinant DNA segments introduced to genetically modified maize (*Zea mays*). *J. Agric. Food Chem.* 50, 2100–2109.

51 Meyer R. (**1995**) Nachweis gentechnologisch veränderter Pflanzen mittels der Polymerase Kettenreaktion (PCR) am Beispiel der Flavr Savr Tomate. *Z. Lebensm. Unter. Forsch.* 201, 583–586.

52 Van Duijn G., van Biert R., Bleeker-Marcelis B., Peppelman H., Hessing M. (**1999**) Detection methods for genetically modified crops. *Food Control* 10, 375–378.

53 Hupfer C., Hotzel H., Sachse K., Engel K.-H. (**1998**) Detection of the genetic modification in heat-treated products of Bt maize by polymerase chain reaction. *Z. Lebensm. Unters. Forsch.* A 206, 203–207.

54 Holck A., Va M., Didierjean L., Rudi K. (**2002**) 5'-Nuclease PCR for quantitative event-specific detection of the genetically modified Mon810 MaisGuard maize. *Eur. Food Res. Technol.* 214, 449–453.

55 Windels P., Taverniers I., Depicker A., van Bockstaele E., de Loose M. (**2001**) Characterization of the Roundup Ready soybean insert. *Eur. Food Res. Technol.* 213, 107–112.

56 Berdal K. G., Holst-Jensen A. (**2001**) Roundup Ready soybean event-specific real time quantitative PCR assay and estimation of the practical detection and quantification limits in GMO analyses. *Eur. Food Res. Technol.* 213, 432–438.

57 Terry C. F., Harris N. (**2001**) Event-specific detection of Roundup Ready soya using two different real time PCR detection chemistries. *Eur. Food Res. Technol.* 213, 425–431.

58 Zimmermann A., Lüthy J., Pauli U. (**2000**) Event-specific transgene detection in Bt11 corn by quantitative PCR at the integration site. *Lebensm. Wiss. und Technol.* 33, 210–216.

59 Golenberg E. M., Bickel A., Weihs P. (**1996**) Effect of highly fragmented DNA on PCR. *Nucleic Acids. Res.* 24, 5026–5033.

60 Hupfer C., Schmitz-Winnenthal J., Engel K.-H. (**1999**) Detection of genetic change in transgenic potatoes during the distillery process in ethanol production. *Lebensmittelchemie* 53, 13–14.

61 Hellebrand M., Nagy M., Mörsel T. (**1998**) Determination of DNA traces in rapeseed oil. *Z. Lebensm. Unters. Forsch.* 206, 237–242.

62 Straub J. A., Hertel C., Hammes W. P. (**1999**) Limits of a PCR-based detection method for genetically modified soya beans in wheat bread production. *Z. Lebensm. Unters. Forsch.* 208, 77–82.

63 Moser M. A., Kniel B., Schmitz-Winnenthal J., Hupfer C., Engel K.-H. (**1999**) Einfluß verfahrenstechnischer Parameter auf den analytischen Nachweis gentechnisch veränderter Zutaten. *Getreide, Mehl, Brot.* Sonderdruck 6/99.

64 Straub J. A., Hertel C., Hammes W. P. (**1999**) The fate of recombinant DNA in thermally treated fermented sausages. *Eur. Food Res. Technol.* 210, 62–67.

65 Hurst C., Knight A., Bruce I. (**1999**) PCR detection of genetically modified soy and maize in foodstuffs. *Molecular Breeding* 5, 579–586.

66 Vaitilingom M., Gendre F., Brignon P. (**1998**) Direct detection of viable bacteria, molds and yeasts by reverse transcriptase PCR in contaminated milk samples after heat treatment. *Appl. Environ. Microbiol.* 64, 1157–1160.

67 Busch U., Mühlbauer B., Böhm U., Liebl B. (**2001**) Gentechnisch veränderter Mais und Soja – qualitativer und quantitativer Nachweis. Deutsche Lebensmittel-Rundschau 97. Jahrgang, Heft 4.

68 ILSI Europe Novel Food Task Force in collaboration with the European Commission's Joint Research Centre (JRC) and ILSI International Food Biotechnology Committee. (2001) Method development in relation to regulatory requirements for the detection of GMOs in the food chain. In: ILSI Europe Report Series, Summary Report of a Joint Workshop held in December 2000; pp. 13.

69 Windels P., Theuns I., Dendauw J., Depicker A., Van Bockstaele E., De Losse M. (1999) Development of a line specific GMO detection method: A case study. *Fac. Landbouwkd. Toegepaste Biol. Wet.* 64(5b), 459–462.

70 Zimmermann K., Mannhalter J. W. (1996) Technical Aspects of Quantitative Competitive PCR. *BioTechniques* 21, 268–279.

71 Reischl U., Kochanowski B. (1995) Quantitative PCR. *Mol. Biology* 3, 55–71.

72 Piatak M. Jr., Luk K.-C., Williams B., Lifson J. D. (1993) Quantitative Competitive Polymerase Chain Reaction for Accurate Quantitation of HIV DNA and RNA Species. *BioTechniques* 14, 70–81.

73 Raeymaekers L. (1993) Quantitative PCR: Theoretical Considerations with Practical Implications. *Anal. Biochem.* 214, 582–585.

74 Gilliland G., Perrin S., Blanchard K., Bunn H. F. (1990) Analysis of cytokine mRNA and DNA: Detection and quantitation by competitive polymerase chain reaction. *Proc. Natl. Acad. Sci. USA* 87, 2725–2729.

75 Hübner P., Studer E., Lüthy J. (1999) Quantitative competitive PCR for the detection of genetically modified organisms in food. *Food Control* 10, 353–358.

76 Hübner P., Studer E., Lüthy J. (1999) Detection of genetically modified organisms in food. *Nature Biotechnol.* 17, 1137–1138.

77 Wurz A., Bluth A., Zelz P., Pfeifer C., Willmund R. (1999) Quantitative analysis of genetically modified organisms (GMO) in processed food by PCR based methods. *Food Control* 10, 385–389.

78 Pietsch K., Waiblinger H.-U. (1999) Kompetitive PCR zur Quantifizierung 'kontentioneller' und transgener Lebensmittelbestandteile. *Lebensmittelchemie* 53, 13.

79 Hardegger M., Brodmann P., Herrmann A. (1999) Quantitative detection of the 35S promoter and the NOS terminator using quantitative competitive PCR. *Eur. Food Res. Technol.* 209, 83–87.

80 Studer E., Rhyner C., Lüthy J., Hübner P. (1998) Quantitative competitive PCR for the detection of genetically modified soybean and maize. *Z. Lebensm. Unters. Forsch.* 270, 207–213.

81 Hupfer C., Hotzel H., Sachse K., Moreano F., Engel K.-H. (2000) PCR-based quantification of genetically modified Bt maize: single-competitive versus dual-competitive approach. *Eur. Food Res. Technol.* 212, 95–99.

82 Matsuoka T., Kubribara H., Akiyama H., Miura H., Goda Y., Kusakabe Y., Isshiki K., Toyoda M., Hino A. (2001) A multiplex PCR method of detecting recombinant DNAs from five lines of genetically modified maize. *J Food Hyg. Soc. Japan* 42, 24–32.

83 Studer E., Dahinden I., Lüthy J., Hübner P. (1997) Nachweis des gentechnisch veränderten Maximizer Mais mittels der Polymerase-Kettenreaktion (PCR). *Mitt. Gebiete Lebensm. Hyg.* 88, 515–524.

84 Krech A. B., Wurz A., Stemmer Ch., Feix G., Grasser K. D. (1999) Structure of genes encoding chromosomal HMG1 proteins from maize. *Gene* 234, 45–50.

85 Hernandez M., Rio A., Esteve T., Prat S., Pla M. (2001) A Rapeseed-Specific Gene, Acetyl-CoA Carboxylase, Can be Used as a Reference for Qualitative and Real-Time Quantitative PCR Detection of Transgenes from Mixed Food Samples. *J. Agric. Food Chem.* 49, 3622–3627.

86 Bustin S. A. (2000) Absolute quantification of mRNA using real-time reverse transcription polymerase chain

reaction assays. *J. Molec. Endocrinol.* 25, 169–193.
87 PE Applied Biosystems. (**1997**) Relative Quantification of Gene Expression. User Bulletin #2.
88 Specht K., Richter T., Müller U., Walch., Werner M., Höfler H. (**2001**) Quantitative Gene Expression Analysis in Microdissected Archival Formalin-Fixed and Paraffin-Embedded tumor Tissue. *Am. J. Pathol.* 158, 419–429.
89 Li X., Wang X. (**2000**) Application of real-time polymerase chain reaction for the quantitation of interleukin-1 mNA upregulation in brain isochemic tolerance. *Brain Res. Protoc.* 5, 211–217.
90 Gut M., Leutenegger Ch., Huder J., Pedersen N., Lutz H. (**1999**) One-tube fluorogenic reverse transcription-polymerase chain reaction for the quantitation of feline coronaviruses. *J. Virol. Methods* 77, 37–46.
91 Höhne M., Santisi C. R., Meyer R. (**2002**) Real-time multiplex PCR: An accurate method for the detection and quantification of 35S-CaMV promoter in genetically modified maize-containing food. *Eur. Food Res. Technol.* 215, 59–64.
92 Taverniers, I., Windels, P., Van Bockstaele, E., De Loose, M. (**2001**) Use of cloned DNA fragments for event-specific quantification of genetically modified organisms in pure and mixed food products. *Eur. Food Res. Technol.* 213, 417–424.
93 Pietsch K., Waiblinger H.-U. (**2001**) Quantification of genetically modified soybeans in food with the LightCycler system. *Rapid Cycle Real-Time PCR* 385–389.
94 Waiblinger H. U., Gutmann M., Hädrich J., Pietsch K. (**2001**) Validierung der Real-time PCR zur Quantifizierung von gentechnisch veränderter Soja. Deutsche Lebensmittel-Rundschau Heft 4.
95 Broll H., Zagon A., Butschke., Grohmann L. (**2002**) GVO Analytik: Validierung und Ringversuche. GMO Analytik heute. Symposium organized by Scil Diagnostics GmbH and GeneScan Europe AG. Frankfurt am Main, Germany, 23rd January 2002.
96 Vaitilingom M., Pijnenburg H., Gendre F., Brignon P. (**1999**) Real-Time PCR detection of genetically modified Maximizer maize and Roundup Ready soybean in some representative foods. *J. Agric. Food Chem.* 47, 5261–5266.
97 Moreano F., Busch, U., Engel K.-H. Distortion of GMO Quantification Results: Influence of Particle Size Compositions as Assayed by Various DNA Extraction Methods. *J. Agric. Food Chem.* Submitted.
98 Moreano F., Busch, U., Engel K.-H. Distortion of GMO Quantification Results: Influence of Heat Induced DNA Degradation as Assayed on Different Maize Milling Fractions. *J. Agric. Food Chem.* Submitted.
99 DIN ISO 21570 Foodstuffs – Methods of Analysis for the Detection of Genetically Modified Organisms and Derived Products. Quantitative Nucleic Acid Based Methods. In preparation.
100 Personal communication. Dr. L. Grohmann, GeneScan Analytics GmbH.

12
Mutations in *Lactococcus lactis*, and their Detection

Jan Kok and Bertus van den Burg

12.1
Introduction

Genetically modified organisms (GMOs) are organisms in which permanent DNA alterations have been introduced through recombinant DNA techniques. Over the past two decades, *Lactococcus lactis* has been made amenable to recombinant DNA technology. Tools to transform *L. lactis*, to introduce and express (foreign) DNA, to secrete (heterologous) proteins and to mutate the chromosome by single and double crossover recombination strategies have all been developed. Constitutive or regulated promoters can be used to drive (foreign) gene expression. The culmination of genetic dissection of *L. lactis* was the determination in 2001 of the entire nucleotide sequence of the chromosome of *L. lactis* subsp. *lactis* IL1403 [1]. All of these advances unlock a wide spectrum of possibilities for the introduction of desirable characteristics in the species or the removal of undesirable traits. For instance, *L. lactis* has already been used to produce proteins from many different prokaryotic and eukaryotic organisms. The genetic methods can be used to alter the strains in such a way that they could be used for entirely new purposes. The major advantage of using recombinant DNA techniques over traditional methods of strain selection is that new traits can be introduced very rapidly and with great accuracy, while in classical mutagenesis the background of the introduced mutation is mostly unknown. For several reasons, it may be necessary to monitor the modified strain, either as such or in the products in which it is used. This review details a number of methods by which this can be done. We will first provide an overview of the genetic changes that occur in this organism naturally, and will then detail to what extent and by which methods the designed mutations can be distinguished from natural mutations and can be traced in food products.

12.2
Composition of the Genome of Lactococcus lactis

The genome of an organism comprises of its chromosome(s) and all extrachromosomal DNA elements. The genome of L. lactis consists of one circular chromosome and plasmid DNA. The nucleotide sequence of the entire chromosome of L. lactis subsp. lactis strain IL1403 has recently been determined and shown to contain 2365×10^6 base pairs [1], incorporating 35.4 % guanosine (G) and cytosine (C) nucleotides. In the chromosome, 2310 open reading frames (ORF) were detected, of which 64 % (1482) could be assigned a biochemical or biological function. There are 465 ORFs (20 %) of which the products show homology to hypothetical proteins of unknown function, while 363 ORF products (~16 %) are, until now, specific for Lactococcus in that they do not show similarity to other proteins. The nucleotide sequence of the chromosome of L. lactis subsp. cremoris is, at the time of writing, almost complete. The nucleotide sequences of the genomes of the two subspecies diverge by less than ~15 % on average.

Generally, strains of L. lactis contain from one to more than 10 different plasmid species, ranging in size from 2 to over 130 kb [2]. Many industrially important traits are encoded by these plasmids; for example, properties such as the capacity to ferment lactose or citrate, to degrade casein, to transport oligopeptides, to produce bacteriocin or exopolysaccharides, or to resist bacteriophage have all been connected to particular plasmids in certain strains. In fact, the relatively frequent loss of some of these traits was a first indication of the location of the corresponding genes on plasmids [2]. Interestingly, many of the functions that are plasmid-encoded in some strains have been found in other strains to be encoded by the chromosome.

Additional genomic features of L. lactis that are important with respect to genetic flexibility and which will be discussed in more detail below are the presence of transposons in certain strains, the wide distribution of a variety of different insertion sequence (IS) elements on the chromosome and on plasmids and the occurrence of prophages and remnants thereof.

12.3
Flexibility in the Genome of Lactococcus lactis

Bacterial genomes are currently considered as relatively flexible structures that are under selective pressure and/or physical constraints to retain their overall organization [3]. Possibly, evolution acts to optimize the genomic organization of bacterial species such that they are best adapted to the niches in which they occur. Although L. lactis is found in different places in the environment, we only consider the dairy strains in this review. Dairy lactococci have small genomes and are optimally geared towards growth in milk. For instance, they possess a large number of genes involved in milk protein and milk sugar degradation, while most dairy strains are multiple amino acid auxotrophs [4–6].

A number of different genetic elements such as temperate bacteriophages, conjugative transposons and ISs are involved in genome rearrangements in a large variety of bacterial species. Transposons are implicated in horizontal gene transfer, while repetitive IS elements and, for that matter, ribosomal (*rrn*) operons provide targets for homologous recombination [7, 8]. IS elements, transposons and prophages have also been shown to add to – or to be the causative agents of – genomic rearrangements in *L. lactis*, and will be discussed below. For the latest details on genome plasticity in *L. lactis*, the reader is referred to the excellent recent review by Campo et al. [9].

A number of natural gene transfer systems have been shown to operate in *L. lactis* that allow for rapid large modifications of the genome, and thus contribute to the genetic (evolutionary) flexibility of the bacterium. As an extensive review of these topics is beyond the scope of this chapter, the reader is referred to some extensive reviews on the subject [10, 11], in which references to the original literature can be found. Here, the transfer systems will be dealt with only from the perspective of their contribution to genetic fluidity in *L. lactis*.

12.3.1
Conjugation

Conjugation of chromosomal markers as well as the transfer of entire plasmids, specifying such traits as lactose fermentation, bacteriocin production and resistance, polysaccharide production, proteinase production and resistance to bacteriophage, has been observed in many strains of *L. lactis*. The mechanism of conjugal transfer in *L. lactis* has been most thoroughly examined for the lactose/proteinase plasmid in the closely related *L. lactis* strains ML3, NCDO712 and C2. Although the initial conjugation frequencies of the plasmid were extremely low (2×10^{-7} transconjugants per recipient), rare colonies appeared with unusual colony morphology. Cells of these strains aggregated in broth cultures, forming large visible clumps of cells. These cells were shown to contain co-integrates of the lactose/proteinase plasmid with a so-called "sex factor", and were able to transfer the traits at very high frequencies of close to 1. The co-integrates are formed by insertion sequence (IS*S1*)-mediated transposition: IS*S1* on the plasmid causes its integration at different positions in the "sex factor" with a duplication of the IS element. A clear difference between the situation in strains ML3 and NCDO712 concerns the fact that the "sex factor" in the former is, in fact, a low-copy number plasmid (pRS01). In NCDO712 (and in ML3) an equivalent (but not exact) copy of pRS01 is located on the chromosome, as was shown by pulse field gel electrophoresis. The co-integrates in the NCDO712 system differ slightly in size and were shown to contain different portions of chromosomal DNA as a result of one-ended transposition. Gasson et al. [12] have described the unidirectional transfer of chromosomal markers between strains, a phenomenon reminiscent of *Escherichia coli* Hfr conjugation.

Aggregation, allowing close physical contact and exchange DNA between cells, is caused by the concomitant expression of the *clu* gene, which is present on the "sex factor" and is activated by the co-integrate formation and a second chromosomal gene, *agg*.

Another form of genetic mobility is provided by a Group II intron located on both pRS01 and the "sex factor" [13]. Group II introns are large self-splicing RNA structures, which often function as mobile genetic elements. They move at low frequency from one site in the genome to a new genetic location. If, however, an intron-less copy of the genetic site occurs in a cell (for instance by conjugation), the intron very rapidly transposes from its original place to the empty site by a process called "intron homing" [14].

12.3.2
Transduction

Transfer of chromosomal as well as plasmid DNA markers by transduction, the erroneous transfer of bacterial DNA from one bacterial cell to another by bacteriophage, has also been described in *L. lactis*. Again, the phenomenon has been most extensively studied in the *L. lactis* strains NCDO712 and C2. Although the mechanism was first described for virulent bacteriophages in the early 1960s [15, 16], in later reports the phenomenon has almost exclusively been analyzed using temperate bacteriophage. Initial low-frequency transduction of the lactose plasmid led to a high frequency of transduction phenomenon, in which an approximately 100-fold increased frequency was caused by deletion of the plasmid to a size that fitted in the bacteriophage head. In the case of the lactose/proteinase plasmid pLP712 of strain NCDO712, inherent plasmid instability resulted in the selection of the smaller variants by the transduction process. The instability of pLP712 could be attributed to the activity of IS elements on the replicon. Plasmids considerably smaller than the size limitation set by the phage head are also efficiently transduced [17]. Plasmid integration into the resident chromosome has been observed after lactose gene transduction and may be caused by IS-mediated transpositional recombination [17–19]. The transduction of chromosomal markers is *rec* dependent: after the erroneous packaging of a fragment of chromosomal DNA in the phage head and transfer it integrates into the proper location in the recipient chromosome by homologous recombination [20]. Although HFT of chromosomal markers does not normally occur, elevated transduction of chromosomal markers, but not of plasmid genes, by the temperate bacteriophage BK5-T was observed when the phage was grown on its lytic host *L. lactis* subsp. *cremoris* H2 [21]. The presence of a number of *pac* sites on the H2 chromosome that showed homology to the BK5-T *pac* site suggested that these facilitated the insertion of chromosomal DNA fragments in the phage head to form transducing particles.

12.3.3
Transformation

Transformation by a natural process has not been observed in *L. lactis* even though the genes for the late competence proteins have been identified in the chromosome [1]. Transformation under specific laboratory conditions is routinely employed. Protoplast transformation has been used in the past, but the general technique is electrotransformation of whole cells [11]. Application of a high-voltage electric field pulse to the cells is thought to transiently permeabilize the cellular membrane, allowing the passage of DNA. Specific growth conditions and treatment of cells, both of which are meant to weaken the thick lactococcal cell walls, as well as the electroporation conditions have been examined to increase the number of genetically transformed cells and transformation frequencies of 10^6 to 10^7 per microgram of plasmid DNA can be routinely obtained with plasmid free laboratory strains. The development of electrotransformation has been instrumental in the rapid advancement of recombinant DNA technology in *L. lactis*.

12.3.4
IS Elements and Transposons

All strains of *L. lactis* contain one or more copies of a number of different insertion sequences (for a review, see Ref. [22]). The number of different IS elements and the number of copies of each IS element in a strain varies widely, but comprises several ten-folds for many strains investigated thus far. The IS elements are located on the chromosome as well as on various plasmids [23]. They are typically between 800 bp and 2500 bp in size, are flanked by inverted repeats, and generally encode their own transpositions functions. Replicative transposition of an active IS element in a plasmid leads to chromosomal integration of the entire vector between two copies of the IS element. IS elements are the causative agents of many genetic rearrangements by providing sites for co-integrate and deletion formation. They are involved in shuffling of genes or blocks of genes, changing their relative order, they can insert in genes and impair gene function, or they can be involved in activation of (silent) genes by providing promoter activity [24–26]. The fact that they are often associated with important dairy functions suggests that they have played an important role in the evolution of *L. lactis* and its adaptation to its current environment. Using probes of the different IS elements it has been feasible to identify strains of *L. lactis* [10, 23].

In the genome sequence of strain IL1403, six different IS elements have been found varying in numbers from 1 (IS*982*) to 15 (IS*983*), together comprising 42 kb of DNA. The nonrandom distribution of two of the IS elements (IS*1077*, always associated by IS*904*, and IS*983*) is suggestive of lateral transfer of a large segment of the genome of one *Lactococcus* donor carrying one of the IS elements to a recipient carrying the other [1].

Certain strains of both subspecies of *L. lactis* carry a chromosomally located transposon encoding sucrose fermenting ability and the genes involved in the biosynthesis, secretion and quorum sensing of the lantibiotic nisin [27, 28]. Insertion of the transposon in the chromosome occurs in an orientation-specific manner at one preferred site, although several secondary sites and multi-copy integration have been observed [29].

12.3.5
Lactococcal Phage as Sources of Genetic Plasticity

Above, we have seen examples of transfer of genetic information through (generalized) transduction by virulent and temperate bacteriophages. The majority of strains of L. lactis are lysogenic and carry one or more prophages [30, 31]. These prophages are relevant for the purpose of this review as they integrate, during the lysogenic cycle, in the bacterial chromosome via homologous recombination and, thus, change the genetic content of the newly lysogenized cell. On the one hand, the prophages increase the metabolic burden on the cell and may even lyse the cell after prophage induction, while on the other hand they may encode functions that increase the fitness of the lysogen. In other words, the host–prophage relationship is an example of a dynamic – albeit very delicate – genetic equilibrium [32]. Another level of plasticity is provided by the Group I intron scrI in one of the sequenced phages, r1t [33]. A gene on scrI specifies an endonuclease that allows the intron to move to an intronless allele by intron homing.

12.3.5.1 An example of natural genetic flexibility: the *Lactococcus lactis* NCDO712 family

Bacteria can be classified into two large groups on the basis of genomic stability. One group contains species with highly flexible genomes, while the genomic organization of the species in the second group, among which L. lactis, has been strongly conserved. Nevertheless, when comparing the physical maps of a number of L. lactis strains, clear differences were observed at the level of the order of the genes on the chromosomes. A large inversion involving almost half of the chromosome was observed when L. lactis subsp. lactis IL1403 was compared with L. lactis subsp. cremoris MG1363 [34]. The latter strain is a plasmid-free derivative of L. lactis subsp. cremoris NCDO712, an industrial strain originally isolated at the National Collection of Dairy Organisms in the United Kingdom in 1954. During the four decades that it has been used by different laboratories, the strain has undergone various spontaneous and/or (chemically) induced mutations. The relationship between all the derivatives and their genetic constitution has been extensively studied [35], and exemplifies the speed at and the extent to which large or smaller rearrangements can occur in the genome of L. lactis. In total, five major chromosomal rearrangements have been observed in different combinations in nine strains belonging to this genetic lineage: a large inversion involving about 50% of the chromosome, excision of an uncharacterized prophage, deletion of the prophage ΦT712, excision of part (30 kb) or the entire sex factor (60 kb), and deletion of some 25 kb of DNA encompassing the *opp* operon for oligopeptide uptake. The natural large inversion of half of the chromosome that is seen in some strains is caused by homologous recombination between two inactive copies of the IS905 element [36]. A labile plasmid structure that is sometimes observed in the plasmid-free NCDO derivative MG1363 is, most probably, the excised "sex factor" and, again, an example of genetic instability.

During construction of the plasmid-free NCDO 712 derivative MG1363, an intermediate strain was obtained that carried only the lactose/proteinase plasmid pLP712 [37]. This plasmid was highly unstable and readily gave rise to various deletion variants, exemplifying yet another level of genetic instability in *L. lactis* that can have far-reaching effects considering the fact that many strains carry many different plasmids species.

12.4
Mutations in *Lactococcus lactis* as a Consequence of Environmental Factors and DNA Metabolism

The genetic material of any organism is repeatedly damaged due to the action of chemicals, thermal stress, and environmental radiation. The concurrent change in the chemical structure of the affected nucleotide can affect its base-pairing ability and can lead to the incorporation of an incorrect nucleotide in DNA. Nucleotides can occur in two tautomeric forms, one of which can base-pair with a nucleotide other than the proper one and can, thus, cause incorrect base incorporation in DNA. Changes in DNA can also take place as a consequence of the normal process of DNA replication. However, due to the specificity of DNA polymerase the error rate during nucleotide incorporation is extremely low while an intrinsic check-up mechanism in DNA polymerase (its $3' \rightarrow 5'$ exonuclease activity) removes incorrect nucleotides should they occur during the polymerization step. Replication mistakes do not always lead to single or point mutations: small deletions or insertions can also happen in DNA, especially in regions containing small repeated sequences. These sequences are responsible for replication slippage, in which the template strand and the newly synthesized DNA strand move relative to each other, and part of the sequence is either copied twice or not at all. Highly specialized proteins inside the cell constantly scrutinize the genetic material, and accurately and very rapidly repair changes occurring in DNA. It is due to all of these processes acting in concert that mutations – permanent changes in DNA – are very rare events. In bacteria and eukaryotic organisms alike, the estimated mutation rate is approximately 1 nucleotide change per 10^9 nucleotides each time a DNA molecule is replicated. Put differently, a single bacterial gene of about 1000 base pairs will suffer from a mutation (not necessarily a deleterious one) once in about 10^6 generations.

12.5
Methods to Mutate the Genome of *Lactococcus lactis*

Methods to mutate the genome of *L. lactis* include treating cultures with ultraviolet (UV) light, base analogues, DNA damaging agents such as the alkylating agent ethylmethane sulfonate (EMS), or the intercalating agent ethidium bromide. Through intelligent screening or selection procedures, the desired mutant has to be identified among the random mutants and the majority of nonmutated cells. A drawback of these random procedures is that, apart from the sought-after mutation, many more may be present elsewhere in the genome of the mutant. Depending on the chemical used, point mutations of a certain type or deletions are introduced in DNA.

Plasmids have been more or less specifically removed from lactococcal strains by growing the cells under suboptimal conditions for growth (starvation, acid conditions, high or low temperatures), by protoplast formation and subsequent regeneration, or by treating the cells with chemicals that interfere with plasmid replication (e. g., ethidium bromide). Under certain conditions, one or more plasmid species can be removed from the cell and, by applying the procedures repetitively, plasmid cured derivatives of several strains of *L. lactis* have been obtained. Plasmid integration and the formation of smaller plasmids by deletion events have also been reported.

All of these mutation strategies are more or less random, and the outcome – that is, the exact nature and location of the mutation – cannot be steered. Over the past two decades, technology has been developed with which it is possible to direct the genetic modification of *L. lactis* in a very precise way. A number of these techniques will now be discussed.

12.5.1
Genetic Engineering of *Lactococcus lactis*

Many tools for the genetic modification of *L. lactis* are now available [38-40]. Plasmid vectors have been constructed for the cloning of genes. Vectors have also been made to select and analyze promoter and terminator structures as well as sequences necessary for the secretion of proteins. Constitutive and regulatable promoters have been isolated and used to develop gene expression vectors with which it is possible to express genes of any origin in the organism [41]. Most of the vector systems developed to date are aimed at studying fundamental questions in the molecular biology of *L. lactis*. These systems make use of traditional antibiotic selection markers and could, thus, never be used in food fermentation processes. For these systems to be applicable in the food industry, they should be devoid of antibiotic resistance genes. So-called "food-grade" plasmids have been designed by using selectable genes derived form *L. lactis*. The nisin resistance determinant *nsr* could be used to select plasmid transformants on specific plates [42]. A plasmid carrying a *lacF* gene served as a dominant selection marker for growth of *L. lactis* YP2-5 on lactose. Strain YP2-5 carries a loss of function mutation in *lacF* in the chromosomal *lac* operon, which is complemented by plasmid-located *lacF* [43]. Another plasmid-strain combination for food-grade cloning and selection was devised on the basis of ochre and amber suppressor genes [44, 45] in combination with the proper stop codon (nonsense) mutations in an essential gene for purine metabolism. As milk is purine free, the mutants cannot grow unless the plasmid carrying the suppressor gene is present. The suppressor genes specify an altered tRNA that recognizes the nonsense mutation as a sense codon and will restore functionality of the mutated gene product. A number of other genes, e. g., *thyA* (for thymidilate synthase) and *alr* (for alanine racemase), have been proposed to serve in the proper genetic backgrounds as dominant markers in food-grade vector systems [46, 47]. An example of a system that makes use of a nonlactococcal selectable gene in *L. lactis* is provided by pLR333 and pLR334. These plasmids are based on the lactococcal theta replicon pWV02 and contain the *scrA/B* genes of *Pediococcus pentosaceus* for positive selection of plasmid-containing cells on sucrose plates [48]. Although these plasmids are not homologous – that is, they were not made by using *L. lactis* DNA exclusively ("self-cloning") – they do consist of DNA from food microorganisms only.

The chromosome of *L. lactis* has been made accessible by the construction of various insertion vectors [49–51]. The most versatile of a number of different insertion vectors, one based on the replicon of pWV01 [40], will be described in some more detail here in order to allow a deeper treatise of the subject of creating calculated chromosomal mutations.

A chromosomal insertion vector was made from the replicon of the cryptic lactococcal plasmid pWV01 (Fig. 12.1). This was done by physically separating the pWV01 origin of replication from the gene encoding the plasmid replication protein, RepA. The latter was placed in the chromosome of *E. coli*, *Bacillus subtilis*, and *L. lactis*. In this way, a set of RepA$^+$ helper strains was created expressing RepA and allowing replication of the pWV01 origin fragment (pORI). The origin fragment, carrying an antibiotic resistance marker and a multiple cloning site, can be used as a chromosomal integration vector when it is endowed with a piece of chromosomal DNA. Single crossover recombination results in the integration of the entire plasmid at a precisely defined genetic locus. In this way genes can either be specifically inactivated or inserted at specific locations in the chromosome. The pORI derivatives pORI13 and pORI19 have been made to allow random mutagenesis of the chromosome [52, 53]. The vector pORI13 carries a promoter-less *lacZ* gene that has been used to probe the chromosome for regions that are under environmental regulation [53]. A vector system for the easy selection of double crossover (replacement) recombination events was also based on pORI: the presence of a constitutively expressed *lacZ* gene in pORI240 and pORI280, which differ only in the antibiotic resistance marker used for selection, allows for easy visualization of the integration and excision events in a plate assay. A further refinement of these integration strategies was their combination with the pG$^+$HOST series of plasmids, which provide the temperature sensitive RepA^{ts} *in trans*. This allowed separating the two more or less rare events of transformation and subsequent integration: the pORI derivative is introduced in a strain carrying the temperature-sensitive

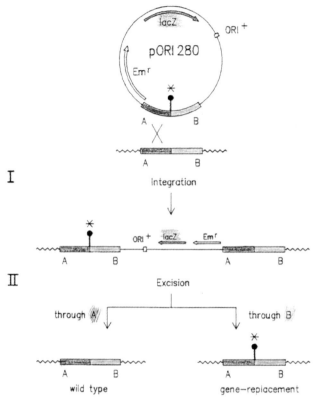

Figure 12.1. Scheme of the two-step procedure to obtain gene replacement (double crossover) recombination. Plasmid pORI280 carries a selectable erythromycin resistance marker (Em^r), a constitutively expressed gene for the *E. coli* enzyme β galactosidase (*lacZ*), and an origin for plasmid replication (ORI$^+$). pORI280 cannot replicate independently but must be propagated in a host strain that expresses the plasmid replication protein RepA. *L. lactis, B. subtilis* and *E. coli* host strains are available for this purpose. The red and taupe area represents the gene(s) of interest. This gene(s) has been taken from the lactococcal chromosome and inserted into pORI280 by general cloning techniques. The lollipop-and-star represents the intended mutation to be introduced in the lactococcal genome. This could be a single nucleotide change or a small or large deletion. It could also be a whole (heterologous) gene(s), to be inserted in the lactococcal chromosome at the junction between the red and taupe DNA sequences. In step I, the plasmid is introduced in *L. lactis* lacking RepA. Using selection on plates containing Em and X-gal (a LacZ substrate that gives a blue color upon hydrolysis by the enzyme), blue colonies are picked up that carry the entire plasmid in their chromosomes. These are the result of recombination (indicated by the cross) either through the red regions of homology between the plasmid and the chromosome, as depicted here, or via the taupe regions (not shown). Subsequently these integrants are grown for a number of generations without Em and plated on media containing X-gal but lacking the antibiotic (step II). Among the majority of blue colonies (the original mixture of the two integrant strains) colorless colonies appear in which resolution of the integrated structure has taken place via the two homologous (red or taupe) DNA regions in the integrant chromosome. In the example given here, excision through the taupe regions results in the intended gene replacement. As is clear from this scheme, no vector DNA is left behind and the mutated strain could, thus, be used in several rounds of gene replacement mutagenesis to create multiple mutations.

plasmid. A rise to the nonpermissive temperature and use of the correct antibiotics removes the pG$^+$HOST plasmid, and thus RepA^{ts}, allowing subsequent selection of the integration event.

Another way to mutate chromosomal genes in a random manner is to use a number of different vectors based on heterologous transposons, or plasmid pVE6007 or one of its derivatives [54]. The latter plasmids carry an active copy of the lactococcal insertion sequence IS*S1* and *repA*^{ts}. After introduction of the plasmid in *L. lactis* and growth for a certain period of time at the permissive temperature, the culture is switched to the nonpermissive temperature. Integrants carrying a copy of the entire plasmid in their chromosome as a result of IS*S1* transposition activity are selected on antibiotic plates and tested for mutant phenotypes. The plasmid on the chromosome of a mutant is present between two copies of IS*S1* – a situation that is stable as long as the strain is kept at the nonpermissive temperature for plasmid replication. When a culture of the integrant is shifted to a temperature that allows replication of the RepA^{ts} plasmid, recombination can occur between the duplicated IS*S1* copies. This process is, in fact, stimulated by rolling-circle replication of the plasmid and leads to excision of the plasmid while leaving a clean copy of IS*S1* behind at the chromosomal location. In other words, the genetic locus is still mutated but now by the IS*S1* element alone [54].

12.6
Strategies to Detect Genetically Modified *Lactococcus lactis*

The detection of GMOs in food products has received major attention following the introduction of new plant varieties. In 1999, more than 40% of all corn and more than 45% of all soybean grown in the US were genetically modified variants [55]. Moreover, more than 60% of all food products in US supermarkets contain GMOs. In recent years, public concern about possible threats related to these products has been growing. Many public debates concerning the environment and public health safety, for example gene flow to other organisms, destruction of agricultural diversity, increased allergenicity and gastrointestinal problems, are ongoing. Although hard evidence for such effects is often still lacking or difficult to obtain, these concerns have spurred the search for procedures to detect even extremely small traces of the GMO. Another reason to have such protocols at one's disposal is for companies to be able to safeguard protected technologies or knowledge.

Several tools are available for GMO detection, in particular related to crop GMOs, although many of them should also be applicable for the detection of modified lactococci.

First, from the above treatise it is clear that a large diversity can be brought about in the genetic material of *Lactococcus* by either natural processes or by a range of genetic engineering techniques. It will be difficult or impossible in many cases to distinguish a lactococcal strain carrying a natural (spontaneous or UV- or chemically induced) mutation or IS insertion from a strain in which the same mutation has been introduced by recombinant DNA technology. The insertion of new genetic material (one or more genes) is, of course, another matter although it could still be difficult if the new DNA is from a very related organism (i.e., a strain from the other subspecies). The most obvious genetic modification that is relatively simple to detect is the insertion of a foreign DNA fragment that leads to the production of a heterologous protein. Such DNA sequences contain unique signatures that allow identification at the DNA level. Furthermore, the heterologous gene product can be the target for detection. A selection of available DNA and protein detection techniques will be discussed below.

12.6.1
Sample Preparation

The possibility of detecting GMOs in food or feed products largely depends on the absolute amount in which the GMO is present. It is not unlikely that in a specific matrix containing lactic acid bacteria only a very small fraction consists of a GMO. Recent EU legislation dictates that foods with a GMO content >1% should be labeled as such [56]. Statistical approaches have shown that the detection of such levels by most procedures currently employed (in particular, PCR-based procedures) is seriously impaired by sampling protocols and errors [57, 58]. For most (if not all) procedures described below the sampling and preparation of the material to be tested is most critical, both with respect to the sensitivity and the reproducibility that can be obtained [58]. The genetically modified lactococci to be detected are often part of products consisting of a complex matrix of proteins and other biomolecules, and contain many other lactococci and/or other microorganisms. During extraction of the nucleic acids extreme care must be taken in order to prevent their degradation. When protein-based procedures are employed, degradation of the target GMO protein by proteases and peptidases must be prevented at all costs.

Many components in food products can interfere with the efficiency of PCR reactions. Examples include proteins, fats, polyphenols, and polysaccharides [61], while agents often applied during nucleic acid preparation (e.g., ethanol, isopropanol) can also reduce PCR sensitivity [60]. Some contaminants can lead to PCR product artefacts that could be mistakenly classified as positives, whereas other contaminants or DNA degradation can give results that are difficult to interpret. To ensure efficient isolation of target nucleic acids, the extraction procedure should be validated extensively for each target and source material. In practice, this means that several extraction procedures need to be compared both with and without added (spiked) target molecules [61].

12.6.2
DNA-based Procedures

The most common techniques to detect modified or foreign DNA are various formats of the polymerase chain reaction (PCR), Southern hybridization and, more recently, DNA-micro-array technology [62–64]. All of these procedures rely on the sequence-specific hybridization of two strands of DNA, and try to enhance and visualize the difference(s) in the nucleotide sequence of the GMO relative to the original strain.

12.6.2.1 Southern hybridization

After transfer of isolated DNA to membranes, hybridization with labeled nucleic acid probes is used to detect homologous sequences. Owing to the availability of novel labels (e. g., fluorescein, digoxigenin, biotin) these procedures have become much faster than with previously applied radioactively labeled (^{32}P) probes. In general, the sensitivity of Southern blot-based procedures is significantly less than that of most PCR-based procedures.

12.6.2.2 PCR

Many PCR-based procedures have been applied in recent years for the detection of low abundance DNA; for example, the demonstration of small numbers of viruses in different biological materials, the quantification of residual DNA in therapeutic products, the diagnosis of infectious bacteria as well as for the identification of GMO products such as Roundup Ready™ soy in food products. Several procedures have recently been discussed in great detail with respect to GMO detection [65], and therefore will be described only briefly here.

Qualitative PCR PCR allows the million-fold amplification of specific DNA fragments. In principle, if a unique fragment can be selected it should be possible to detect a single GMO in a pool of 10^5 to 10^7 nonGMOs. After PCR, the amplified DNA molecules can be visualized using agarose gel electrophoresis, high-performance liquid chromatography (HPLC) or capillary electrophoresis, and their amount determined. Often, the DNA fragment is treated with restriction endonucleases, hybridized with a specific probe, or sequenced to confirm its identity. In addition, nested PCR procedures, in which two additional primers that hybridize specifically to the amplified DNA fragment are used in a second round of PCR, have been shown to yield reliable results [66].

The sensitivity of these PCR methods is strongly affected by the procedure used to prepare the starting material (see above). Poor template preparation – that is, the isolation of GMO DNA to be amplified together with the DNA from the majority of nonGMO strains – often results in much lower detection levels than those expected on theoretical calculations.

Quantitative PCR (end-point and real-time) In general, quantitative PCR procedures employ the inclusion of a known concentration of an internal standard that is co-amplified with the target DNA fragment of the GMO. After quantitative end-point PCR, both DNA fragments are separated and quantified. The starting concentration of the GMO can then be calculated by means of regression analysis. An additional advantage of this procedure is that the negative effects of contaminating agents on PCR efficiency can be seen directly by its effect on the amplification of the internal standard DNA. Quantitative real-time PCR has a major advantage since the concentration of amplified DNA in these procedures is proportional to PCR cycle number during the exponential phase of the reaction [67]. Real-time PCR is dependent on the availability of techniques to monitor and determine PCR product formation during the PCR and, moreover, to differentiate between specific and nonspecific PCR products. Several sensitive detection reagents and tools have been developed recently, including the exploitation of the double-strand DNA binding dye SYBR Green I, hybridization of probes or fluorescence resonance energy transfer probes (FRET), hydrolysis probes (also known as TaqMan® technology; [68, 69]), and molecular beacons [70].

Exhaustive limiting dilution PCR This procedure does not use the co-amplification of externally added reference DNA. On the contrary, the procedure must be optimized to the extent that the amplification of an endogenous gene occurs in an all-or-nothing fashion. Furthermore, it must be assumed that one or more of the targets to be detected will give positive signals. The quantification is performed after comparison of many replicates of serial dilutions of the target material. At the largest dilutions where only some samples yield positive results, Poisson statistics are exploited to determine the absolute number of targets in the sample [71]. As in the case of most other PCR procedures, exhaustive limiting dilution PCR is sensitive to contaminating agents.

12.6.3
Nucleotide Sequence-based Procedures

Determination of the nucleotide sequence can be the ultimate tool to detect the presence of small alterations at the DNA level. For example, the occurrence of single base substitutions or small deletions can be demonstrated unambiguously. Significant progress has been made in this field spurred, as it has been, by the importance of single-nucleotide polymorphism (SNP) analysis in genetic studies of diseases. Tools for large-scale, high-throughput sequencing and the accompanying analyses of the huge quantity of obtained data are available. Thus far, these tools have not been exploited for the characterization of genetic diversity in lactic acid bacteria, but there does not seem to be a major technological restriction. Since SNP analysis technology requires the amplification by PCR of DNA fragments to be sequenced, quantification is often hampered.

12.6.3.1 **Micro-arrays**
DNA micro-arrays have proven extremely powerful and sensitive tools for the analysis of gene expression. Many gene-specific oligonucleotides or amplicons (the probes) can be coupled to different surfaces (often a glass support). Specific hybridization to the gene probes can be monitored after hybridization with, most often, fluorescently labeled target nucleic acids. Hybridization conditions can be chosen such that even single nucleotide mismatches can be detected. In most cases the

nucleic acids are multiplied by PCR procedures, prior to hybridization, in order to yield sufficient material to allow detection. Consequently, quantification of the abundance of the gene(s) of interest is often difficult, as with other PCR-based procedures. This technology has already been used successfully for the detection of pathogenic bacteria by specific identification of their virulence factors [72]. Conceivably, this technology will also prove extremely suitable for the detection of genetic diversity brought about by genetic engineering.

12.6.4
Protein-based Procedures

Most protein-based procedures to detect GMOs available to date exploit antibodies. Antibodies can be highly specific; for example, monoclonal antibodies recognize small and unique epitopes and, therefore, are extremely suited for sensitive protein detection methodologies. Since monoclonal antibodies generally interact with short peptide sequences (approximately nine amino acids) they can also be used to detect the presence of (partially) degraded GMO protein. On the other hand, polyclonal antibodies are often much more sensitive, which makes them more suited if small amounts of GMO protein are expected [73].

12.6.4.1 Western hybridization
For Western hybridization procedures, proteins from the sample of interest are first separated by sodium dodecyl sulfate–polyacrylamide gel electrophoresis (SDS–PAGE) and subsequently transferred to nitrocellulose or polyvinyl membranes. Interaction of antibodies specific for the protein of interest is accomplished, and bound antibody is detected by exploiting an enzyme covalently attached to the antibody (e.g., horseradish peroxidase or alkaline phosphatase), by applying a secondary antibody directed towards the first (GMO protein-specific) antibody, or by using *Staphylococcus aureus* protein A coupled to an easily assayable enzyme. These procedures are extremely specific and qualitative. Quantification is possible to a limited extent only if the detection limit of the GMO protein-specific antibody has been determined beforehand. A major disadvantage of Western hybridization procedures is the time needed to obtain results.

12.6.4.2 ELISA
Enzyme-linked immunosorbent assay (ELISA) procedures are much faster than Western hybridization, are very sensitive, and are much more suited for quantitative purposes. Furthermore, when performed in microtiter plates, ELISAs can easily be applied in high-throughput settings. A variation on the ELISA theme makes use of nitrocellulose strips in which a colored GMO protein-specific antibody is incorporated in a region of the strip that is dipped into the sample solution. Upon reaction of the antibody with the GMO protein the colored complex diffuses through the strip matrix until it is bound in a region of the strip that carries a sec-

ond GMO protein-specific antibody, forming a colored zone. A second zone in the strip containing an antibody against the color reagent provides a control for strip quality. It is to be expected that more and improved of these lateral flow strip-based procedures will become available in due time. In particular, improvements of the sensitivity are to be expected, for example by using fluorescence signal detection. One of the major advantages of strip-based procedures is their suitability for application in the field.

12.6.4.3 SPR

Surface plasmon resonance (SPR) occurs when light is reflected under certain conditions from a conducting film at the interface between two media of different refractive index. In most commercially available systems the media are the sample and the glass of the sensor chip, while the conducting film is a thin layer of gold on the chip surface. SPR causes a reduction in the intensity of reflected light at a specific angle of reflection. This angle varies with the refractive index close to the surface on the side opposite from the reflected light. When molecules in the sample bind to the sensor surface, the concentration and therefore the refractive index at the surface changes and an SPR response is detected (see www.biacore.com for additional technical background information).

SPR could provide a means for sensitive detection and quantification of GMO proteins in samples of diverse nature. In brief, receptor biomolecules (e. g., specific antibodies) are immobilized on the surface of the sensor chip, and the ligand (GMO molecule) is introduced in the flow system that is applied to the surface. Specific binding of the ligand can be monitored due to the change in refractive index caused the mass change resulting from the complex formation between receptor and ligand.

12.6.4.4 Two-dimensional gel electrophoresis and mass spectrometry

Two-dimensional (2D) gel electrophoresis techniques are used extensively to analyze and compare the protein content of biological samples. After separation of the proteins on the basis of their charge (first dimension), the proteins are separated on the basis of their molecular weights (second dimension) by means of SDS–PAGE. After staining, a pattern of unique spots is obtained in which each spot represents a specific protein. In principle, this procedure can be exploited to screen for the occurrence of novel protein (variants) resulting from genetic modification. A major drawback of the procedure is the large variation in protein content that results from the differential expression of endogenous genes caused by environmental stimuli. Moreover, 2D gel electrophoresis is still laborious and time-consuming. It is of utmost importance to standardize the procedures for sampling and sample preparation. Recently, capillary electrophoresis techniques have been developed that can assist in automation of protein separation [74]. In combination with the protein separation by (2D) SDS–PAGE or capillary electrophoresis, mass spectrometry techniques have been developed that enable automated determination of the molecular weight of large numbers of proteins in a particular sample. After treatment of the proteins by specific proteases prior to mass spectrometric molecular weight determination, the amino acid content and/or amino acid sequence can be assigned unambiguously. In view of the rather high costs of these procedures it is not likely that these technologies will be exploited for GMO detection on a large scale, unless further improvements with respect to selectivity and automation are achieved.

12.7
Conclusions

By using the genetic engineering technology described above, it is now possible to insert any DNA fragment into the chromosome of *L. lactis*. Conversely, entire genes or operons can be deleted from the genome using the same strategies. By using replacement recombination, genetic changes as small as single nucleotide mutations can be introduced without leaving any trace of vector DNA used to make the genetic changes. For many of these changes, apart from the insertion of heterologous genes, it is impossible to distinguish whether they are the result of a natural mutation or of a genetic engineering event. Of course, if this were to be requested by industry, legislative bodies or the consumer, any mutation made by engineering technology could be labeled as such by introducing, together with the mutation, a genetic "bar code" – much in the way as was done to label yeast mutants [75]. Such a genetic label would identify the mutation, could be used to identify the producer of the mutant, and would provide additional possibilities to design specific PCR identification strategies.

Although at present many tools are available that can be exploited to detect changes in the genome of lactic acid bacteria, it is clear that the selection of the most suitable tool will be case-dependent. In view of the fact that most procedures are technically complex, much care should be taken in interpreting the outcome. Extensive validation studies will be needed which, in our opinion, should be carried out by independent and highly specialized research institutions or laboratories.

References

1 A. Bolotin, P. Wincker, S. Mauger, O. Jaillon, K. Malarme, J. Weissenbach, S. D. Ehrlich, A. Sorokin. *Genome Research* **2001**, *1*, 1–23.
2 L. L. McKay. *Antonie van Leeuwenhoek* **1983**, *49*, 259–274.
3 S. Casjens. *Annu. Rev. Genet.* **1998**, *32*, 339–377.
4 W. M. de Vos. *Lactic acid bacteria, genetics, metabolism and application.* Kluwer Academic Publishers, the Netherlands, **1996**.
5 J. Kok, W. M. de Vos. *Genetics and Biotechnology of lactic acid bacteria.* Blackie Academic and Professional, UK, **1994**.
6 E. R. S. Kunji, I. Mierau, A. Hagting, B. Poolman, W. N. Konings. *Lactic acid bacteria, genetics, metabolism and application*, Kluwer Academic Publishers, the Netherlands, **1996**.
7 M. Syvanen. *Bacterial Genomes: Physical Structure and Analysis.* Chapman & Hall, New York, **1997**.
8 S. L. Liu, K. E. Sanderson. *Proc. Natl. Acad. Sci. USA* **1996**, *93*, 10303–10308.
9 N. Campo, M. Dias, M.-L. Daveran-Mingot, P. Ritzenthaler, P. Le Bourgeois. *Lactic Acid Bacteria, Genetics, Metabolism and Application.* Kluwer Academic Publishers, the Netherlands, **2002**.
10 M. J. Gasson. *FEMS Microbiol. Rev.* **1990**, *87*, 43–60.
11 M. J. Gasson, G. F. Fitzgerald. *Genetics and Biotechnology of Lactic Acid Bacteria.* Blackie Academic and Professional, UK, **1994**.
12 M. J. Gasson, J.-J. Godon, C. Pillidge, T. J. Eaton, K. Jury, C. A. Shearman. *Int. Dairy J.* **1995**, *5*, 757–762.
13 G. M. Dunny, L. L. McKay. *Lactic Acid Bacteria, Genetics, Metabolism and Application.* Kluwer Academic Publishers, the Netherlands, **1999**.
14 M. Belfort, P. S. Perlman. *J. Biol. Chem.* **1995**, *270*, 30237–30240.
15 W. E. Sandine, P. R. Elliker, L. K. Allen, W. C. Brown. *J. Dairy Sci.* **1962**, *45*, 1266–1271.
16 L. K. Allen, W. E. Sandine, P. R. Elliker. *J. Dairy Sci.* **1963**, *30*, 351–357.
17 G. F. Fitzgerald, M. J. Gasson. *Biochimie* **1980**, *70*, 489–502.
18 L. L. McKay, K. A. Baldwin. *Appl. Environ. Microbiol.* **1978**, *36*, 360–367.
19 M. J. Gasson, S. Swindell, S. Maeda, H. M. Dodd. *Mol. Microbiol.* **1992**, *6*, 3213–3223.
20 D. G. Anderson, L. L. Mckay. *J. Bacteriol.* **1983**, *155*, 930–932.
21 B. E. Davidson, I. B. Powell, A. J. Hillier. *FEMS Microbiol. Rev.* **1990**, *87*, 79–90.
22 D. A. Romero, T. R. Klaenhammer. *J. Dairy Sci.* **1993**, *76*, 1–19.
23 A. Schäfer, A. Jahns, A. Geis, M. Teuber. *FEMS Microbiol. Lett.* **1991**, *80*, 311–318.
24 M. Shimizu-Kadota, M. Kirki, H. Hirokawa, N. Tsuchida. *Mol. Gen. Genet.* **1985**, *200*, 193–198.
25 P.-J. Cluzel, A. Chopin, S. D. Ehrlich, M.-C. Chopin. *Appl. Environ. Microbiol.* **1991**, *57*, 3547–3551.

26 H. M. Dodd, N. Horn, Z. Hao, M. J. Gasson. *Appl. Environ. Microbiol.* **1992**, *58*, 3683–3693.
27 M. J. Gasson. *FEMS Microbiol. Lett.* **1984**, *21*, 7–10.
28 N. Horn, S. Swindell, H. M. Dodd, M. J. Gasson. *Mol. Gen. Genet.* **1991**, *228*, 129–135.
29 P. J. G. Rauch, W. M. de Vos. *J. Bacteriol.* **1992**, *174*, 1280–1287.
30 B. E. Davidson, I. B. Powell, A. J. Hillier. *FEMS Microbiol. Rev.* **1990**, *87*, 79–90.
31 T. R. Klaenhammer, G. F. Fitzgerald. *Genetics and Biotechnology of Lactic Acid Bacteria*. Blackie Academic and Professional, UK, **1994**.
32 F. Desiere, S. Lucchini, C. Canchaya, M. Ventura, H. Brüssow. *Lactic Acid Bacteria, Genetics, Metabolism and Application*. Kluwer Academic Publishers, the Netherlands, **2002**.
33 A. Nauta. PhD Thesis, University of Groningen, Groningen, the Netherlands, **1997**.
34 P. Le Bourgeois, M. Lautier, L. van den Berghe, M. J. Gasson, P. Ritzenthaler. *J. Bacteriol.* **1995**, *177*, 2840–2850.
35 P. Le Bourgeois, M.-L. Daveran-Mingot, P. Ritzenthaler. *J. Bacteriol.* **2000**, *182*, 2481–2491.
36 M.-L. Daveran-Mingot, N, Campo, P. Ritzenthaler, P. Le Bourgeois. *J. Bacteriol.* **1998**, *180*, 4834–4842.
37 M. J. Gasson. *J. Bacteriol.* **1983**, *154*, 1–9.
38 W. M. de Vos, G. F. M. Simons. *Genetics and Biotechnology of Lactic Acid Bacteria*. Blackie Academic and Professional, UK, **1994**.
39 K. Leenhouts, G. Venema. *Plasmids: A Practical Approach*. Oxford University Press, UK, **1993**.
40 K. Leenhouts, G. Venema, J. Kok. *Methods Cell Sci.* **1998**, *20*, 35–50.
41 W. M. de Vos. *Curr. Opin. Microbiol.* **1999**, *2*, 289–295
42 B. R. Froseth, L. L. McKay. *J. Dairy Sci.* **1991**, *74*, 1445–1453.
43 W. M. de Vos, P. Vos, G. Simons, S. David. *J. Dairy Sci.* **1989**, *72*, 3398–3409.
44 F. Dickely, D. Nilsson, E. Bech Hansen, E. Johansen. *Mol. Microbiol.* **1995**, *15*, 839–847.
45 K. I. Sorensen, R. Larsen, A. Kibenich, M. P. Junge, E. Johansen. *Appl. Environ. Microbiol.* **2000**, *66*, 1253–1258.
46 P. Ross, F. O'Gara, S. Condon. *Appl. Environ. Microbiol.* **1990**, *56*, 2164–2169
47 P. Bron, S. Hoffer, J. Lambert, J. Delcour, W. M. de Vos, M. Kleerebezem, P. Hols. *Streptococcal Genetics*, **2002**, P. 33.
48 R. Kiewiet. Thesis, University of Groningen, Groningen, the Netherlands, **1996**.
49 K. Leenhouts, J. Kok, G. Venema. *Appl. Environ. Microbiol.* **1989**, *55*, 394–400.
50 M. C. Chopin, A. Chopin, A. Rouault, N. Galleron. *Appl. Environ. Microbiol.* **1989**, *55*, 1769–1774.
51 H. Israelsen, E. B. Hansen. *Appl. Environ. Microbiol.* **1993**, *59*, 21–26.
52 J. Law, G. Buist, A. Haandrikman, J. Kok, G. Venema, K. Leenhouts. *J. Bacteriol.* **1995**, *177*, 7011–7018.
53 J. W. Sanders, G. Venema, J. Kok, K. Leenhouts. *Mol. Gen. Genet.* **1998**, *257*, 681–685.
54 E. Maguin, H. Prevost, S. D. Ehrlich, A. Gruss. *J. Bacteriol.* **1996**, *178*, 931–935.
55 R. N. Beachy. *Science* **1999**, *285*, 335.
56 European Commission. **2000** *Official Journal* L 006, 13–14.
57 S. Kay, G. van den Eede. *Nature Biotechnol.* **2001**, *19*, 405.
58 J. Gilbert. *Food Control* **1999**, *10*, 363–365.
59 A. Gasch, in: *Foods produced by Modern Genetic Engineering* (Schreiber, F. A., Bögle, K. W., eds), 2nd Status Report, **1997**, BgVV-hefte, pp. 90–97.
60 Qiagen, Critical factors for successful PCR. *Practical Guidelines Technical Literature*, **1999**.
61 A. Lovatt, Applications of quantitative PCR in the biosafety and genetic stability assessment of biotechnology products. *Rev. Mol. Biotechnol.* **2002**, *82*, 279–300.
62 J. Sambrook, D. Russel. *Molecular Cloning: A Laboratory Manual* (3rd edition), Cold Spring Harbor Laboratory Press, **2000**.

63 F. E. Ahmed. *J. Environ. Sci. Health* **1995**, *C11*, 1–51.
64 M. Hertzberg, M. Sievertzon, H. Aspeborg, P. Nilsson, G. Sandberg, J. Lundeberg. *Plant J.* **2001**, *25*, 585–591.
65 F. E. Ahmed. *Trends Biotechnol.* **2002**, *20*, 215–223.
66 A. M. A. van Hoef, E. J. Kok, E. Bouw, H. A. Kuiper, J. Keijer. *Food Addit. Contam.* **1998**, *15*, 767–774.
67 F. E. Ahmed. *Environ. Sci Health* **2000**, *C18*, 75–125.
68 K. J. Livak, S. J. A. Flood, J. Marmaro, W. Giusti, K. Deetz. *PCR Methods Appl.* **1995**, *4*, 357–362.
69 C. A. Heid, J. Stevens, K. J. Livak, P. M. Williams. *Genome Res.* **1996**, *6*, 986–996.
70 J. A. Vet, B. J. Van der Rijt, H. J. Blom. *Expert Rev. Mol. Diagn.* **2002**, *2*, 77–86.
71 P. J. Sykes, in: *The PCR Technique: Quantitative PCR* (Larrick, J. W., ed.), **1998**, Eaton Publishing Co., pp. 81–93.
72 V. Chizhikov, A. Rasooly, K. Chumakov, D. D. Levy. *Appl. Environ. Microbiol.* **2001**, *67*, 3258–3263.
73 G. M. Brett, S. J. Chambers, L. Huang, M. R. A. Morgan. *Food Control* **1999**, *20*, 401–406.
74 I. V. Kourkine, C. N. Hestekin, B. A. Buchholz, A. E. Barron. *Anal. Biochem.* **2002**, *74*, 2565–2572.
75 D. D. Shoemaker, D. A. Morris, M. Mittmann. *Nature Genet.* **1996**, *14*, 450–456.

13
Detection Methods for Genetically Modified Microorganisms used in Food Fermentation Processes

Walter P. Hammes, Christian Hertel, and Torsten Bauer

13.1
Introduction

Detection methods for genetically modified plants have been developed and their reliability has been established [1–4]. These methods cannot be applied without modifications to microorganisms for several reasons. For example, as crops are already on the market, the methods can be adapted to the specific modification and validated. In plants, elements such as the marker genes, promoter sequences and terminators used for the construction are well defined and still rather limited in their numbers. Furthermore, the phenotype of the genetically modified plants can usually easily be detected, for example resistance to insects or virus. On the other hand, no genetically modified microorganisms have gained a status that its admission for release and placing on the market has been applied for, and thus, real examples are missing. Therefore, validated data do not exist for detection of legally approved genetically modified microorganisms (GMM). In general, the detection relies on the available experience for tracing of traits on the genomic as well as the phenotypic level. A genetically modified *Lactobacillus curvatus* can serve as a paradigm of a case treated in such a manner. This GMM harbors a gene from *L. sakei* coding for a catalase [5], and was used to establish a method in the official collection of methods according to § 35 of the German Food Act for detection of GMM in a food (fermented sausage). This method is described in greater detail below. Of great importance for detection is also the special nature of microorganisms, and especially of prokaryotes, which needs to be taken into consideration when targeting specific traits, and this aspect will be discussed in the following sections.

In order to change properties such as the regulation of gene expression, self-cloning is a rather simple method that leads to a rearrangement of the genome without introducing genetic DNA from another strain or species. The use of these modified organisms in a contained environment is not within the scope of genetic modification according to Directive 90/219 EC. On the other hand, it falls under the definition of genetic modification upon release of the organisms.

A genetically modified yeast used for the production of bread provides an example of an organism which was approved by the ACNFP [6]. No corresponding self-cloned crops can be produced with the methods presently available. Clearly, the proof of an illegal modification by self-cloning is beyond practicability; that is, the modification can be traced but not unambiguously related to gene technology. In a close connection to self-cloning stands the intermediary genetic modification of a strain followed by the reconstitution of its wild-type condition. An example was provided by Curic et al. [7], who described the introduction of a plasmid encoding the *ilv* genes for the synthesis of branched-chain amino acids rendering the constructs auxotrophic. The construct provides a means to facilitate the screening for mutants of the Ald$^-$ (acetolactate decarboxylase) phenotype. The mutants were finally selected and cured from the plasmid. The resulting strain is a suitable starter organism for the production of fermented dairy products with the enhanced flavor of diacetyl. Again, this type of genetic modification escapes from detectability.

13.2 Properties of Microorganisms

Some of the properties of microorganisms are encoded on extra-chromosomal elements. These so-called plasmids occur in greater numbers, may encode properties of technological or health-related importance, and may contribute to high levels of expressed gene products. Plasmids can be exchanged between different organisms by *conjugation*. During the course of that horizontal gene transfer (HGT), genetic information spreads within microbial associations. In addition, two further mechanisms contribute to HGT. First, *transduction* is mediated by bacteriophages which, upon disintegration of the host bacteria, co-transfer the genetic material of the virus and parts of the host's DNA to a new host. The nucleic acid content of the phage particle may become integrated into the genome of the new host bacterium, which then becomes a lysogenic strain carrying a prophage together with its recently acquired DNA sequences. Finally, by *transformation*, free DNA is taken up by a bacterium and becomes integrated into the genome and expressed. These microbe-specific HGT events must be borne in mind when the presence of a genetic modification has to be detected in a microorganism and related to a genetic modification event. First, the specific sequence may have been acquired by a natural HGT event; and second, the once-performed modification may be subject to rearrangements in the genome, thereby interfering with detectability of genetic modification. These rearrangements take place especially frequently in bacteria and are mediated by elements, such as insertion elements and transposons. Altogether, the frequent HGT processes and genomic rearrangement processes cause a plasticity of the bacterial genome, which has to be seen as an inherent property of microorganisms.

To detect a specific property on the genotype or phenotype level, it has to be known that this property is unambiguously the result of a genetic modification.

As shown in Fig. 13.1, the levels to which the detection of the GMM can be directed differ in their unambiguousness. It is highest on the nucleic acid level. PCR-based or hybridization techniques can provide unambiguous results, as far as it is possible with microorganisms. On the other hand, on the phenotype level – and especially upon application of technical or biological processes on the food – the unambiguousness of detection decreases. In the focus of methods used for detection of GMM in food are, therefore, those targeting the genomic level. However, in foods the positive detection of a transgene may need careful interpretation as, for example, fermentation processes are usually performed in nonsterile environments. These may contain numerous additional microorganisms, including also unknown or uncultivable species, the properties and genetic endowment of which is unknown and may serve as a source for new properties that might subsequently be detected in a fermentation organism. This property might even be identical with any that has been incorporated by genetic modification.

When GMMs are used for the production of food, four categories can be differentiated with regard to their detectability:

I. The product is free of any DNA and impurities, which are indicative for a GMM (e.g., highly purified food ingredients or additives).
II. The product contains DNA but no GMM (e.g., liquid products such as beer or wine, that were subjected to separation processes such as filtration or centrifugation).
III. Products that contain the dead GMM with its DNA (e.g., a pasteurized yogurt or baked goods).
IV. Products containing the living GMM (e.g., a nonpasteurized yogurt, cheese, beer, sauerkraut, fermented sausage, etc.).

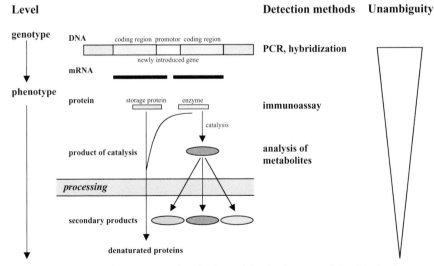

Figure 13.1. The unambiguousness of methods used for the detection of GMO in food.

It is clear that the involvement of a GMM in the production of category IV food is easy to detect, as not only the DNA is readily available and extractable, and the modification can be traced in detail, but the organisms can also be cultured and their physiological and biochemical properties investigated. Finally, the organisms can be identified down to the strain level; this may be important in cases where the identity of the GMM as a constituent of a starter culture, probiotic food or protective culture has to be shown. In fact, for that purpose all methods can be applied that are commonly used for identification at the genotype or phenotype level. In cases where culturing is not needed or is not possible – for example when the organisms acquire a status of "viable but not culturable" (VBNC) – it is possible to deduce whether live or dead cells are present by the use of fluorescent staining techniques followed by microscopy. Alternatively, when applicable, the technique of flow cytometric analysis can be used [8, 9]. Among category III foods, there is an overlap with foods of category II. In cases where a GMM is heat-killed and immediately frozen, the DNA and most of the components remain accessible for analysis, despite being in a partially denatured form. However, when the cells lyse – either spontaneously or when lysis is induced by means of genetic modification – the conditions for detection of the GMM are similar to those used for category II food. Induced lysis has been the subject of several studies aimed at an accelerated ripening of cheese [10, 11] that is brought about by hydrolytic (proteolytic) activity released from the GM-starter cultures. In general, the detectability of a genetic modification in a dead GMM contained in food is comparatively straightforward, provided that the DNA is retained within the cell compartment. It has been shown that physico-chemical as well as enzymatic DNA-degrading effects are greatly reduced in dead cells when compared with free DNA [12]. With regard to the detection of DNA, the molecule must be extracted, and the strength of DNA inclusion must be taken into account. Generally, DNA is more easily extracted from Gram-negative bacteria and yeasts than from Gram-positive bacteria, and attention must also be paid to components of the food that might interfere with detectability (see below). Isolation of transgenic DNA from the food may, therefore, require extensive purification which must be adjusted to the location of the transgene – that is, chromosomal or plasmid integration. Finally, the target of detection – the DNA, protein or metabolite – may undergo extensive degradation or denaturation depending on the processing conditions to which a food is subjected. The effects of factors such as heat, pH, and enzymatic attack in food have been studied (T. Bauer, unpublished results), while those generally exerted by shearing forces have also been described [13]. These processes, together with separation techniques and deliberate enzyme application, may finally lead to category I food. For these, the detection of a GMM or its effect on the food is not feasible.

13.3
Current Methods for Detection of GMM

Methods for the detection of GMM in sausages and yogurt have been established and were published during the late 1990s upon validation in ring trials [14, 15]. The procedure of analysis for the detection of GMM in foods is depicted in Fig. 13.2. In this example, fermented sausage had been produced with the recombinant strain *L. curvatus* LTH 4002 used as starter. This GMM harbors plasmid pLSC300 encoding the katalase gene *katA* of *L. sakei* LTH 677 [5]. These sausages were then subjected to analysis [14]. Total DNA was isolated from homogenized sausages using a universal phenol/chloroform extraction method, with the efficiency of extraction being checked using agarose gel electrophoresis. For the unambiguous detection of the recombinant strain, primers were constructed which targeted against *katA* and its border sequence within the plasmid, respectively. This approach takes into account that most strains of *L. sakei* contain *katA* and that this species can be present in the indigenous ripening flora of fermented sausages. The specific primers are used for the amplification of a 1321 bp fragment of pLSC300 by PCR. In parallel, the amplificability of the isolated DNA was checked by amplification of 1025 bp fragment of the 16S rDNA using universal primers. The lengths of the amplified DNA fragments were determined by agarose gel electrophoresis. In addition, the amplified sequence was validated by Southern hybridization using a nonradioactive labelled oligonucleotide probe targeted against the nested sequence.

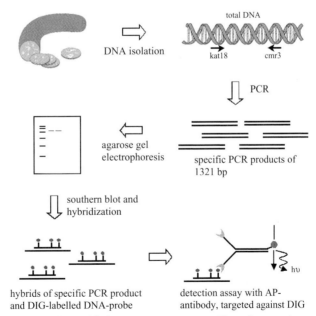

Figure 13.2. Detection of the genetically modified strain *L. curvatus* LTH 4002 in fermented sausages according to the official method of the German Food Act LMBG § 35. AP: Alkaline phosphatase; DIG: Digoxigenin.

This procedure of analysis corresponds closely to the validated method for a detection of genetically modified strain of *Streptococcus thermophilus* in yogurt [15], with the exception that the DNA extraction method was adapted to this special food matrix. The GMM had been endowed with the *cat* gene encoding resistance to chloramphenicol. Again, the specific primers were targeted against the recombinant gene (*cat*) and its border sequence within the chromosome (*lacZ* gene) to unambiguously identify the recombinant nature of the organism.

13.3.1
DNA Isolation

The isolation of DNA is the first step in an analysis to detect GMM use in food by molecular methods. In cases where the cells are not lysed, the GMMs may be separated from the food matrix by first homogenizing the food and then separating the cells by filtration and centrifugation. Subsequently, the DNA can be isolated according to published protocols. To optimize the yield of isolated DNA, the method needs to be adapted according to the taxonomic position of the GMM.

For DNA extraction, the case described above [14] can serve again as an example. In this situation, a universal phenol/chloroform extraction procedure was used to isolate total DNA from fermented sausage containing living cells of the GMM. The applicability of this method to obtain high-quality DNA that is suitable for detection by PCR has been demonstrated for heat-treated fermented sausages [12] and cream [16]. The method was also adapted for the isolation of total DNA from yogurt, including the bacterial genomic DNA of *Streptococcus thermophilus* [15]. These two methods were validated in a ring trial and added to the official collection of methods according to German Food Act LMBG § 35.

In a situation where the GMM is killed during the processing of fermented foods and its DNA is released into the food matrix, the free DNA must be isolated directly from the food matrix. As most of the approved GMOs are plants, numerous data are available for the isolation of DNA from plant material. Zimmermann et al. [17] compared nine different extraction methods including CTAB, Wizard, DNeasy and others. The isolation of DNA using commercial kits such as Wizard or DNeasy that are based on the specific binding of DNA to resins resulted in high-quality DNA, though the yield was poor compared with that achieved for other methods. However, the resin-based isolation method was recommended as the quality of the DNA was seen to be the most important factor.

The DNA-based detection of GMM may be hampered by the presence in the food matrix of inhibitors that interfere with cell lysis, degrade DNA during isolation, or directly interfere with the PCR itself. Numerous compounds or substances such as organic or phenolic compounds, glycogen or milk proteins were already identified to inhibit PCR [18]. Especially in the case of meat and milk products, heme and Ca^{2+} were found to be strong PCR-inhibitors [19, 20]. Inhibition of the PCR can also occur due to the presence of contaminants in reagents or may even be caused by the powder used on disposable gloves. In order to exclude false negative results, external or internal standards for PCR-based detection methods can be applied which monitor the efficacy of the PCR [21]. With regard to foods, heterologous internal standards (so-called mimics) were successfully used to reliably detect the food pathogens *Yersinia enterocolitica* [22] and *Campylobacter coli/Campylobacter jejuni* [23].

13.3.2
DNA Stability

The persistence of recombinant DNA in fermented foods is of major concern for the detection of GMM by DNA-based methods. While the DNA is contained in cells, it is naturally protected against degradation. In this respect, Straub et al. [12] showed that the DNA from genetically modified *Lactobacillus curvatus* strains used as starter organism to produce summer sausages was heavily protected against either enzymatic or physical attack, even when contained in dead cells. On the other hand, DNA that is released from cells into the food matrix – as it occurs during food processing – undergoes physical, chemical, and enzymatic degradation. There are numerous ecological factors which prevail during the production and storage of foods which may either enhance the degradation of free DNA or contribute to its protection against degradation. It has been shown that free plasmid DNA in summer sausages remained detectable after nine weeks of storage [12]. These results, together with those obtained in challenge experiments performed to investigate the effect of added DNase I in sausages, indicate that the meat matrix exerted a protective effect on the free plasmid DNA. Further information of DNA stability in milk has been presented [24], showing that free bacterial DNA remained detectable for 12 days in ultra high-temperature-treated milk stored at 4 °C, whereas storage at 20 °C resulted in rapid degradation of DNA within a few days.

The processing of fermented foods can also include a final heat treatment step in order to pasteurize the product; examples include summer sausages, yogurt, and beer. In summer sausages, neither the temperature employed for the heat treatment nor the fat content or pH significantly affected the detectability of free recombinant DNA in the product [12]. A combination of heat treatment with other degrading factors may result in additive or synergistic effects. For example, Klein et al. [25] reported that the different steps applied in the process of sugar production contribute to a strong degradation of free plasmid DNA. The overall efficacy of DNA elimination can be calculated as ca. 10^{14}. By contrast, Hotzel et al. [26] showed that residual DNA from yeasts used in the production of beer can still be detected in the final product (by PCR), despite the prevailing acidic conditions (pH ca. 4.5).

13.3.3
Organism-specific Detection of GMM

The safety assessment of foods containing viable GMM raises a number of unique issues such as pathogenicity, toxigenicity, genetic stability and gene transfer that may not be relevant when assessing foods containing nonviable GMM or products thereof. Guidelines and recommendations have been published to provide a structured approach to the safety assessment of GMMs [27, 28]. It is generally required that, with regard to the host organism, the full taxonomic profile is provided, down to strain level where appropriate. Thus, there is a need to determine the taxonomic position of the GMM as well as to identify the strain. Furthermore, to ensure that the detected recombinant DNA is contained in the very host strain that had been used for construction of the GMM, the detection of the recombinant DNA must be combined with the identification of the host organism. This measure takes into account the possibility of horizontal transfer of recombinant DNA between different strain, species or even less-related organisms. Ludwig et al. [29] presented an overview about DNA-based methods which can be used especially for this type of combined GMM detection approach. The methods are based on specific DNA probe hybridization and on diagnostic DNA *in vitro* amplification. Taking into account the contaminating or indigenous microbiota – which may be present together with the GMM in foods – it is evident that living cells are required for the combined use of probes and/or PCR primers specific for the recombinant DNA and the microorganism. The available methods are highly sensitive, but are applicable to living GMM exclusively.

The ribosomal RNAs (rRNA) have been proven to be excellent phylogenetic markers [30], and can be used as targets for the taxonomic identification of GMM. The molecules contain evolutionary highly as well as less conserved sequence positions and regions, which reflect earlier and more recent events in the evolution of the organism, respectively. These regions can contain taxonomic (phylogenetic) relevant signature positions, which are diagnostic for groups of phylogenetically related organisms, for example family, genus, or species. They often provide useful target sites for taxonomic (phylogenetic) probes and primers, which can be used in combination with a variety of hybridization and PCR techniques [31, 32, 29]. Numerous oligonucleotide probes and primers for specific detection of the species of fermentation organisms have already been published, and details are freely available from literature sources. The availability of comprehensive 16S rRNA sequence databases and software packages such as ARB facilitate the rapid rational design and *in-silico* specificity profiling of such probes and primers [33]. As far as pure cultures of the GMMs are available, various hybridization (e. g., dot/slot blot or colony hybridization) and diagnostic PCR techniques can be applied. For example, colony hybridization was used successfully detect a genetically modified *Lactococcus lactis* subsp. *lactis* in pure or mixed cultures and fecal samples [34, 35]. The proteinase-negative strain was modified by introducing the proteinase gene *prtP*. Unambiguous detection of the GMM was ensured by the combined use of probes specific for the recombinant gene and the species

Lactococcus lactis. In addition, the recombinant DNA was reliably detected applying a diagnostic PCR system.

The primary structures of the rRNAs are usually too extensively conserved to provide strain-specific probe or primer target sites. However, techniques based on the detection of strain-specific DNA sequences by applying hybridization or PCR techniques with specific probes or primers are available. These unique sequences can be derived, for example, from RAPD fragments [36] or may be obtained by the subtraction hybridization technique [37, 38]. The applicability of the latter method to develop strain-specific PCR-based detection systems for food-fermenting organisms has been demonstrated [39, 40]. Bunte et al. [40] used such a PCR system to monitor the strain *Lactobacillus paracasei* LTH 2579 in fermenting sausages and human fecal samples. Minor sequence variations between closely related strains of the species *L. paracasei* were identified by subtraction hybridization using the genomic DNA of the target strain and five subtracter strains. The principle of subtraction hybridization is depicted in Fig. 13.3. Upon hybridization, a nonhybridizing DNA fragment of 235 bp was obtained; this can either be used as DNA probe for hybridization or to develop a diagnostic PCR system. The specificity of the PCR system was validated using numerous strains of the species *L. paracasei*.

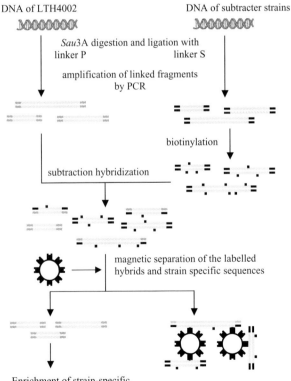

Figure 13.3. Schematic illustration of subtraction hybridization. (Modified from Ref. [39].)

13.4
Conclusion

The plasticity of the bacterial genome, the frequency of horizontal gene transfer, and the application of self-cloning complicate the detection of GMM in foods, especially when recombinant DNA is naturally present within a rich indigenous or contaminating flora. For an unambiguous detection, the detection of recombinant DNA combined with the identification of the host organism is required, though this approach needs intact cells of the GMM to be present. For this purpose, specific and sensitive molecular methods are available. As no GMM have been admitted for release and the placement on the market of products created by their use, validated data for the detection of legally approved GMM do not currently exist. Nevertheless, detection methods have been established for model foods. Thus, with certain limitations, it is possible to detect GMM in food as long as their genetic background is well described – as is legally required in the course of the application procedure.

References

1 ILSI, *Detection Methods For Novel Foods Derived From Genetically Modified Organisms*, ILSI Europe Report Series. Brussels, Belgium, **1999**.
2 R. Meyer, *Food Control* **1999**, *10*, 391–399.
3 E. Gachet, G. G. Martin, F. Vigneau, G. Meyer, *Trends Food Sci. Technol.* **1998**, *9*, 380–388.
4 A. Pasqualone, *Industrie Alimentari*. **1999**, *39*, 444–451.
5 C. Hertel, G. Schmidt, M. Fischer, K. Oellers, W. P. Hammes, *Appl. Environ. Microbiol.* **1998**, *64*, 1359–1365.
6 ACNFP, *Annual Report 1990*, Department of Health, Ministry of Agriculture, Fisheries and Food, UK, **1990**.
7 M. Curic, B. Stuer-Lauridsen, P. Renault, D. Nilson, *Appl. Environ. Microbiol.* **1999**, *3*, 1202–1206.
8 J. S. Miller, J. M. Quarles, *Cytometry* **1990**, *11*, 667–675.
9 G. Nebe-von-Caron, P. J. Stephens, C. J. Hewitt, J. R. Powell, R. A. Badley, *J. Microbiol. Methods* **2000**, *42*, 97–114.
10 B. A. Law, *Int. Dairy J.* **2001**, *11*, 383–398.
11 L. U. Guldfeldt, K. I. Sorensen, P. Stroman, H. Behrndt, D. Williams, E. Johansen, *Int. Dairy J.* **2001**, *11*, 373–382.
12 J. A. Straub, C. Hertel, W. P. Hammes, *Z. Lebensm. Unters. Forsch.* **1999**, *210*, 62–67.
13 C. S. Lengsfeld, T. J. Anchordoquy, *J. Pharmaceut. Sci.* **2002**, *91*, 1581–1589.
14 Anonymous, L 08.00-44. Untersuchung von Lebensmitteln: Nachweis einer gentechnischen Veränderung in Rohwurst durch Amplifizierung der veränderten DNA-Sequenz mit Hilfe der PCR und Hybridisierung des PCR-Produktes mit einer DNA-Sonde. In: Amtliche Sammlung von Untersuchungsverfahren nach § 35 LMBG, BgVV, Lose Blattsammlung, Beuth-Verlag, Berlin/Köln, **1996**.
15 Anonymous, L 02.02-4. Untersuchung von Lebensmitteln: Nachweis einer gentechnischen Veränderung von Streptococcus thermophilus in Joghurt durch Amplifizierung der veränderten DNA-Sequenz mit Hilfe der PCR und Hybridisierung des PCR-Produktes mit einer DNA-Sonde. In: Amtliche Sammlung von Untersuchungsverfahren nach § 35 LMBG, BgVV, Lose Blattsammlung, Beuth-Verlag, Berlin/Köln, **1997**.
16 J. A. Straub, C. Hertel, W. P. Hammes, *J. Food Prot.* **1999**, *62*, 1150–1156.
17 A. Zimmermann, J. Luthy, U. Pauli, *Z. Lebensm. Unters. Forsch.* **1998**, *207*, 81–90.
18 I. G. Wilson, *Appl. Environ. Microbiol.* **1997**, *63*, 3741–3751.
19 A. Akane, K. Matsubara, H. Nakamura, S Takahashi, K. Kimura, *J. Forensic Sci.* **1994**, *39*, 362–372.
20 J. Bickley, J. K. Short, D. G. McDowell, H. C. Parkes, *Lett. Appl. Microbiol.* **1996**, *22*, 153–158.
21 B. C. Trapnell, *Am. J. Physiol.* **1993**, *264*, L199–L212.
22 S. Thisted Lambertz, A. Ballagi-Pordany, R. Lindquist, *Lett. Appl. Microbiol.* **1998**, *26*, 9–11.

23 J. A. Straub, C. Hertel, D. Made, W. P. Hammes, *Z. Lebensm. Unters. Forsch.* **1999**, *209*, 180–184.
24 M. Braeutigam, C. Hertel, W. P. Hammes, *FEMS Microbiol. Lett.* **1997**, *155*, 93–98.
25 J. Klein, J. Altenbuchner, R. Mattes, *J. Biotechnol.* **1998**, *60*, 145–153.
26 H. Hotzel, W. Müller, K. Sachse, *Eur. Food Res. Technol.* **1999**, *209*, 192–196.
27 SCF, *Commission Recommendation 97/618/EEC concerning the scientific aspects of the presentation of information necessary to support applications for the placing on the market of novel foods and novel food ingredients and the preparation of initial assessment reports under regulation (EC) No 258/97 of the European Parliament and of the Council*, Official Journal of the European Communities, L253, Brussels, **1997**.
28 ILSI, *Safety assessment of viable genetically modified micro-organisms used in food*, ILSI Europe Report Series, Brussels, Belgium, **1999**.
29 W. Ludwig, E. Brockmann, C. Beimfohr, C. Hertel, B. Jacobsen, K. H. Schleifer, *Syst. Appl. Microbiol.* **1995**, *18*, 477–485.
30 W. Ludwig, K. H. Schleifer, *FEMS Microbiol. Rev..* **1994**, *15*, 155–173.
31 K. H. Schleifer, W. Ludwig, R. Amann, Nucleic acid probes. In: M. Goodfellow, O. McDonnell (eds.) *Handbook of New Bacterial Systematics*, pp. 463–510, Academic Press, London-New York, **1993**.
32 R. Amann, W. Ludwig, K. H. Schleifer, Phylogenetic identification and *in situ* detection of individual microbial cells without cultivation. *Microbiol. Rev.* **1995**, *59*, 143–169.
33 R. Amann, W. Ludwig, *FEMS Microbiol. Rev.* **2000**, *24*, 555–565.
34 C. Hertel, W. Ludwig, K. H. Schleifer, *Syst. Appl. Microbiol.* **1992**, *15*, 447–452.
35 E. Brockmann, B. L. Jacobsen, C. Hertel, W. Ludwig, K. H. Schleifer, *Syst. Appl. Microbiol.* **1996**, *19*, 203–212.
36 A. Tilsala-Timisjärvi, T. Alatossava, *Appl. Environ. Microbiol.* **1998**, *64*, 4816–4819.
37 A. J. Bjoursen, J. E. Cooper. *Appl. Environ. Microbiol.* **1988**, *54*, 2852–2855.
38 V. G. Matheson, J. Munakata-Marr, G. D. Hopkins, P. L. McCarty, J. M. Tiedje, L. J. Forney, *Appl. Environ. Microbiol.* **1997**, *63*, 2863–2869.
39 L. Wassill, W. Ludwig, K. H. Schleifer, *FEMS Microbiol. Lett.* **1998**, *95*, 87–94.
40 C. Bunte, C. Hertel, W. P. Hammes, *Syst. Appl. Microbiol.* **2000**, *23*, 260–266.

Index

a
aad 34, 36f
AAT 75
AbiQ 106
abortive infection (Abi) system 106
abuse principle 124
acetaldehyde 109
Acetobacter 64
α-acetolactate 109
α-acetolactate decarboxylase 74, 109
acetolactate reductoisomerase 74
acid tolerance 110
acidophilic 103
Agaricus bisporus 86
agarose gel electrophoresis 197
Agrobacterium rhizogenes
– hairy root disease 26
Agrobacterium tumefaciens 69f
– crown gall tumor 26
– T-DNA 26
– tumor-inducing plasmid 26
agronomic property
– generation time 46
– sprouting time 46
aguardente 64
alaD 109
alanine dehydrogenase 109
alanine racemase 109
ALD6 66, 75
ALD7 66, 75
ALDC 74
alfalfa 42
allergic reaction 101
all-fish-gene cassette 174, 180
all-salmon gene cassette 181
alpha-amylase inhibitor 40
alr 109
als 34
Amaranthus hypochondriacus 47
AmdS 66, 94

amino acid composition 47
ampicillin 29
AMY1 73
α-amylase 89
amylase 89f, 92
amyloglycosidase 89
angiotensin-I-converting-enzyme 109
Anguilla spp. 185
animal health 15
animals 3–19
antibiotic resistance 8, 32, 34, 36–39, 93
antibiotic resistance gene
– *aadA* gene 29
– *bla* gene 29
– *hpt* gene 29
– *nptII* gene 29
antifreeze protein 12f
antihypertensive effect 109
antinutritive factor 48
antisense-RNA 107
AOX 66
apple 40, 42
aquaculture 12, 174
Arabidopsis 44ff, 52
arctic charr 176, 178
arginine deiminase 110
ARO4 66, 76
ARS-based plasmid 67
Arxula adeninivorans 68
Aspergillus 78
Aspergillus aculeatus 87
Aspergillus awamori 70, 87, 92, 94
Aspergillus foetidus 87
Aspergillus fumigatus 47
Aspergillus japonicus 87
Aspergillus nidulans 66, 69, 94
Aspergillus niger 87, 90, 92, 94
Aspergillus oryzae 64, 76, 86f, 90, 92, 94
Aspergillus sake 86
Aspergillus sp. 88, 91

Aspergillus spp. 64
ATF1 66, 75
atlantic salmon 176, 179
– growth enhancement 180
ATPase 110
auxotrophic marker 93, 95

b

Bacillus thurigiensis 40
Bacillus thuringiensis toxin 29
bacterial artificial chromosome 8
bacterial resistance 42
– attacins 43
– bacteriophage T4 lysozyme 43
– cecropin 43
– hydrogen peroxide 43
– sarcotoxin 43
bacteriocin 103f, 105
bacteriophage 105, 234
– Group I intron 236
bacteriophage resistance 106
– lactic acid bacteria 105
baker's yeast 72f
– biomass yield 77
– bread 76
– catabolite repression 76
– genetically modified 78
– maltase 78
– maltose permease 78
– melibiase 77
bakery product 102
baking quality 49
banana 50
bar 32, 34f, 38
barley 45, 49f, 64
barnase 38
barstar 38
basis Regulation
– Regulation (EC) No 178/2002 141
Basta 30
BayTE 39
beer 64
– flavor 73
– flavor stability 74
– maturation time 74
– Reinheitsgebot 72
– stale flavor 74
bialaphos 30
bioconservation 105
biogenic amine 109
bioluminescence 30
biomass 96
biosafety 19
bird 7

bla 32, 33, 35, 38f
Bollgard 194
bottom-fermenting yeast 72
bovine spongiform encephalitis 16
Brassica napus 52
Brassica rapa 34
bread 64, 76
brewer's yeast
– carbohydrate utilization 72
– flocculation 73
– genetically modified 78
– glucoamylase 78
– hydrogen sulfide 74
Bt 176 maize 221f
butter 102, 109
buttermilk 102
bxn 34

c

cabbage 103
cachaca 64
CaMV 35S promoter 200
Candida maltosa 68
Candida spp. 68
capillary electrophoresis
– detection 246
CAR1 66, 76
β-carotene 47
carotenoid content 46
carp 12, 176, 178, 183
carrot 43
Cartagena-Protocol 144
casein degradation
– cheese 108
– fermented dairy product 108
cassava 48, 102
catalase 110
catfish 12
cattle 4, 6, 11f, 16
cell lysis 96
cell surface 111f
cephalosporin 29
cereal 64
Certified Reference Material 190, 219
cetyl-trimethylammonium bromide 191
champagne 64
channel catfish 176, 178
cheese 102f, 105, 109
chicken 12
chicory 38, 195
chinook salmon 176, 179
chocolate 64
chymosin 89
Cichlidae 182

citric acid 87
citrus tree 46
Clostridia 105
Clostridium spp. 105
Clostridium tetani
– toxin C-fragment 112
CO_2 104
cocoa 64
coconut 64
cod 185
Codes Alimentarius Commission 130
codon usage 29
co-electroporation 106
coho salmon 176, 179, 181
cold shock 110
Colletotrichum lindemuthanium 70
colony hybridization 258
CommProp GMFF 139f, 145
– genetically modified food and feed 141ff
– genetically modified food or feed 144
– integral approach 144
– labeling 143
– objective and definitions 141
– procedure-related labeling 143
– unavoidable contamination 143
CommProp Traceability 145
– feedstuff 144
– flavoring 144
– food additive 144
– foods and food ingredients 144
competitive PCR
– Bt 176 maize 216
– cryIA(b) gene 218
– ivr1 gene 218
– maize-specific invertase (ivr1) gene 217
– Roundup Ready soybean 216
– soy-specific lectin (le1) gene 217
composed food 206
– immunoassay 210
congenital immunization 15f
conjugation 106, 233
conjugative transposon 233
corn rootworm 40
cotton 31, 34, 36, 40, 194
CP4 EPSPS 32–35
cracker 102
CRE/lox system 31
creA 94
cross-over
– double 68
– single 68
cryIA(b) 195
cryIAb 32, 35, 40
cryIAc 35f, 40

cryIFa2 35
cryIIAb 40
cryIIIA 35ff
cryIIIB 40
cryIIIBb 40
cryoprotection 110
CTAB 256
CTAB-extraction 221
cucumber 41f
cutthroat trout 176, 179
cyanide toxicity 48
Cyprinus carpio 176, 178, 183

d

dairy product 102
dam 38
Debaryomyces hansenii 63
dehydration-stress tolerance
– arginine decarboxylase 44
– Mn superoxide dismutase 44
– osmoprotectant 44
delayed fruit ripening 37, 39
Deliberate Release Directive 130
– Directive No 90/220/EEC 136
detection assay
– food-grade modification 153
– target 150–153
– unknown modification 154
detection strategy
– construct specific 195f
– event specific 195f
– screening 195f
detection
– capillary electrophoresis 246
– DNA-based 242
– ELISA 245
– genetically modified microorganism 251, 253, 255, 257, 259
– micro-array 244
– nucleotide sequence-based 244
– organism-specific 258
– PCR 243
– protein-based 245
– Southern hybridization 243
– subtraction hybridization 259
– surface plasmon resonance 246
– two-dimensional gel electrophoresis 246
– unambiguity 253
– Western hybridization 245
dextran 103
diacetyl 73f, 103f, 109
dicotyledonous plant 27f
direct gene transfer
– particle bombardment 27

- transformation of protoplast 27
Directive 2000/13/EC, Labeling Directive 143
Directive 2001/18/EC 139ff, 144
Directive 70/457/EEC 140
Directive 70/458/EEC 140
Directive 79/112/EEC 123, 139f
Directive 82/471/EEC 140
Directive 90/219/EEC 153
Directive 90/220/EEC 123, 127, 140, 154
Directive No 90/220/EEC
- Deliberate Release Directive 136
disease resistance 15f
- bacterial resistance 42
- fungal resistance 41
- nematode resistance 43
- virus resistance 41
distiller's yeast 72
DNA
- amplicon-specific 161
- analysis 191
- chemical stability 155, 208
- degradation 212, 214f, 217, 254, 257
- depurination 208
- detection 156, 158–161, 164, 166, 168, 223
- detection strategy 192
- extraction 157f, 191, 211, 220f
- fragmentation 208f
- heat-treated 217f
- hybridization 158
- isolation 192, 256
- isothermic amplification 164
- linear 68
- mass spectrometry 166
- methylation 8
- micro-array 165
- microinjection 4, 7f, 11
- nontranscribed 152
- phenol/chloroform extraction 256
- photon-driven monotoring 166
- purification 157f, 192
- quantification 222
- quantitation 221
- random integration 4, 11
- sequencing 160
- single-stranded 159
- stability 257
- thermal cycling 164
- transformation 65, 69f
- vaccine 15
DNA quantitation 223
DNA-based method, PCR 210
DNeasy 256
downstream processing 96

DtpP 108
DtpT 108
durum wheat 49

e

ECA 76
90/220 EEC 31
eel 185
EFSA 142ff
eggplant 40, 49
electroporation 7, 11, 111
electrospray-MS 166
electro-transformation 106
ELISA
- detection 245
emulsifier 107
endogenous retrovirus 5
environmental stress
- dehydration stress 44
- iron stress 45
- salt stress 45
enzymatic hydrolysis 87
enzyme 88
enzyme production 95
enzyme-immunoassay detection 164
enzyme-linked immunosorbent assay 210
EPS-operon 108
EPSPS 195
Erwinia carotovora atroseptica 43
ethidium bromide 160, 191
ethyl carbamate 76
excision system 30f
exopolymer synthesis 103
exopolysaccharide (EPS) 107f
- lactic acid bacteria 108
explosive expression 107
expression vector 92, 94
extracellular immunization 16

f

farm animal 3–19
fat-replacer 107
fatty acid content 37
fermentation 97
- batch 96
- cheese 106
- continuous 94, 96
- dairy 106
- fed-batch 94, 96
- seed 96
- submerged 96, 105
- surface 96
fermented food 62–79
- animal origin 63

– plant origin   63f
fermented meat product
– preservation   63
fermented sausage   102
fermenter   96f
ferritin   47
fertility restoration   38
filamentous fungi   62f, 66, 79, 86–97, 101
– autonomously replicating vector   70
– auxotrophic marker   91
– ectopic integration   70
– electroporation   69
– expression vector   92
– gene disruption   90
– homologous integration   70
– metabolic engineering   71
– nonhomologous integration   70
– protoplast fusion   78
– protoplast   69
– rice-wine   78
– selection marker   69
– soy sauce   78
fish   7, 12, 16, 18
– gene transfer   177
– genetic engineering   174
flavor compound   109
Flavr Savr   39, 194
FlavrSavr-Tomato   124
– labeling   128
flax   34
FLO1   66
flocculation   73
5-fluoro-orotic acid   91f
fluorescence correlation spectroscopy   166
fluorescence resonance energy transfer   161, 163
– probe   244
folic acid   109
food additive   86
– baking industry   89
– brewing industry   89
– dairy industry   89
– starch industry   89
– wine industry   90
food fermentation   100–112
food processing   257
– aid   89f
– degradation of DNA   208f, 212
– degradation of proteins   207f
food quality   46–49
food-grade
– concept   153
– dominant marker   239
– gene delivery   106f

– gene transfer   106
– organism   153
– plasmid   238f
– selection marker   238
– starter strain   107
– vector   111
fructan   48
fruit mash   104
functional food   101f
fungal resistance   41
– endochitinase   42
– human lysozyme gene   42
– hypersensitive response   42
Fusarium venenatum   86, 88, 90, 92

g
Gadus morhua   185
GAM1   73
gelling agent   107
gene cassette   176
gene disruption   90ff, 95
gene replacement   95, 240
gene silencing   27, 41
gene stacking   224
gene targeting   16
gene transfer   9, 18f, 233
– additive   8, 15
– efficiency   4
– food-grade   106
– germ-line   15
– in fish   7, 12
– in mammals   3ff
– in poultry   7
– microinjection   177
– somatic   11, 15
genetic information
– conversion   151
– degeneration   152
– flux   150
genetic modification
– detection   149
genetically engineered enzyme   88
genetically engineered food
– legal situation   121–145
genetically modified crop   199
– detection   188–202
– qualitative detection   197
– quantification   198
– sample preparation   190
– sampling   189
– screening   193
– specific detection   195
– verification   198
genetically modified *Lactococcus lactis*

– detection   241
genetically modified microorganism
– detection   251, 253, 255, 257, 259
– food fermentation   253
– *Lactobacillus curvatus* LTH 4002   255
– pLSC300   255
– *Streptococcus thermophilus*   256
– *Lactobacillus curvatus*   251
genetically modified organism
– labeling   144, 206
– traceability   144. 206
genetically modified plant   26–52
– herbicide-resistant cotton   31
– herbicide-resistant maize   31
– herbicide-resistant soybean   31
– insect-resistant cotton   31
– insect-resistant maize   31
– molecular requirement   28
– Roundup Ready soybean   31
– tissue requirement   28
geneticin   29
genotype   149
gin   64, 77
β-glucan   49, 73
β-glucanase   49, 90
glucoamylase   87
Gluconacetobacter xylinus   64
glucose isomerase   89
β-glucuronidase (GUS)   30
gluten   49
glutenin   49
glycoamylase   92
glycosyltransferase   108
GmFad2-1   39
goat   6, 11
gox   32–35
GPD1   66, 75
Gram-negative bacteria   29
grape   40, 64
grape vine   42
grapefruit   40f
GRAS   79, 88, 90
growth hormone   11f
gus   37, 33, 39

h

Hansenula polymorpha   68, 77, 86, 95
HAP4   66
heat shock response   110
herbicide resistance   32ff, 36, 38
herbicide resistance gene
– *bar* gene   30
– *pat* gene   30
heterofermentative   103

HIS3   66f
homofermentative   103
homologous integration   67
homologous recombination   6f, 16, 68, 70, 134, 233, 236f
horizontal gene transfer
– conjugation   252
– transduction   252
– transformation   252
HSP30   66
HXT   66
HybProbes   161f
hybridization   158, 160, 162, 165, 198, 218, 244
hydrogen peroxide   104
hygromycin B   29
hygromycin phosphotransferase   29

i

Ictalurus punctatus   176, 178
identity preserved   188
IGF-1   17
ILV5   66
industrial enzyme   96
insect resistance   32, 35ff
– alpha-amylase inhibitor   40
– B.t.-toxin   40
– lectin   40
– proteinase inhibitor   40
insertion vector
– pG$^+$HOST   239, 241
– pOR1   239f
– pVE6007   241
interlaboratory study   200
intracellular immunization   15
intron   28
– homing   234, 236
iron level   47
IS element   233, 235
isopentenyltransferase (IPT)   30
ISS1 transposition   241

k

kanamycin   29
katA   255
katalase gene
– *Lactobacillus sakei* LTH 677   255
keratin   17
Kloeckera   63
Kluyveromyces lactis   68, 77, 86, 95
Kluyveromyces marxianus   68
Kluyveromyces spp.   68
koji   78
kombucha   64

l

Labeling Directive
– Directive 2000/13/EC 143
Lab-on-a-chip technology 166
laccase 90
lactase 89
lactic acid bacteria 102f, 109
– amino acid auxotrophy 108
– bacteriophage resistance 105
– bioconservation 104
– exopolysaccharide (EPS) 107f
– heterofermentative 101
– homofermentative 101
– proteolysis 108
– stress response 110
– vector system 111
lacticin 3147 105
Lactobacillus 76
Lactobacillus acidophilus 102
Lactobacillus amylolyticus 102
Lactobacillus casei 102
Lactobacillus confusus 64
Lactobacillus curvatus 102, 110, 257
Lactobacillus delbrueckii subsp. *bulgaricus* 102
Lactobacillus fermentum 102
Lactobacillus helveticus 102
Lactobacillus paracasei 259
Lactobacillus plantarum 102
Lactobacillus reuteri 102
Lactobacillus sake 102, 110
Lactobacillus sanfrancisco 102
Lactobacillus spp. 64, 107
Lactococcus lactis 106f, 109, 111f, 231, 240f, 243, 247
– conjugation 233
– DPC 3147 105
– genetic element 233
– genetic engineering 238
– genetically modified 258
– genome 232
– genomic rearrangement 233
– Group II intron 234
– insertion vector 239
– IS element 235
– mutation rate 237
– prophage 236
– sex factor 233
– subsp. *cremoris* 101f, 232
– subsp. *cremoris* MG1363 236
– subsp. *cremoris* NCDO712 236
– subsp. *lactis*
– supsp. *lactis* IL1403 101f, 236
– subsp. *lactis* var. *diacetylactis* 102
– transduction 234
– transformation 235
– transposon 235, 236
lactose intolerance 101
L-alanine 109
lantibiotic 104f
lettuce 40, 50
LEU2 66f
Leuconostoc lactis 102
Leuconostoc mesenteroides
– subsp. *cremoris* 102f
– subsp. *mesenteroides* 103
Leuconostoc spp. 103
levan 103
LibertyLink 193f
ligase chain reaction 164
LightCycler 161
lipase 89
Listeria monocytogenes 105
Listeria spp. 105
LMBG 124
low iron tolerance
– Fe(III) chelate reduction 45
– siderophore 45
low-calorie sweetener 48
low-carbohydrate beer 73
luciferase 30
lycopene 47
lysostaphin 16

m

Macrozoarces americanus 174
MAE1 73, 75
MAE2 73, 75
maize 31ff, 35, 38, 40, 64, 135, 137, 193, 198
– milling product 220
maize Bt11
– primer 200
maize EPSPS 33
maize event 176 (Maximizer)
– primer 200
maize event
– crylA(b) 213
– EPSPS 213
maize kernel
– sample preparation 190
maize Mon810 (Yield Gard)
– primer 200
maize Roundup Ready (GA21)
– primer 200
maize StarLink (CBH 351)
– antibody 200
– primer 200
maize T25 (Liberty Link)
– primer 200

MAL 66
malate 73
MALDI-TOF 166
male sterility 32, 37f
malolactic conversion 104
MalR 76
MalS 76
MalT 76
malting quality 49
mammalian artificial chromosome 8f
marker gene 18
marker rescue 165, 167
mastitis 105
maximizer 193
meat and fish product 102
MEL1 66, 73, 77
melibiose 73, 103
mesophilic 101
MET2 66, 74
MET3 66, 72, 74
MET10 66, 74
MET14 66, 74
MET25 66, 74
metabolic engineering 71, 89, 95, 109
metallothionein-like protein 47
mice 4, 11f, 15f, 51
micro-array 165
– detection 244
microhabitat product 63
MIG1 66, 76f
MIG2 66, 77
milk 103, 108
milk composition
– α-casein 14
– β-casein 13f
– α-lactalbumin 13f
– β-lactoglobulin 13f
– acetyl CoA carboxylase 14
– antigen 14
– human lactoferrin 14
– human lysozyme 14
– immunoglobulin 14
– k-casein 14
– lactase 14
– lactose 13
– phe-free α-lactalbumin 14
millet 64
– beer 77
Misgurnus mizolepis 176, 178
miso 64
mitotic stability 70
modified fatty acid 39
molecular beacon 162, 218
monoclonal antibody 16

monocotyledonous plant 27f
mosaicism 5–8
– transgenic fish 177
mozzarella 103
Mucor 78
Mucor spp. 64
mud loach 176, 178
Mx1 gene 15
myogenic differentiation 12
myostatin 12

n
NADH-oxidase 109
negative labeling
– inevitable trace 137
– maximum contamination permitted 137
– Novel Food Regulation 129f
nematode resistance, proteinase inhibitor 43
neo 29
neomycin 29
– phosphotransferase (NPTII) 29
Neurospora crassa 69
Neurospora sitophila 64
neutral proteinase
– *Bacillus subtilis* 109
NewLeaf 194
niaD 66, 70, 92
Nisin 105
– inducer 111
NisK 111
NisR 111
nitrate reductase 92
nonantibiotic 104
nonstarter lactic acid microflora 103
NOS-terminator
– primer 200
Novel Food 31, 79
Novel Food Regulation 143, 205
– additives and flavorings 128f
– competent authorities in the member states 130–134
– compliance with world trade law 130
– history 121f
– labeling 126ff
– negative labeling 129f
– Replacement Regulation 135
– requirement for novel foods and food ingredients 124
– sanction 129
– scope of application 123f
– Supplementary Regulation 135
– the authorization procedure 125f
– the elements of the label 127
– the notification procedure 126

novel food   18
novel food and food ingredients   122
– labeling   126
– substantially equivalent   125
Novel Foods and Food Ingredients Instrument   136
– criminal offences and misdemeanors   139
– general rules for novel foods   137
– labeling   138
– negative labeling   137f
– rules on labeling   137
nptII   29, 32–39, 193ff
nuclear transfer   6ff
nucleic acid sequence-based amplification   164
nutraceutical   14, 18, 112

o

oats   41
ocean pout   174, 177
Oenococcus oeni   104
Oenococcus spp.   104
Official Journal of the European Communities   126
ogogoro   64
oilseed rape   31–34, 38ff, 42, 52
Olestra   124
oligonucleotide   159, 161f, 165
oligopeptide   108
olive   102
oncom   64
Oncorhynchus clarki   176, 179
Oncorhynchus kisutch   176, 179, 181
Oncorhynchus mykiss   176, 178f
Oncorhynchus nerka   177
Oncorhynchus tshawytscha   174, 176, 179
open reading frame   152
Opp   108
orange   42
Oreochromis hornorum   176, 178, 182
Oreochromis niloticum   182
Oreochromis niloticus   176, 178
orotidine-5'-phosphate decarboxylase   91

p

pacific chinook salmon   174
pacific salmon
– growth enhancement   181
palm juice   64, 77
papaya   37, 195
papaya 55-1, 66-1   200
– primer   200
parthenocarpic fruit   48
pat   32ff

pat/bar   195
PCR   156, 158, 166f, 177, 183, 245, 255
– bioinformatic consideration   163
– competitive   216ff
– cross border sequence   197
– detection   243
– differential display reverse transcription   168
– efficiency   212
– exhaustive limiting dilution   244
– general principle   159
– inhibition   211, 256
– multiplex   160
– nested   192, 197
– plant species-specific   189
– qualitative   211f, 214, 243
– quantitation kit   221
– quantitation system   222
– quantitative   160, 215f, 243
– real-time   161, 168, 192, 198, 218, 220, 222, 243f
– sensitivity   211, 215
– specificity   211, 216
– target sequence   196
– universal   189
– validation   199
PCR-ELISA   160, 192
PCR-LCR-EIA   164
pea   41
peanut   40f, 64
pectinase   90
Pediococcus acidilactici   102f
Pediococcus pentosaceus   102f
Pediococcus spp.   103
penicillin   29
Penicillium   78
Penicillium camemberti   63, 86
Penicillium chrysogenum   69
Penicillium nalgiovense   63
Penicillium roqueforti   63, 86
Pep   108
pepper   49
peptidas   109
PER (phage encoded resistance)   107
peroxidase   110
PGK1   66
PGU1   66, 75
phage artificial chromosome   8
phage resistance   103
phage-defense system   105f
pharmaceutical trait   50
pharmaceutical
– erythropoietin   51
– glucocerebrosidase   51f

- hemoglobin 51
- hirudin 51f
- interferon 51
phenotype 149
phenylethyl alcohol 76
phosphinotricin 30
phosphomannose isomerase (PMI) 30
Photinus pyralis 30
phylogenetic marker 258
phytase 17
phytic acid 47
phytoalexin 42
Phytophthora infestans 43
Pichia methanolica 86
Pichia pastoris 66, 68, 77, 86, 95
Pichia spp. 68
Pico Green 191
pig 4, 6, 11, 14, 16f
pineapple 43
pinII 35
pito 64
plant product 102
Plasmodium falciparum, merocoite stage surface antigen 112
pLP712 237
- instability 234
- integration 234
plum tree 41
pluripotent stem cell 6
Podospora anserina 70
polymerase chain reaction see PCR
pombe 64
post-translational modification 51
potato 35, 37, 41ff, 46f, 50f, 194
potato New Leaf Y
- primer 201
potato NewLeaf Plus
- primer 201
primary transcript 152
primer 159, 161
prion protein 16f
probiotic bacteria 101f
procedure-related labeling 154
- CommProp GMFF 143
processed food 206, 215, 217, 223
prohibition principle 124, 142
promoter 11
- β-galactosidase 95
- 35S 193ff
- 35S-CaMV 196, 198
- alcohol oxidase 1 95
- amylase 94
- amyloglycosidase 94
- antifreeze protein 174, 177
- carbon catabolite-repressed 94
- carp β-actin
- cauliflower mosaic virus 35S RNA 28
- cellobiohydrolase I 94
- constitutive 28, 111
- constitutive 35S 51
- constitutive PGK1 75
- Cytomegalovirus 178f
- formate dehydrogenase 95
- generic 92
- giant silk moth cecropin B 178
- growth-phase-dependent 105
- histone H3 177
- HSP30 73
- inducible 28
- maize ubiquitin 1 gene (*ubi1*) 28
- MeGa™ 52
- metallothionein 177
- metallothionein-B 181
- methanol oxidase 95
- mouse metallothionein 178, 183
- mud loach β-actin 178
- neutral amylase 94
- nopaline synthase (nos) 28
- ocean pout antifreeze protein 178f
- ocean pout antifreeze protein gene 180
- octopine synthase (ocs) 28
- ovule-specific 49
- patatin 46, 51
- phase-specific stationary 73
- pH-dependent 105
- rice actin 1 gene (*act1*) 28
- rice Glu-B1 47
- rous sarcoma virus long terminal repeat 178
- salt-dependent 105
- sockeye salmon histone 3 178f
- sockeye salmon metallothionein 178f
- sockeye salmon metallothionein-B 179
- phage inducible 107
protease 89, 90, 92, 95, 109
proteinbased detection method 156
protein-based method
- Bt maize 210
- herbicide-tolerant soybean 210
proteolysis
- lactic acid bacteria 108
proteolytic system 109
- cell-wall-bound serine-protease 108
- peptidase 108
- peptide transporter 108
protoplast transformation
- electroporation 27
- PEG-transformation 27

provitamin A biosynthesis   46
pRS01   233f
PrtP   108
pufferfish   185
pullulanase   50
pWV01   239
pyrG   91f, 94f

q
qualitative PCR   211
– ADP-glucose pyrophosphorylase (AGPase) gene   212
– hygromicin phosphotransferase gene   212
– insect-resistant Bt 176 maize   212
– patatin gene   212
quantitative PCR   215
Quorn™   86

r
rabbit   6, 16
raffinose   103
rainbow trout   176, 178f
ramification amplification   164
rapeseed   194
real-time PCR   161, 218, 222f
– Bt 176 maize   220
– commercially available kit   220
– genetically modified soybean   219
– Roundup Ready soybean   220
– transgenic maize   219
– transgenic soy   219
redfish   185
reference material   157
Regulation (258/97/EC)   154
Regulation (EC) 1139/98
– Bt-176 maize   205
– mandatory labeling   205
– Roundup Ready soybean   205
Regulation (EC) 50/2000
– additive   205
– flavoring   205
Regulation (EC) No 1139/98   123, 139f
Regulation (EC) No 178/2002
– Basis Regulation   141
Regulation (EC) No 1813/97   123
Regulation (EC) No 258/97   139f, 205
– Novel Food Regulation   122
Regulation (EC) No 49/2000   140
Regulation (EC) No 50/2000   128, 140
Regulation (EC)
– threshold   205
rennet   89
replacement recombination   247

Replacement Regulation   137f
– labeling   136
– threshold value   135
reporter gene   29, 34, 36f, 39, 180
– *gusA* (uidA) gene   30
– *luc* gene   30
restorer of fertility   32
restriction fragment length polymorphism   183
restriction fragment-length polymorphism see RFLP
restriction-modification (R/M) system   106
retroviral vector   7
– cloning capacity   5
– lentiviral vector   5
Rhizopus   78
Rhizopus oligosporus   64
Rhizopus spp.   64
rice   34, 40–47, 52, 64
RNA   149
– messenger RNA (mRNA)   149
– ribosomal RNA (rRNA)   149
– transfer RNA (tRNA)   149
rolling cycle amplification   164
rotavirus-induced diarrhea   101
Roundup Ready   193f
ruminant   13

s
Saccharomyces bayanus   64, 72, 74
Saccharomyces carlsbergensis   72
Saccharomyces cerevisiae   64–67, 72, 74–77, 86
Saccharomyces monacensis   72
Saccharomyces pastorianus   72
Saccharomyces spp.   64
sake   64
– flavor   76
– phenylethyl alcohol   76
– toxic substance   76
Salmo salar   176, 179
salmonid   12
salt-stress tolerance
– HAL1 gene   45
– Na^+/H^+ antiport   45
Salvelinus alpinus   176, 178
SAM   77
sam-K   39
sampling   156f
– guideline   190
Sauerkraut   102
sausage   103
Schizosaccharomyces pombe   64, 68, 75, 86
– de-acidification of wines   77
– pompe   77

– rum 77
– ogogoro 77
Schwanniomyces occidentalis 68
scrapie 16
screening method
– genetically modified crop 193
Sebastes spp. 185
seedless fruit 48
selectable marker 8f, 29, 90
selection marker 94, 106
– β-lactamase gene 93
– *amdS* gene 69
– bleomycin 95
– drug resistance 69
– food-grade 238
– hygromycin 95
self-cloning 153f, 239, 252, 260
serine hydroxymethyltransferase 109
sheep 6, 11, 16f
shoyu 64
silage 103
single cell protein 86
single molecule detection 166
single-nucleotide polymorphism 244
small ruminant 4
sockeye salmon 177
somatoliberin 11
somatostatin 11
somatotropin 11
sorghum 64
sour maize gruel 64
sourdough 76, 102
Southern blot 158, 198
Southern hybridization 255
– detection 243
soy bean 31ff, 39, 40f, 64, 193
– sample preparation 190
soy milk 103
soy sauce 64, 77, 102
soya 198
soya bean 135, 137
soya drink 190
soya flake 190
soya flour 190
soya lectin gene Le1 197
soya or maize flour
– sample preparation 190
soya or maize grit
– sample preparation 190
soya Roundup Ready
– antibody 201
– primer 201
soya sausage 190
spectinomycin 29

SPS-Agreement 130
squash 37, 194
STA2 66, 78
stabilizer 107
Standing Committee for Foodstuffs 126, 129
Staphylococcus aureus protein A 245
Staphylococcus spp. 105
starch and dextrin 73
StarLink 193
starter bacteria 106
starter culture 63, 78, 101, 103, 112
stationary phase 110
strand displacement amplification 164
Strephtococcus spp. 105
Streptococcus thermophilus 102f, 107, 109
Streptomyces hygroscopicus 30
Streptomyces viridichromogenes 30
streptomycin 29
stress response
– lactic acid bacteria 110
submerged fermentation 87f
subtraction hybridization
– detection 259
subtractive hybridization 181
sucrose 103
sufu 64
sugar beet 34, 48
sugarcane 40f, 64
suicide system 107
sunflower 40
surface plasmon resonance 165, 167
– detection 246
sweet potato 41
swine 15
Swiss-type cheese 103
SYBR-green 161, 244

t
Takadiastase™ 87
tandem integration 93
Taq-DNA-polymerase 159, 218
TaqMan 161
TaqMan Probes 161
Taq-polymerase 161
TBT-Agreement 130
tea extract 64
TEM-1 β-lactamase 29
temocillin 29
tempe 64
temperate bacteriophage 134, 233
terminator
– amyloglycosidase 94
– Bovine growth hormone polyA site 178
– *nos 3′* 193–196

- ocean pout antifreeze protein 3' region 178
- ocean pout antifreeze protein gene 3' region 180
- simian virus 40 poly A site 178
- sockeye salmon growth hormone 1 178f
- translational 93
terpenoid 75
thickener 107
threshold regulation 188
tilapia 12, 176, 178
- growth acceleration 182
- growth enhancement 182
tobacco 41–44, 49ff
- guttation fluid 52
- post-harvest expression system 52
tofu 64, 190
tomato 36, 39, 41ff, 45, 47, 49, 194
top-fermenting ale yeast 72
transcription 149
transduction 234
transformation method
- *Agrobacterium* transformation 26f
- direct gene transfer 27
transformation
- frequency 235
- polyethylene glycol (PEG-induced) 111
- protoplast 111
transgene 9, 18f
- chinook salmon growth hormone 178f
- coho salmon growth hormone 178
- expression 11f, 15, 28, 50
- gene construct 8
- giant silk moth cecropin B 178
- heterochromatin formation 8, 10
- human glucose transporter type 1 178f
- human growth hormone 1 178
- rainbow trout growth hormone 1 178
- rat gulonolactone oxidase 179
- rat hexokinase type II 178f
- silencing 8
- sockeye salmon growth hormone 1 178f
- stability 50
- tilapia growth hormone 178
- winter flounder antifreeze protein 179
transgenesis 3
- nuclear transfer 6
- pluripotent stem cell 6
- pronuclear DNA microinjection 4, 9
- retroviral vector 5, 9
transgenic animal 3–19
- agricultural trait 10, 16
- biochemical pathway 10, 17

- biosafety 18
- carcass composition 11
- disease resistance 10
- feed efficiency 10f
- gene farming 3
- growth rate 10f
- milk composition 13
- models for human diseases 3
- wool production 17
- xenotransplantation 3
transgenic cell 6, 10
- identification 8
transgenic fish 182
- antifreeze protein 176, 180
- benefits expected 175
- biosynthesis of ascorbic acid 176
- carbohydrate utilization 176
- detection 183
- development 175
- disease resistance 176
- enhanced growth 176
- food safety 183f
- gene cassette 176, 178f
- growth hormone 180
- mosaicism 177
- pleiotropic effect 184
- production 175
translation 149
transposon 235
- Tn5 29
Trichoderma 78
Trichoderma harzianum 42
Trichoderma longibrachiatum 75, 88
Trichoderma reesei 69, 88, 90, 92
Trichosporon 63
TRP1 66f
two-dimensional gel electrophoresis, detection 246

u
URA3 66f, 95
urease 110

v
vaccine delivery system 111
vaccine
- cholera toxin B subunit gene (CTB) 51
- edible plant-based 50
- enterotoxin (LT-B subunit gene) 51
- oral immunization 50
- virus-like particle (VLP) 50
vector system 111
vector
- extrachromosomal 67

– integrative 67f
– replacement cassette 68
vegetable 103
virus resistance 36f
– ribosome-inactivating protein 41
– viral coat protein 41
– viral replicase gene 41

w

walnut 40
watermelon 49
Western hybridization
– detection 245
wheat 41, 42, 47, 49, 52, 64
White Paper on Food Safety 139
wine 64, 74
– flavor characteristic 75
– processing efficiency 75
wine flavor
– β-1,4-glucanase 75
– acetyltransferase 75
– glycerol 75
– malate permease 75
– malic enzyme 75
wine yeast 72f
– fermentation performance 74
Wizard 256
wizard-extraction 221
World Trade Organization (WTO) 130
WTO 145

x

Xanthomonas oryzae 42
xylanase 89

y

Yarrowia lipolytica 68, 86
yeast 62ff, 79, 101
– auxotrophic marker 67
– centromeric plasmid 67
– Crabtree-effect 68
– dominant markers 67
– electroporation 66
– episomal plasmid 67
– hyperglycosylation 68
– metabolic engineering 71
– methylotrophic 68, 95
– nonconventional 68
– replicating plasmid 67
– spheroplast method 65
– substrate range 73
– transformation 66
– vector 67
yeast artificial chromosome 4, 8
Yieldgard 193
yogurt 102f

z

Zygosaccharomyces rouxii 64, 68
– miso 77
– soy sauce 77